Proteases in Apoptosis: Pathways, Protocols and Translational Advances

Kakoli Bose
Editor

Proteases in Apoptosis: Pathways, Protocols and Translational Advances

Editor
Kakoli Bose
Integrated Biophysics and Structural Biology Lab
Advanced Centre for Treatment,
Research and Education in Cancer (ACTREC),
Tata Memorial Centre
Navi Mumbai, Maharashtra, India

ISBN 978-3-319-19496-7 ISBN 978-3-319-19497-4 (eBook)
DOI 10.1007/978-3-319-19497-4

Library of Congress Control Number: 2015944824

Springer Cham Heidelberg New York Dordrecht London

Printed on acid-free paper

Springer International Publishing AG Switzerland is part of Springer Science+Business Media (www.springer.com)

To my mother, Rekha Ghosh
for whom, I am who I am
&
my daughter Roshnee
who has been a constant support and
a source of inspiration all along this
journey . . . .

Apoptosis: An Oncologist's Perspective . . .

Programmed cell death (PCD) that allows human body to dispose of billions of *de trop* cells poses an important impediment against cancer. Therefore, being the head of one of the most prestigious cancer hospitals in Asia, I envisage the immense potential of the apoptotic molecules as therapeutic targets against cancer. From ongoing studies over the years, it is clear that the effective treatment of cancer needs multi-targeted intervention, and a major breakthrough in managing and curing the disease is only possible through innovative search of novel targets and drugs.

Modern medical science greatly acknowledges the contribution of fundamental research through conceiving and nurturing the notion – *Interpretation precedes Intervention*. Keeping that essence in mind, *Proteases in Apoptosis: Pathways, Protocols and Translational Advances* has made a conscientious effort to provide a resplendent account of intricate apoptotic pathways and the key modulators involved. It is an extremely well-written book with vivid descriptions of the family of proteases involved in programmed cell death. Each and every chapter graphically describes the complex apoptotic network with a very simple and lucid flow of information. Moreover, the last two chapters that elaborately discuss experimental protocols, translational advances and therapeutic possibilities fulfill its *bench to bedside* mandate.

Specific targeting of malignant cells is the major challenge of cancer drug design. Drugs that would lead to disruption of genetic and molecular machinery of tumours therefore need to be explored and developed in the coming years. Therapeutic molecules targeting apoptotic pathways and signal transduction cascades offer the possibility of greater efficacy and lower toxicity. The perfect synchronization of this generic approach with an advanced individualized treatment modality will definitely lead towards a successful cancer management program. In cancer treatment, where precise prognosis, intervention and elimination of disease still remains a far-fetched dream, a whole-hearted and concerted effort from the research as well as medical communities can turn this unrealized aspiration into reality.

Director, Tata Memorial Centre
Professor and Chief, Breast Unit
Mumbai, India

Dr Rajendra A. Badwe, MD.

Foreword

Apoptosis or programmed cell death is a highly regulated process that plays an important role in tissue homeostasis, developmental processes and the immune system. It is a process that comes in force from early embryonic development and continues in the adult life to maintain a perfect balance between cell proliferation, cell differentiation and cell death. Several molecules and signalling pathways regulate the complex process of apoptosis. Understanding the relevance of apoptosis in several pathological conditions is important to get better insight into the pathogenesis of the disease and would provide leads/cues to develop novel therapeutic approaches for the treatment of diseases like cancer and Alzheimer's.

It gives me pleasure to introduce to the readers. The book on *Proteases in Apoptosis: Pathways, protocols and translational advances* that has been edited and written by Dr. Kakoli Bose who has been working in the field of apoptosis and cancer for the past 15 years.

Dr. Bose is Principal Investigator and Assistant Professor at ACTREC which is a premier organization of the country dedicated to cancer research and patient care. Dr. Bose completed her graduate studies at North Carolina State University, Raleigh, and postdoctoral training at Tufts New England Medical Centre, Boston. Dr. Bose's research interest focuses on non-classical mechanisms of programmed cell death with emphasis on understanding structure-function relationship of proteins involved in novel adapter-independent extrinsic pathways and caspase-independent apoptosis with the aim of targeting them for disease intervention.

Dr. Bose has carefully crafted the six chapters presented in this book that covers various aspects of apoptosis ranging from basic molecular mechanisms to preclinical and clinical implications.

As a consequence of these features, the readership will be broad from the novice to experts in the field. Contributors to this book are both national and international experts, and I am confident that the book would be of interest to research scientists and clinicians who are interested in deciphering the mechanisms that regulate cell death processes.

Director, ACTREC
Prof. & Head, Tumor Immunology Group,
ACTREC, Tata Memorial Centre
Navi Mumbai, India

Dr. S.V. Chiplunkar

Preface

Death proclaims creation, although sounds utterly oxymoronic, is the most befitting description of programmed cell death *aka* apoptosis. With almost 50–60 billion ageing and physiologically impaired cells being replaced daily in a normal adult human, apoptosis makes way for healthy tissues to rebuild and regenerate, thus preserving the positive force of life. It's been a long journey since 1950s when the concept of programmed cell death began to take shape which later burgeoned exponentially so as to become an integral part of biomedical research. Groundbreaking discoveries in the early 1990s brought into forefront the role of proteases mainly caspases in the complex network of apoptosis activation process. This family along with other lesser known proteases, such as granzymes, calpains, cathepsins and HtrAs, initiate, activate, execute and modulate the entire apoptotic cascade through coordinated and precise mechanisms thus maintaining a delicate balance between cell survival and death. Perturbation in this equilibrium leads to several diseases of major medical significance such as cancer and neurodegenerative disorders which underscores the potential of these enzymes as current and future therapeutic targets.

In this book, we have described the role of proteases in programmed cell death. Apart from providing a broad overview on these proteases, this book also annotates the recent developments in various methodologies for studying their role under normal and diseased conditions. It also discusses significant contributions of these proteins in translational research and their future prospects in therapeutic intervention.

The book is organized in six chapters. The first chapter provides an introductory note on apoptosis in general, different pathways, molecules involved in these critical pathways and diseases associated with its deregulation. The second chapter introduces the reader to the caspase family of cysteinyl proteases. It focuses on structural and functional classification of caspases, their mechanisms of activation, substrates, inhibitors and role in different apoptotic pathways.

Molecules associated with alternate cell death mechanisms are emerging as potential therapeutic targets especially in cases where traditional pathway fails to activate and are covered in detail in the third and fourth chapters. The fifth chapter elaborates structural and functional assays on mechanism of these proteases and their involvement in apoptosis. This chapter brings together a wide array of complementary techniques that have been developed for the specific detection and analysis of these proteases and their activities. Finally, the sixth chapter concludes with a vivid description and review of animal models and non-invasive imaging modalities in developmental therapeutics targeting proteases in apoptotic pathways. Taken together, the different chapters of the book deal with important aspects of the proteases associated with programmed cell death along with challenges and recent advancements in the field of research.

This book aims at providing up-to-date information on proteases associated with different cell death pathways with thorough discussions on current and potential preclinical and clinical applications. It also intends to inform and inspire undergraduate and graduate students alike and stimulate them towards pursuing biomedical research.

I am grateful to all the contributing authors for providing their expertise and thoughtful insights on proapoptotic proteases. I also thank all my students, lab members and friends who spent considerable amount of time proofreading and giving invaluable comments.

Navi Mumbai, India Kakoli Bose

Contents

Contributors

Saujanya Acharya Integrated Biophysics and Structural Biology Lab, Advanced Centre for Treatment, Research and Education in Cancer (ACTREC), Tata Memorial Centre, Navi Mumbai, Maharashtra, India

Kakoli Bose Integrated Biophysics and Structural Biology Lab, Advanced Centre for Treatment, Research and Education in Cancer (ACTREC), Tata Memorial Centre, Navi Mumbai, Maharashtra, India

Christine E. Cade Department of Molecular and Structural Biochemistry, North Carolina State University, Raleigh, NC, USA

Lalith K. Chaganti Integrated Biophysics and Structural Biology Lab, Advanced Centre for Treatment, Research and Education in Cancer (ACTREC), Tata Memorial Centre, Navi Mumbai, Maharashtra, India

Pradip Chaudhari Small Animal Imaging Facility, Advanced Centre for Treatment, Research and Education in Cancer (ACTREC), Tata Memorial Centre, Navi Mumbai, Maharashtra, India

A. Clay Clark Department of Molecular and Structural Biochemistry and Center for Comparative Medicine and Translational Research, North Carolina State University, Raleigh, NC, USA

Present affiliation: Professor and Chair, Department of Biology, University of Texas at Arlington, Arlington, TX, USA

Raja Reddy Kuppili Integrated Biophysics and Structural Biology Lab, Advanced Centre for Treatment, Research and Education in Cancer (ACTREC), Tata Memorial Centre, Navi Mumbai, Maharashtra, India

Nitu Singh Integrated Biophysics and Structural Biology Lab, Advanced Centre for Treatment, Research and Education in Cancer (ACTREC), Tata Memorial Centre, Navi Mumbai, Maharashtra, India

Chapter 1
Apoptosis: Pathways, Molecules and Beyond

Nitu Singh and Kakoli Bose

Abstract Programmed cell death or apoptosis manifests itself through a complex network of biochemical pathways and distinct morphological signatures. It is a natural phenomenon in multicellular organisms required to maintain tissue homeostasis through selective removal of ageing and unwanted cells. Impairment of this tightly regulated cellular process leads to various pathophysiological conditions including neurodegenerative disorders, ischemic damage, acquired immunodeficiency syndrome and cancer. Recognizing its immense therapeutic potential, a plethora of research endeavors has been undertaken in the past two decades that target molecules involved in apoptosis. Caspases, a conserved family of cysteinyl proteases that initiate and execute programmed cell death through extrinsic and intrinsic pathways are major focus of apoptosis research. However, study of molecules associated with lesser-known caspase-independent cell death is slowly gaining prominence, especially in cases where the traditional pathways fail to activate apoptosis. The goal of this chapter is to provide a broad overview of different apoptotic pathways, molecules involved and their crosstalk with special emphasis on proteases. This chapter also discusses different diseases associated with deregulation of apoptosis, current status on pre-clinical and clinical trials, their limitations and future prospects.

Keywords Apoptosis • Protease • Caspase • Tissue homeostasis • Caspase-independent cell death • Cancer • Neurodegenerative diseases

Apoptosis

'Programmed cell death', which is also referred to as 'apoptosis' (in Greek meaning falling of leaves or petals from flowers), is an evolutionary conserved phenomenon observed in multicellular organisms [1] essential for selective removal of ageing,

N. Singh • K. Bose (✉)
Integrated Biophysics and Structural Biology Lab, Advanced Centre for Treatment, Research and Education in Cancer (ACTREC), Tata Memorial Centre, Navi Mumbai, Maharashtra 410210, India
e-mail: kbose@actrec.gov.in

K. Bose (ed.), *Proteases in Apoptosis: Pathways, Protocols and Translational Advances*, DOI 10.1007/978-3-319-19497-4_1

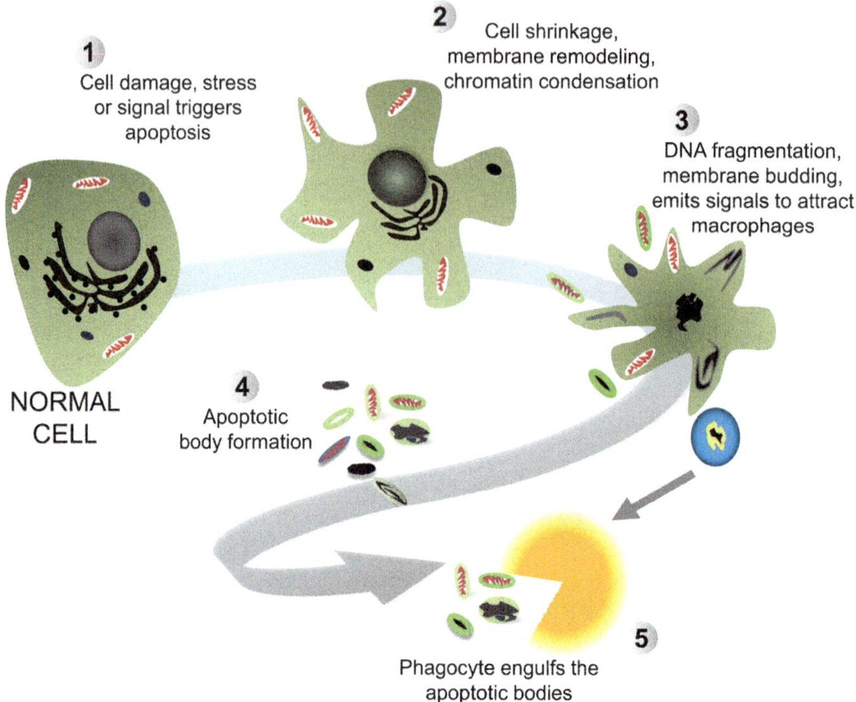

Fig. 1.1 Diagrammatic representation of various cellular and morphological changes observed during apoptosis. Cell death by apoptosis is normal and energy dependent process initiated by cellular damage, stress or number of endogenous and extracellular stimuli. It is accompanied by decrease in cell volume, nuclear changes with chromatin condensation followed by DNA fragmentation and membrane blebbing. This results in the formation of small apoptotic bodies surrounded by cell membrane which are later cleared by the process of phagocytosis in the extracellular milieu avoiding any inflammatory reaction

unwanted and impaired cells. During apoptosis, shrinkage and membrane blebbing alters cellular morphology finally leading to its disintegration into smaller apoptotic bodies (Fig. 1.1). Unlike necrosis, where the cell dies by swelling and releasing its contents thus causing an inflammatory response, apoptosis is a controlled process where the contents are strictly maintained within the cell membrane [2]. As its name suggests, apoptosis is a natural form of cell death that is required for immune response as well as normal development of tissues and organs such as differentiation of toes and fingers in embryos [3–9]. Tight regulation of apoptosis is also critical for proper development of placenta and healthy pregnancy, impairment of which might lead to conditions such as preeclampsia [10]. Apoptosis is thus essential for maintaining cellular homeostasis i.e. a fine balance between cell proliferation and death. Disruption in this balance however, is responsible for several pathological conditions including autoimmune diseases, neurodegenerative disorders, stroke, acquired immunodeficiency syndrome (AIDS) and cancer [4, 11–16].

1 History

Although programmed cell death was first observed by Carl Vogt way back in the early nineteenth century, research on this cellular process was initiated more than a century later when apoptosis was observed through electron microscopy and the formation of apoptotic bodies were reported by John F. R. Kerr [17]. Since then plenty of research ventures in this field have catered much of our understanding to the complexities of the pathways involved, including interactions and crosstalk of critical molecules with myriads of other cellular components.

Proteases play a key role in apoptosis through cleavage of several proteins that are critical for normal functioning of the cellular system. A major class of proteases called caspases regulates the apoptotic machinery, although several other less prominent players are also involved such as granzymes, calpains, cathepsins and HtrAs [18]. Upon receiving specific apoptotic signals, these proteases (either present or are released in the cytosol) get activated and initiate cascade of events leading to proteolytic cleavage of key proteins essential for normal cellular functions such as cytoskeletal proteins, DNA repair enzymes and nuclear proteins. This proteolytic cascade in turn activates degradative enzymes, for instance, DNases that cleave nuclear DNA and complete the cell death process. Finally, these cells are packaged into apoptotic bodies and get eliminated by phagocytosis [19, 20].

2 Apoptotic Pathways

Classical apoptotic cascades follow two major pathways, extrinsic and intrinsic. Although, the extrinsic pathway is mainly cytoplasmic and intrinsic pathway originates in the mitochondria, several molecules in these pathways crosstalk with each other thus adding to their complexity toward specific therapeutic intervention.

2.1 Extrinsic or Receptor-Mediated Cell Death Pathway

The extrinsic cell death pathway is often referred to as 'receptor-mediated pathway' where extracellular cell death ligands trigger activation of cell-surface death receptors [21, 22]. Given the importance of the cellular process, multiple membrane **d**eath **r**eceptors (DRs) that belong to the tumor necrosis receptor super-family, are found to be involved such as TNF-R1 (Tumor Necrosis Factor Receptor 1), Fas (also known as CD95 or APO-1), DR3, DR4 (TRAILR1), DR5 (TRAILR2) and DR6 [23]. Among these, extensive studies on TNF-R1, Fas and DR4/DR5 have been done, while a lot is yet to be known about the other receptor-mediated pathways. These transmembrane proteins have conserved 'cysteine-rich' extracellular tertiary structural fold for recognition of their specific ligands. They also comprise an intracellular stretch of amino acids (~80 residues) known as 'death-fold domains' that are essential for activation of the pathway through homotypic protein-protein interactions [24, 25].

In the classical pathway, death signal brings the ligands to the vicinity of the extracellular region of monomeric receptors and induces their trimerization. The trimeric receptors and their cytosolic adaptor molecules then interact with each other through their respective **d**eath **d**omains (DD). This interaction recruits upstream zymogens of cysteinyl proteases or procaspases (mainly procaspase-8) to these preassembled adaptor molecules where another level of interaction occurs. The upstream or initiator procaspases have a large N-terminal protein-protein interaction domain called 'prodomain'. A death-fold domain called death effector domain or DED that resides within the prodomain of procaspase-8 undergoes homotypic interaction with C-terminal DED of the adaptor molecule with subsequent formation of a huge protein complex called **d**eath **i**nducing **s**ignaling **c**omplex (DISC). It thereby increases the local concentration of these monomeric proenzymes through proximity induced by the adaptor molecules [26]. This subsequently leads to their dimerization, removal of prodomain and subunit rearrangements to form active initiator caspases. This active caspase-8 in turn cleaves and activates the downstream or executioner procaspases, mainly procaspase-3 and 7. Upon its activation, executioner caspase-3, cleaves **X**-linked **i**nhibitor of **a**poptosis **p**rotein (XIAP) creating a positive feedback loop of self-activation thereby synergistically enhancing the cell death process [27, 28]. Proteolytic cleavage of important cellular proteins [29] by these effector caspases finally rings the death bell of the cells. The entire pathway is tightly regulated at every single level by several molecules such as decoy proteins and inhibitors [25, 30, 31].

2.1.1 TNF-mediated Pathway

The tumor necrosis factor (TNF) is a moonlighting pro-inflammatory cytokine that is involved in important cellular signaling processes such as NF-κB pathway, and is associated with the progression of several critical diseases such as neurodegenerative disorders, autoimmune diseases and cancer [21]. Two major receptors of TNF are TNF-R1 and R2. Although TNF-R2 is found mainly in the immune system, TNF-R1 has ubiquitous expression and is involved in most of the signaling pathways that TNF modulates. There have been controversies in the literature upon TNF receptor assembly and ligand binding. On one hand, literature reports ligand dependent or independent activation of the monomeric receptor [32, 33] while, on the other side, it has been shown that ligation of TNF occurs on the preassembled receptors. According to the latter mechanism, an extracellular receptor region known as **p**re-**l**igand **a**ssembly **d**omain (PLAD) is required for efficient ligand docking and receptor mediated caspase activation. Literature suggests that PLAD either induces a conformational change or higher order oligomerization in the receptor making it capable of relaying apoptotic signal onto the downstream molecules [34, 35]. Although, in the former mechanism, receptor activation demands ligand docking, in the latter, the activation of preassembled receptor is blocked by a complex called **s**ilencer **o**f **d**eath **d**omain (SODD) [36]. TNF binding to the receptor complex releases SODD followed by binding of adaptor molecule TRADD (**TNF-R** **a**ssociated **d**eath **d**omain protein) through their death domains. Recruitment of TRADD at the receptor site is followed by its binding to another adaptor molecule FADD (**F**as **a**ssociated **d**eath **d**omain) via respective death fold domains. Through homotypic interaction via death effector domains (DED) of free N-terminal region of FADD and N-terminus of initiator procaspase-8 or -10, cell death signal is relayed to activator caspases-3, -6 and -7 eventually leading to cell death (Fig. 1.2).

Unlike other death ligand mediated pathways such as Fas and TRAIL, TNF regulated extrinsic pathway is not spontaneous as it is limited by pro-survival signal that is generated upon NF-κB activation [37]. Thus proapoptotic property of TNF can be manifested only upon NF-κB inactivation during transcription or inhibition of protein synthesis. Since NF-κB promotes activation of antiapoptotic molecules such as **TNF** **r**eceptor **a**ssociated **f**actors, TRAF-1 and -2 (inhibit caspase-8) and **i**nhibitor of **a**poptosis **p**roteins, cIAP-1 and -2 (inhibit caspases-3, -7 and -9), its inhibition by transcription or protein synthesis inhibitors might accentuate proapoptotic properties of TNF-R1. However, more in-depth studies are required to gain insight into this complex mechanism and answer some of the intriguing questions such as how the preassembled trimeric receptor is inactive under normal cellular conditions and in that case what is the role of SODD as well as activating ligands. Answer to these paradoxical questions might help further our understanding of TNF pathway in apoptosis and hence will help in targeting the pathway more efficiently.

2.1.2 Fas-mediated Pathway

Fas play a key role in physiological regulation of apoptosis in the cell. It has also been found to be involved with several diseases including immune system disorders [38]. Upon apoptotic signal, Fas ligand binds to the extracellular region of its receptor, leading to receptor trimerization and increase in intracellular concentration of its C-terminal death domain. This close assemblage of DD recruits homologous DD of the adaptor molecule FADD forming a functional DISC [39–41]. The free N-terminal DED of FADD then binds to the long prodomain of procaspase-8 or -10 leading to oligomerization, zymogen processing and subsequent activation of the apoptotic pathway (Fig. 1.2). Catalytically active initiator caspases then activate downstream procaspases (procaspase-3, -6 and -7) by cleaving at their respective recognition sequences. Removal of prodomain and activation of downstream caspases creates a positive feedback loop as they in turn activate the upstream proteases thus initiating the caspase cascade [42]. Finally, effector caspases induce apoptosis by cleaving important cellular components that are critical for its survival and efficient functioning. However, it is interesting to note that although caspase-10 has been found to be another initiator caspase like caspase-8, it has distinct substrates in death receptor pathways or other cellular processes [43], the reason for which is yet to be deciphered.

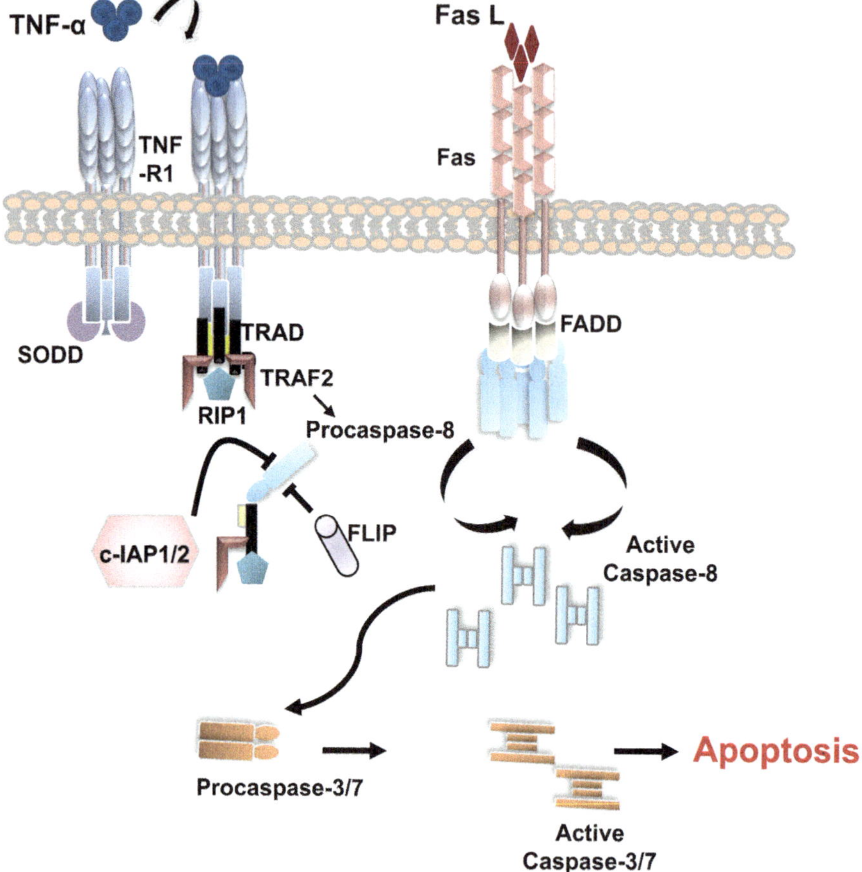

Fig. 1.2 Receptor-mediated extrinsic cell death pathway. In the extrinsic pathway, death ligands, such as FasL, TNF or TRAIL engage their cognate receptors Fas, TNFR or DR5 respectively. Engagement of death receptors with their ligands initiate the recruitment of adaptor proteins such as FADD or TRADD, which in turn recruit zymogenic initiator procaspase-8 to form the death-inducing signaling complex (DISC). DISC recruitment leads to auto-processing and thus activation of caspase 8. Active caspase-8 then proteolytically cleaves and activates executioner caspases 3, -6 and -7 that further culminates in substrate proteolysis and hence cell death. The pathway is tightly regulated with several inhibitory molecules at various stages of the signaling. In TNFR-induced mechanism, the activation of preassembled receptor is blocked by a complex called silencer of death domain (SODD). Fas mediated pathway has a negative regulator molecule FLIP (FLICE-like inhibitory protein). FLIP interferes with DISC functions by preventing the formation of catalytically active caspase-8

Like TNF-mediated signaling, this pathway is also regulated by several modulator molecules. FLIP (**F**LICE-**l**ike **i**nhibitory **p**rotein), a component of DISC is one such negative regulator of the Fas-mediated apoptosis. The two isoforms namely $FLIP_S$ (**s**hort) and $FLIP_L$ (**l**ong) have been found to structurally mimic part or whole of procaspase-8 respectively and hence act as decoy proteins. While $FLIP_S$

comprises tandem DED domains similar to procaspase-8, the longer version also has the catalytically inactive protease domain. It has been suggested that FLIP interferes with DISC functions by preventing formation of mature caspase-8 [44]. However, its role in extrinsic cell death pathway is controversial as it has also been shown to promote NF-κB activation and hence prevent cell death [45]. In addition, FLIP$^{-/-}$ mice exhibited phenotypes similar to caspase-8, and FADD knockout mice suggesting that it might be involved in promoting apoptosis [46]. Moreover, studies have found FLIP$_L$ to act as an activator of procaspase-8 via heterodimer formation [47–49]. Further insights are required to resolve these ambiguities and elucidate the exact role of FLIP in cell death and other pathways.

Other than the death receptor, adaptor protein, caspase-8 and FLIP, several other molecules have also been reported to be part of DISC such as Daxx, RIP and PIDD. Daxx aids Fas to mediate apoptosis through JNK signaling pathway, RIP (**r**eceptor **i**nteracting **p**rotein), a serine-threonine kinase with a death domain, interacts with TRADD and FADD to mediate apoptosis via formation of huge signaling complex called ripoptosome [50–52], while PIDD (**p**53-**i**nduced **d**eath **d**omain protein) is associated with the PIDDosome formation. Although the mechanism of its mode of action so far has not been elucidated, the interactions with RAIDD (**RI**P-**a**ssociated **I**ch-1/Ced-3-homologue protein with a **d**eath **d**omain) and caspase-2 have been implicated [51, 53].

2.1.3 TRAIL-mediated Pathway

TRAIL receptor was identified due to its sequence homology with the other two receptors, Fas and TNF [54, 55]. The role of Apo2L/TRAIL in various biological pathways is not very well delineated. However, like Fas, it has also been implicated in tumor immune response [56, 57] and knockout studies in mice confirm its anti-tumor and anti metastatic roles [58, 59]. The mechanism by which TRAIL induces apoptosis is quite intriguing. Out of five known TRAIL receptors (TRAIL R1-R4 and osteoprotegerin or OPG) [60–65], the first two promote cell death in different tumor cells, while TRAIL-R3 and -R4 that are devoid of cytoplasmic death domains (DD) act as 'decoy' proteins with significantly reduced proapoptotic activities. OPG, however, with a lower binding affinity for TRAIL-1 and -2 ligands has limited effect at physiological temperature [66]. A recent groundbreaking study demonstrates importance of post-translational modification on TRAIL sensitivity. Microarray analysis of more than 100 human tumor cell lines identified a few critical enzyme-encoding genes which are upregulated in TRAIL-sensitive cell lines. These enzymes (*GALNT14*, *GALNT3*, *FUT6* and *FUT3*) post-translationally modify TRAIL through O-glycosylation and trigger receptor trimerization and subsequent DISC formation [67].

TRAIL is capable of inducing apoptosis through both extrinsic and intrinsic pathways. In the former, ligand binding to TRAIL-R1 and -R2 triggers receptor trimerization followed by DISC assembly and subsequent caspase activation and initiation of cell death. On the other hand, TRAIL mediated intrinsic pathway gets activated when mitochondrial proapoptotic molecules Smac/DIABLO or HtrA2 restore caspase-3 activity by relieving inhibitory effect of XIAP [68, 69]. TRAIL also weakly induces NF-κB pathway with the help of adaptor proteins Rip and TRAF-2 [70].

2.1.3.1 TRAIL Based Therapy: Where Are We Now?

Extrinsic cell death pathways are excellent therapeutic targets as they have the ability to trigger cell death irrespective of p53 expression, which is mostly inactivated in several cancers. The therapeutic ability of other death receptors got thwarted due to their toxicity and severe side effects, while TRAIL-R with very mild toxicity became the molecule of interest [71]. It has been observed that surface expression of death receptors resulted in enhanced sensitivity of cancer cells to TRAIL-R induced apoptosis [72]. Although activations of TRAIL-R and the 'decoy' protein FLIP in cancer cells have been found to be inversely proportional to each other, the topic is still debatable and requires further studies for unequivocally establishing the role of FLIP in TRAIL-R inhibition [48, 72, 73]. TRAIL-R mediated apoptosis can be augmented by introduction of proapoptotic Smac/DIABLO or HtrA2, both of which are mitochondrial molecules that are released into the cytosol upon apoptotic induction. Although work has been done on effect of Smac mimetics on TRAIL-R induced apoptosis [68, 74], serine protease HtrA2 that holds a great promise in cancer therapeutics needs to be explored more effectively. Despite role of TRAIL-R in the intrinsic apoptotic pathway has been controversial, its effect nonetheless has been seen to be amplified in presence of tumor suppressor p53. Thus, understanding of TRAIL-mediated pathway in detail and how it can be sensitized in cancer cells to trigger apoptosis will lead toward design of molecules that can effectively target the pathway in combination with conventional and other therapies such as irradiation, proteasome inhibitors, Smac mimetics etc. [75].

2.2 Intrinsic Pathway of Programmed Cell Death

The 'intrinsic' or 'mitochondrial' pathway of apoptosis as its name suggests, originates within the cell as an effect of UV radiation, γ-irradiation, heat, altered redox potential, oncoproteins, viral virulence factor, chemotherapeutic agents, mutation in DNA, changes in the rate of cellular metabolism and intracellular damage [21, 76]. Although it is primarily initiated in the mitochondria, **e**ndoplasmic **r**eticulum (ER) sometimes plays a role in this pathway as well [21].

2.2.1 Mitochondrial Pathway and BCL-2 Family Proteins

It is an established fact that apart from the caspases, BCL-2 (B-cell lymphoma 2) family members are also distinctively involved in the apoptotic pathway. BCL-2

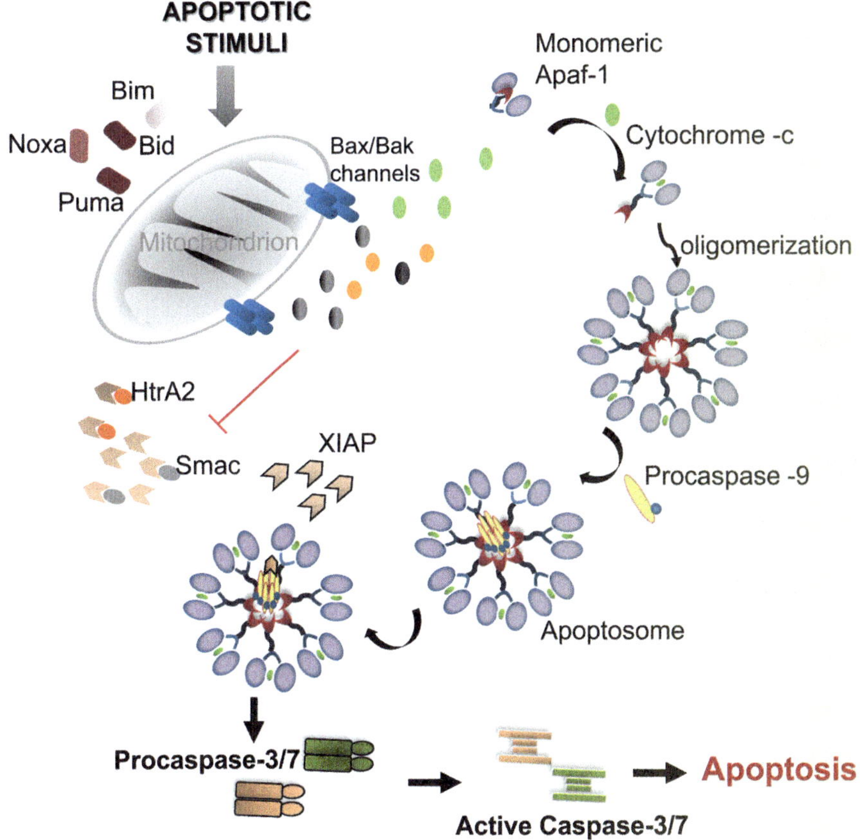

Fig. 1.3 Mitochondrion mediated intrinsic pathway of apoptosis. Cellular stress activates p53, a cell-cycle check-point protein. p53 in turn initiates the intrinsic pathway by up regulating BCL-2 family proteins, Puma and Noxa, which further activates BAX, BAK. Oligomerization of BAX-BAK complex at the outer membrane of the mitochondria results into membrane permeabilization, thereby resulting in efflux of cytochrome-c and other anti-apoptotic proteins such as HtrA2, Smac/Diablo. The released cytochrome-c binds an apoptotic protease-activating factor 1 (Apaf-1) monomer, leading to its oligomerization into a heptameric wheel-like structure called the apoptosome that later recruits and activates initiator procaspase 9. Active caspase-9 exposes the ATPF motif that binds to X-linked inhibitor of apoptosis protein (XIAP) preventing further activation. The mitochondrial protein Smac/DIABLO augments apoptosis by binding XIAP and reversing their grip on active caspase-9. Activated caspase-9 then cleaves and activates effector caspases 3, 7 to trigger apoptosis

proteins are categorized under three sub-groups: anti-apoptotic (BCL-2, BCL-W, BCL-xL, a-1 and MCL-1), pro-apoptotic effectors (BAX, BAK, and BOX) and pro-apoptotic BH3 only proteins (BID, BIM, BAD, etc.) as shown in Fig. 1.3. All of these proteins comprise at least one conserved **B**CL-2 **H**omology domain (BH) (Fig. 1.4). Proapoptotic BCL-2 family proteins can be further categorized based on

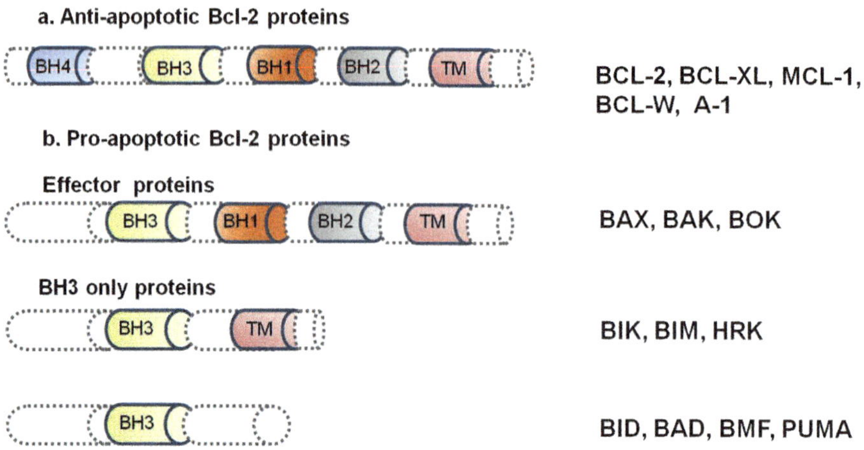

a. Anti-apoptotic Bcl-2 proteins

BH4 BH3 BH1 BH2 TM — BCL-2, BCL-XL, MCL-1, BCL-W, A-1

b. Pro-apoptotic Bcl-2 proteins

Effector proteins

BH3 BH1 BH2 TM — BAX, BAK, BOK

BH3 only proteins

BH3 TM — BIK, BIM, HRK

BH3 — BID, BAD, BMF, PUMA

Fig. 1.4 Domain organization of BCL-2 family of proteins. The B-cell lymphoma 2 (BCL-2) family of proteins are characterized by the presence of up to four regions of sequence homology, commonly known as BCL-2 homology (BH) domains. Based on the number of BH3 domain and their function, they are categorized into three sub-groups: anti-apoptotic, pro-apoptotic effector (the proteins that cause mitochondrial outer membrane permeabilization (MOMP)) and BH3 only (the proteins that transfer the apoptotic signal to the effectors). *BCL-2* B-cell lymphoma protein 2, *BCL-XL* BCL2 related protein, long, *MCL-1* myeloid leukemia cell differentiation protein, *BAX* BCL-2-associated X protein, *BAK* BCL-2 antagonist or killer, *BOK* BCL-2-related ovarian killer protein, *BIK* BCL-2-interacting killer, *BIM* BCL-2-interacting mediator of cell death, *HRK* harakiri, *BID* BH3-interacting domain death agonist, *BAD* BCL-2 antagonist of cell death, *BMF* BCL-2-modifying factor, *PUMA* p53 upregulated modulator of apoptosis, *TM* transmembrane. Based upon the number of BH3 domain and their function they are categorized into three sub-groups, anti-apoptotic (**a**), pro-apoptotic effector (the proteins that cause mitochondrial outer membrane permeabilization (MOMP)) and BH3 only (the proteins that transfer the apoptotic signal to the effectors)(**b**)

the number of ‘BH’ domains they possess. BAX (**B**CL-2 **a**ssociated **X** protein), BAK (**B**CL-2 **h**omologous **a**ntagonist **k**iller) and BOK (**B**CL-2 related **o**varian **k**iller) are structurally similar to BCL-2 with three ‘BH’ domains (BH1-3), whereas BID, BIM and BAD require a single ‘BH3’ domain for their functions (Table 1.1) [77, 80]. Many of these proteins such as BCL-2 and BCL-xL are regulated by p53 either through direct interaction or are transcriptionally activated such as BAX [81, 82]. While de-phosphorylation activates BAD [83], functional form of BID (**B**H3 **i**nteracting **d**omain) is obtained only after its cleavage to a truncated form (t-BID) by caspase-8 of the extrinsic pathway which exposes its BH3 domain [84, 85]. With myriads of pro- and antiapoptotic BCL-2 proteins being involved, an extremely precise balance between their expression levels is necessary during normal conditions as a slight tip of this balance is sufficient to set-off the intrinsic cell death pathway.

Moreover, apart from being a component of lysosomal autophagy, cathepsins, a group of proteases, are also involved in mitochondrial cell death pathway.

Table 1.1 Extrinsic and intrinsic pathway proteins and their alternate nomenclature

Protein name	Acronym	Alternate nomenclature	Pathway	Group
Apo2 ligand	Apo2L	TRAIL/TNFSF10	Extrinsic	Extracellular ligands
Apo3 ligand	Apo3L	TWEAK/TNFSF12/DR3LG	Extrinsic	
Fatty acid synthetase ligand	FasL	Fas ligand, TNFSF6, Apo1, apoptosis antigen ligand 1, CD95L, CD178, APT1LG1	Extrinsic	
Tumor necrosis factor alpha	TNF-α	TNF ligand, TNFA, cachectin	Extrinsic	
Death receptor 3	DR3	TNFRSF12, Apo3, WSL-1, TRAMP, LARD, DDR3	Extrinsic	Receptors
Death receptor 4	DR4	TNFRSF10A, TRAILR1, APO2, KILLER, ZTNFR9	Extrinsic	
Death receptor 5	DR5	TNFRS10B, TRAIL-R2, TRICK2, CHE1	Extrinsic	
Fatty acid synthetase receptor	FasR	Fas receptor, TNFRSF6, APT1, CD95	Extrinsic	
Tumor necrosis factor receptor 1	TNFR1	TNF receptor, TNFRSF1A, p55 TNFR, CD120a	Extrinsic	
Fas-associated death domain	FADD	MORT1	Extrinsic	Adaptor proteins
TNF receptor-associated death domain	TRADD	TNFRSF1A associated via death domain	Extrinsic	
Receptor-interacting protein	RIP	RIPK1	Extrinsic	
Cysteinyl aspartic acid protease-8	caspase-8	MACH-1, MCH5, ICE-LAP6, Mch6, Apaf-3	Extrinsic	Initiator caspase
FLICE-inhibitory protein	c-FLIP	Casper, I-FLICE, FLAME-1, CASH, CLARP, MRIT	Extrinsic	Decoy protein
B-cell lymphoma protein 2	Bcl-2	Apoptosis regulator Bcl-2	Intrinsic	Bcl-2 family members
BCL2 like 1	Bcl-x	BCL2 related protein	Intrinsic	
BCL2 related protein, long isoform	Bcl-XL	BCL2L protein, long form of Bcl-x	Intrinsic	
BCL2 related protein, short isoform	Bcl-XS	–	Intrinsic	
BCL2 like 2 protein	Bcl-w	Apoptosis regulator BclW	Intrinsic	
Myeloid leukemia cell differentiation protein	Mcl-1	Bcl-2-like protein 3, Bcl-2-related protein, EAT/mcl1 mcl1/EAT	Intrinsic	
BCL2 associated athanogene	BAG	BAG family molecular chaperone regulator	Intrinsic	

(continued)

Table 1.1 (continued)

Protein name	Acronym	Alternate nomenclature	Pathway	Group
BCL2 associated X protein	BAX	Apoptosis regulator BAX	Intrinsic	
BCL2 antagonist killer 1	BAK	BCL2L7, cell death inhibitor 1	Intrinsic	
BCL2 antagonist of cell death	BAD	BCL2 binding protein, BCL2L8, BCL2 binding component 6, BBC6, Bcl XL/Bcl-2 associated death promoter	Intrinsic	
BCL2 interacting protein	BIM	BCL2 like 11	Intrinsic	
B-cell lymphoma protein 10	Bcl-10	mE10, CARMEN, CLAP, CIPER	Intrinsic	
Bik-like killer protein	Blk B	B lymphoid tyrosine kinase, p55-BLK, PUMA/JFY1, p53 upregulated modulator of apoptosis	Intrinsic	
Apoptotic protease activating factor	Apaf-1	APAF1	Intrinsic	
Apoptosis inducing factor	AIF	Apaf-3/ICE-LAP6/Mch6	Intrinsic	
Inhibitor of apoptosis proteins	IAP	XIAP, API3, ILP, HILP, HIAP2, cIAP1, API1, MIHB, NFR2-TRAF signaling complex protein	Intrinsic	
High-temperature requirement	HtrA2/Omi	Omi stress regulated endoprotease, serine protease Omi protein A2	Intrinsic	IAP-antagonists
Second mitochondrial activator of caspases/ direct IAP binding protein with low PI	Smac/ DIABLO	–	Intrinsic	
Cysteinyl aspartic acid-protease-9	Caspase-9	ICE-LAP6, Mch6, Apaf-3	Intrinsic	Initiator caspase
Cysteinyl aspartic acid-protease-3	Caspase-3	CPP32, Yama, Apopain, SCA-1, LICE	–	Executioner caspases
Cysteinyl aspartic acid-protease-6	Caspase-6	Mch-2	–	
Cysteinyl aspartic acid-protease-7	Caspase-7	Mch-3, ICE-LAP-3, CMH-1	–	

Modified from Elmore [174]

Upon lysosomal destabilization due to membrane permeabilization, cathepsins are released into the cytosol where they trigger apoptosis by cleaving BID thus initiating degradation of antiapoptotic BCL-2 proteins.

2.2.1.1 The Gateway: Mitochondrial Membrane Permeabilization

The mitochondria accommodates a variety of pro- and antiapoptotic signals such as BCL-2 family of proteins, Ca^{2+} overload, **r**eactive **o**xygen **s**pecies (ROS) etc. The main event that triggers intrinsic apoptotic pathway is when the equilibrium between life-sustaining and destroying signals is tilted towards the latter leading to **m**itochondrial **o**uter **m**embrane **p**ermeabilization (MOMP). Under normal scenario, in healthy cells, the outer mitochondrial membrane allows molecules (such as small proteins and metabolites) upto $\sim$5 kDa to be released through protein channels. However, upon apoptotic signal, the mitochondrial membrane potential ($\Delta\Psi_m$) drops, thereby significantly increasing membrane permeability leading to release of larger proapoptotic molecules from mitochondrial inter-membrane space (IMS) into the cytosol. Currently, two different mechanisms of MOMP are in vogue, which although distinct, do overlap with each other.

According to one, permeabilization initiates at the mitochondrial **i**nner **m**embrane (IM). In healthy cells, the interface between the **i**nter **m**embrane **s**pace (IMS) and the mitochondrial matrix builds up the transmembrane potential ($\Delta\Psi_m$) which is tightly regulated by a flexible large multiprotein complex called **p**ermeability **t**ransition **p**ore (PTP) at the junction of **i**nner and **o**uter **m**embranes (IM/OM). During apoptotic signal, the pore suddenly expands allowing entry of water into the mitochondria thus leading to swelling and subsequent rupture of mitochondrial OM. Thus in this model, drop in $\Delta\Psi_m$ precedes membrane permeabilization [86–88]. On the other hand, the second hypothesis suggests $\Delta\Psi_m$ loss is a consequence of formation of large pores at the OM, releasing apoptotic factors in the cytosol [89, 90]. However, the precise mechanism of this complex coordinated process involving OM rupture, enlargement of PTP and release of mitochondrial factors in the cytosol is yet to be delineated and currently is one of the major foci of basic and translational research.

BCL-2 family members mainly BAX and BAK are the key players in mediating MOMP. Efficient execution of their functions relies upon their interaction with other family members, both pro- and anti apoptotic [91]. Once activated, BAX and BAK undergo significant conformational changes leading to their oligomerization and mitochondrial targeting of BAX [92–94]. Interaction of BAX with proapoptotic t-BID, further leads to its membrane insertion and liposome permeabilization thus supporting a hypothesis that emphasizes requirement of BH3-only proteins for BAX/BAK activation [95]. Although a substantial amount of information is available on BAX/BAK mediated activation of intrinsic apoptotic pathway, many critical questions still remain unanswered. For example, there are debatable reports on the level of oligomerization required for formation of functional BAX/BAK molecules. One of the models suggests formation of dimer chains which is driven by hydrophobic interactions between two molecules of either BAK or BAX. However, the minimum number of molecules required is still controversial with reports spanning a range from four to hundreds of these protein molecules [96–101].

Despite a lot of effort to obtain a deep insight into BAX/BAK mediated MOMP has been taken, the complexities in the mechanism itself makes it quite challenging to comprehend. Live cell imaging studies with GFP-labeled cytochrome-c demonstrate BAX/BAK mediated MOMP occurs within 5 min of its initiation [102]. Several independent studies in this field however led to different theories that describe how MOMP is formed, pores are generated and enlarged followed by release of IMS proteins. Based on structural similarities of BAX/BAK with bacterial pore forming toxins, one of the theories proposes that they are responsible for creating a channel for release of mitochondrial proteins post MOMP [103, 104]. These pores, which increase in size with time most probably due to oligomerization of BAX and BAK, are termed as **m**itochondrial **a**poptosis induced **c**hannels (MAC) [105]. An alternative to this theory is that BAX and BAK interact with lipid of the OM leading to membrane bending and hence formation of transient lipid pores for molecules to exit mitochondria [106–108].

Although the release of protein molecules cannot be regulated by membrane permeabilization, selectivity is either controlled by their interaction with or by remodeling of the inner mitochondrial membrane. The former hypothesis is strengthened by the fact that a mitochondrial inter membrane flavoprotein called **A**poptosis **I**nducing **F**actor (AIF) is released into the cytosol at a later apoptotic stage only after its cleavage, possibly by cytosolic protease calpain I [109]. AIF then migrates to nucleus where it fragments DNA and condenses nuclear chromatin [110]. However, on the other hand, literature suggests that cytochrome-c release is mediated by the conformational changes and dynamics at the inner mitochondrial membrane thus supporting the alternative hypothesis. It has been proposed that the IMS plasticity is primarily controlled by the mitochondrial cristae remodeling which alters its size and thus release of cytochrome-c. Two mitochondrial proteins have been implicated in the process viz. **o**ptic **a**trophy **p**rotein 1 (OPA1) and **p**resenilins-**a**ssociated **r**homboid **l**ike protein (PARL). Cleavage of OPA1 by PARL might lead to closure of cristae junctions thus preventing exit of cytochrome-c [111, 112]. However, the mechanism of release of protein molecules post-MOMP is still controversial with lots of contrasting observations [113, 114]. Detailed studies of IMS dynamics and kinetics of the protein molecules released will help in understanding the mechanism with greater precision.

2.2.1.2 Induction of Apoptosis: Release of Mitochondrial Macromolecules

Upon MOMP, there is release of two classes of apoptotic molecules that are involved either in caspase-dependent or independent pathway. Firstly, cytochrome-c is released in the cytosol leading to **m**itochondrial **p**ermeability **t**ransition (MPT) and hence subsequent drop in membrane potential. An alternate pathway demonstrates release of cytochrome-c via **v**oltage-**d**ependent **a**nion **c**hannel (VDAC) which is opened up by BCL-2 family proteins [115]. Upon release in the cytosol, cytochrome-c binds to the WD-40 repeat domain [116] of **a**poptotic **p**rotease **a**ctivating **f**actor-1 protein (Apaf-1). Prior to binding cytochrome-c, Apaf-1 exists in the cytoplasm as an inactive coiled monomer comprising an N-terminal **ca**spase **r**ecruitment **d**omain (CARD), ATPase domain and several WD-40 repeat domains. Binding of cytochrome-c in presence of ATP induces significant conformational changes in Apaf-1 thus leading to the formation of a functional heptameric wheel-like complex known as apoptosome [117, 118]. Although high resolution structural data is not available, cryo-electron microscopy at 9.5 Å resolution unambiguously demonstrates the positions of different domains of Apaf-1 and relative orientation of the two proteins in the complex [119]. Apoptosome formation leads to complete 'unlocking' of Apaf-1 structure which then recruits and activates initiator procaspase-9 via homotypic CARD-CARD interactions [114]. Activation of caspase-9 is followed by cleavage of executioner procaspases (caspase-3, -6 and -7) thus setting off the caspase cascade and ultimately cell death (Fig. 1.3) [120]. However, whether procaspase-9 activation by apoptosome occurs prior to its dimerization is still debatable and further studies are required to understand the exact mechanism of its action [121, 122]. This complex process is perhaps evolutionarily conserved to protect cells against accidental death upon small amount of cytochrome-c release in the cytoplasm [118].

The second set of proteins released from the mitochondria upon apoptotic signal is the ones that relieve the inhibitory effect of 'Inhibitor of Apoptosis Proteins' on caspases through competitive binding [123, 124]. Mature Smac/DIABLO and HtrA2/Omi that are released from the mitochondria have an N-terminal IAP-binding tetrapeptide motif (AVPI and AVPS respectively). Although not much information is available, another enzyme, GSPT1/eRF3 (eukaryotic class II polypeptide chain release factor) binds survivin through a similar conserved motif AKPF and activates caspases [125, 126]. These motifs have been found to bind BIR3 domains of their targets such as XIAP and survivin thereby releasing their inhibition on active caspases.

2.2.2 ER Mediated Intrinsic Apoptotic Pathway

Though less prominent, apoptosis through intrinsic pathway also involves endoplasmic reticulum (ER). Disruption in ER homeostasis due to stress factors including excessive Ca^{2+} intake, oxidative stress and chemical toxicity lead to 'unfolded protein response' which reduces further protein synthesis and increases expression of chaperones. Apoptosis or necrosis occurs when all these attempts to neutralize the effect of the induced stress fail [127]. In certain disease conditions such as Alzheimer's and stroke, release of a small amount of cytochrome-c from mitochondria and its subsequent entry into the ER and binding to $InsP_3$ receptor (**ino**sitol trisphosphate **r**eceptor that acts as a calcium channel) induces calcium release from ER [128]. This relays a signal to the mitochondria resulting in huge exit of cytochrome-c to the cytoplasm thus creating a positive feedback loop, which is followed by dramatic caspase activation. Some other novel molecules such as CHO, Valosin containing protein (ER stress regulator), Bap31(ER protein involved in ER export of transmembrane proteins) have also been implicated in ER although the mechanism of apoptotic induction by these molecules require further studies [129–132].

Caspase-12 is the only caspase so far that has been associated with ER mediated cell death. Upon activation, it is released in the cytosol and cleaves procaspase-9. However, caspase-12 has only been identified in mouse and rats in its functional form and existence of their human counterpart is still controversial [133]. ER stress mediated apoptosis is predominantly found in neurodegenerative diseases including Alzheimer's, Huntington's, Parkinson's and prion protein related diseases [134].

2.3 Cross Talk Between Extrinsic and Intrinsic Pathways of Cell Death

The extrinsic or 'death receptor mediated' and the intrinsic or mitochondrial apoptotic pathways have different triggering factors that originate in two different cellular compartments. However, although they mostly involve distinct molecules with functions independent of each other, often the common factors do intersect at various junctions with final convergence at the executioner caspases [135]. In some cell types, where activation of proteins related to extrinsic pathway is insufficient, the mitochondrial pathway is triggered by t-BID promoting initiation of the intrinsic pathway [136]. Although it was earlier hypothesized that a weak extrinsic apoptotic pathway in a cell is compensated with enhanced activity of caspase-9 and -3, later it was observed that cells deficient in both the proteins are also sensitive to apoptosis. This apparent anomaly was solved after the discoveries of Smac/DIABLO and HtrA2/Omi which act as 'missing links' between the intrinsic and extrinsic pathways. Crosstalk between these pathways through inhibition of IAPs by Smac and HtrA2 with subsequent activation of caspase-9 and -3 promotes apoptosis. Caspase-6 that requires caspase-9 for its maturation further activates caspase-8 of the extrinsic pathway thus fostering the interaction between these two pathways [137]. Also, upon DNA damage, p53 up-regulates proteins involved in both the pathways such as BAX, Puma, FasL and killer/DR5 [138]. Understanding the mechanism of crosstalk and the molecules involved is important as it will help develop common therapeutic strategies against different diseases that are caused by dysregulation of either or both apoptotic pathways.

2.4 Caspase Independent Cell Death (CICD)

Apart from the two major pathways, a comparatively slower process which does not involve caspases also exists. CICD is initiated when apoptotic signals fail to trigger caspase activation. CICD occurs generally with intrinsic signals post MOMP, while extrinsic pathway leads to a kind of cell death called 'necroptosis' [139]. Although CICD follows some classical apoptotic traits, it exhibits some phenotypic variability as well based on the stimulus and cell type. In CICD, typical chromatin condensation is replaced by some other distinct features such as cytoplasmic vacuolization and nuclear condensation on the periphery [140]. Interestingly, although present in higher order eukaryotes, CICD has not been clearly identified in two model systems, *Caenorhabditis elegans* and *Drosophila melanogaster* which might be due to lack of MOMP in these organisms [141, 142]. In higher order organisms, it was observed that cells lacking Apaf-1 or overexpressing BCL-2 (blocks MOMP) do not undergo CICD [143] suggesting that MOMP might be a prerequisite for this alternative apoptotic pathway. However, it was observed that CICD can only partly compensate for caspase dependent cell death as mice deficient in Apaf-1, caspase-9 and a non-functional cytochrome-c mutant lead to severe lethality. However, this has been found to be comparatively less intense but with phenotypic abnormalities when downstream caspases were inactivated [144–147]. These aberrations in growth might be due to slower pace at which CICD occurs [148]. In physiological setting, neurons and cardiomyocytes that have lower levels of Apaf-1 expression are capable of undergoing CICD [149, 150]. Although not much details of this form of cell death is available till date, its slower rate of occurrence might be utilized in understanding its mechanism by following the fates of fluorescently labelled proteins involved in this pathway [151].

2.4.1 Necroptosis: Following the Death Receptor Pathway

Activation of the TNF receptor also leads to rapid cell death by CICD [152]. This might be due to presence of a significantly higher level of reactive oxygen species produced by phospholipase A_2 (PLA_2). PLA_2, which is otherwise cleaved and inactivated by caspases, is unable to initiate CICD pathway under normal conditions when caspases are present. It has been demonstrated by Holler and co-workers [153] that adaptor molecule FADD is also associated with necroptosis post Fas activation. An intracellular kinase RIP-1 has been found to be involved in signalling necroptosis [52]. It has been assumed that it performs this function by abrogating the interaction between adenine nucleotide translocase (ANT) and cyclophilin D thus leading to deregulation of mitochondrial functions [154].

2.4.2 Mitochondrial CICD

Although necroptosis is limited to certain conditions and cell types, CICD associated with mitochondria is more universal. Post MOMP, under normal conditions, caspase-3 cleaves p75 subunit of mitochondrial respiratory subunit I and as a consequence dysregulation of electron transport, loss of membrane potential and gradual decline in ATP synthesis occur [155]. However, under caspase inhibitory conditions, although no p75 cleavage takes place, cell death is observed which might be due to loss of ATP generation post MOMP after slow release of respiratory chain components from mitochondria [156]. This hypothesis does not hold true for transformed cells where ATP generation is dependent on glycolysis [157]. Thus, the mechanistic details of MOMP mediated CICD needs further elucidation.

2.4.3 Non-caspase Proteases in CICD

Granzymes belong to a family of serine proteases that are produced by **c**ytotoxic **T** **l**ymphocyte (CTL) and **n**atural **k**iller (NK) cells [158]. In humans there are 5 granzymes (A, B, H, K and M), amongst which the most well studied ones are proapoptotic Granzyme A and B. It has been found that Granzyme-A (GrA) mediates apoptosis in a caspase-independent manner. On its delivery to the target cell cytosol through either Ca^{2+}-dependent or perforin-mediated pores, GrA triggers a pathway primarily characterized by formation of single-stranded DNA breaks and appearance of apoptotic morphology. The endonuclease, GAAD (**G**rA-**a**ctivated **D**NAse) is involved in the formation of these DNA strand breaks. GADD activity is inhibited by its specific inhibitor assembly, commonly known as the SET complex which contains an inhibitor of **p**rotein **p**hosphatase 2A (pp32), nucleosome assembly protein SET, high-mobility group protein 2 (HMG2) and

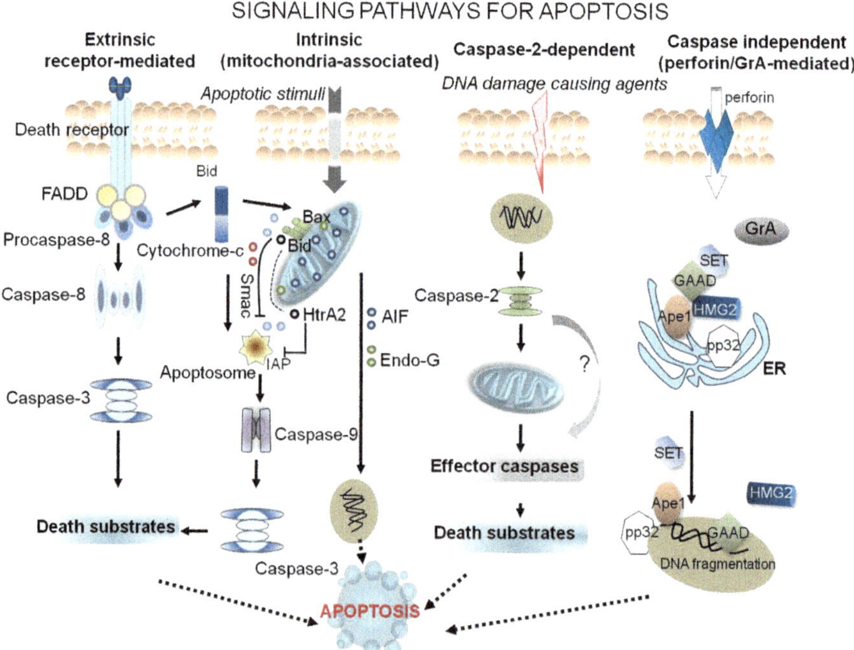

Fig. 1.5 Signaling pathways that regulate cell death in mammalian cells (Adapted from Orrenius et al. [159])

apurinic/apyrimidinc endonuclease 1(Ape1). Granzyme-A cleaves SET, HMG2 and ApeI, but not pp32 to release and activate GADD, which later translocates to the nucleus and creates DNA strand nicks (Fig. 1.5). Like its predecessor, granzyme-B is also involved in alternate cell death pathway. This serine protease has the unusual property of cleaving substrates at aspartic acid residues [160]. Granzyme-B has been found to cleave many caspases (-3, -7, -8 and -10) directly suggesting involvement of this protease in CTL-induced killing [161–168].

Another prominent non-caspase protease in the apoptotic pathway is calpain which is a cysteine, non-lysosomal endoprotease requiring calcium for its activation. There are two isoforms, Calpain I and II which differ in their requirement of Ca^{+2} for apoptosis induction [169, 170]. Calpains are 110 kDa heterodimers that are implicated primarily in neuronal apoptosis where it is triggered as a consequence of increase in intracellular Ca^{+2} levels. Autocatalytic processing activates calpains which in turn cleaves several important cytoskeletal (α-fodrin) as well as nucleoskeletal proteins (lamins A and B). Apart from these, calpains proteolyze a variety of neurofilaments, ion channels, growth factors and enzymes thus leading to complete disruption of cellular integrity and hence cell death [171–173].

Several mitochondrial IMS proteins such as HtrA2/Omi, EndoG etc., initiate CICD after their release from the mitochondria. Upon release in the cytosol, it translocates to nucleus where it is implicated in DNA degradation. HtrA2, a serine protease, can also induce apoptosis independently by cleaving important cellular proteins such as anti-apoptototic protein Pea-15, cytoskeletal associated proteins viz. α-, β-tubulin, vimentin and proteins related to translational machinery [175].

3 Diseases Associated with Deregulation of Apoptosis, and Therapeutic Possibilities

Both the major apoptotic pathways intersect downstream where executioner caspases come into play. So, although a variety of diseases are associated exclusively with a particular pathway, overlap of pathogenesis due to this convergence would also automatically exist. The way a cell builds its intricate apoptotic machinery, similarly, it adopts ways to regulate its normal functions by nurturing a precise balance between different agonists and antagonists thus maintaining homeostasis. A little imbalance therefore can have detrimental effect on the normal functioning of a healthy cell. For example, in the extrinsic pathway, death decoy receptors 1 and 2 (DcR1 and 2) that mimic TNF-R lacking DED domain antagonize the proapoptotic functions of the transmembrane receptor [176]. Also, later in the apoptotic pathway, decoy protein FLIP stalls apoptosis by preferentially binding procaspase-8 instead of the adaptor protein FADD [177]. The intrinsic pathway is tightly regulated by antiapoptotic proteins from BCL-2 family where they prevent release of cytochrome-c to the cytosol from mitochondria [178].

Restoring normal apoptosis in different disease conditions is one of the major challenges of current biomedical research. Diseases either associated with or that emanate from this imbalance of cellular homeostasis vary from neurodegenerative disorders to cancer. Enhanced level of apoptosis might cause neurodegeneration such as in Alzheimer's or Parkinson's disease (AD and PD) [179–181]. Moreover, increase in the rate of apoptosis could result in loss of neurons in the brain due to restricted blood flow as in ischemic stroke. Similarly, AIDS is associated with sharp decline of helper T cells as they commit suicide [182, 183]. The other extreme condition, where due to some inhibitory effect there is less or no apoptosis, leads to accumulation of 'un-dead' cells resulting in tumorigenesis or cancer. Mutations, altered expressions, presence or absence of several apoptosis related genes such as p53 tumor–suppressor, DNA methylation and alteration in mRNA stability are some major hallmarks of cancer. Mutations in p53 gene can result in multidrug resistance and interestingly reintroduction of the wildtype p53 can reverse the disease condition and bring about chemosensitivity [184–186]. In all these cases, therapeutic interventions employing molecules involved in the apoptotic pathways although may seem promising are sometimes limited by potential side effects. For example, caspase and calpain inhibitors have been identified as targets for AD and PD. However, risk of using protease inhibitors for long-term treatment, lack of proper delivery system and complexity of biochemical pathways demand proper attention [187–190]. In neurodegenerative disorders, application of caspase inhibitors is limited by the side-effects that they may cause due to their interference with other cysteine proteases [191]. Moreover, caspase inhibitors might not have a long-term effect on rescued neurons since AIF and other pro-apoptotic factors might also be released [192–194]. Similarly, calpain activation leads to neurofibrillary pathology, due to aberrant APP (**a**myloid **p**recursor **p**rotein) processing, loss of synapse and subsequent cell death [195, 196]. Despite this the calpain inhibitors such as calpeptin, MDL-28170 and PD150606 that can prevent neuronal death and revive cognitive function in AD models are still not in clinical trials [109, 170, 197–200]. Other than memantine (uncompetitive NMDA receptor antagonist) [201–204], a few antiapoptotic drugs that have entered phase I and II trials did not show encouraging results most probably due to redundancy in apoptotic pathways and absence of an ideal animal model system which can mimic the intricacies of human neuronal system. However, early diagnosis and combinatory therapeutic intervention might provide better results in patients suffering from neurodegenerative disorders.

Caspase activators (both peptidomimetics and small molecules) are being designed to specifically target cancer cells. Inducible caspases such as caspase-9 under the control of prostate-specific promoter was found to specifically target prostate cancer cells [204]. Gene therapeutic targeting of caspases to cancer cells has been another popular approach in combating tumorigenesis [205]. Protein engineering tools have been employed to design several caspase chimeras such as immunocaspase-3 and -6 comprising single-chain anti-erbB2/HER2 antibody (e23sFv) and the translocation domain of *Pseudomonas* exotoxin-A fused to an active caspase [206, 207]. These chimeras have shown promising results in breast, ovarian, endometrial, gastric, bladder, prostate, or lung cancer and are currently under evaluation (specifically caspase-3) for clinical intervention. Disruption of inter and intra-molecular bonds at the intersubunit linker of dormant procaspase-3 through design of peptidomimetics such as RGD tripeptide has potential in anticancer therapy [208]. Possibilities of other proteases such as proapoptotic HtrA2 in targeting cancer cells is also being explored [209]. However, although the possibilities seem boundless and the current research scenario looks promising, certain barriers need to be crossed (such as lowering the IC_{50} of current target molecules and enhancing the gene delivery system) before bringing these therapeutic strategies to clinics.

References

1. Duque-Parra J (2005) Note on the origin and history of the term "apoptosis". Anat Rec B New Anat 283:2–4
2. Raff M (1998) Cell suicide for beginners. Nature 396:119–122
3. Lawen A (2003) Apoptosis-an introduction. Bioessays 25:888–896
4. Ozoren N, El-Deiry W (2003) Cell surface Death Receptor signaling in normal and cancer cells. Semin Cancer Biol 13:135–147
5. Thorburn A (2004) Death receptor-induced cell killing. Cell Signal 16:139–144
6. Peter M, Krammer P (1998) Mechanisms of CD95 (APO-1/Fas)-mediated apoptosis. Curr Opin Immunol 10:545–551
7. Strasser A, O'Connor L, Dixit V (2000) Apoptosis signaling. Annu Rev Biochem 69:217–245
8. Degli Esposti M (1999) To die or not to die—the quest of the TRAIL receptors. J Leukoc Biol 65:535–542
9. Abe K, Kurakin A, Mohseni-Maybodi M, Kay B, Khosravi-Far R (2000) The complexity of TNFrelated apoptosis-inducing ligand. Ann N Y Acad Sci 926:52–63
10. Sharp A, Heazell A, Crocker I, Mor G (2010) Placental apoptosis in health and disease. Am J Reprod Immunol 64:159–169
11. Zornig M, Hueber A, Baum W, Evan G (2001) Apoptosis regulators and their role in tumorigenesis. Biochim Biophys Acta 1551:F1–F37
12. Daniel P, Wieder T, Sturm I, Schulze-Osthoff K (2001) The kiss of death: promises and failures of death receptors and ligands in cancer therapy. Leukemia 15: 1022–1032
13. Green D, Evan G (2002) A matter of life and death. Cancer Cell 1:19–30
14. Thompson C (1995) Apoptosis in the pathogenesis and treatment of disease. Science 267:1456–1462
15. Sheikh M, Huang Y (2004) Death receptors as targets of cancer therapeutics. Curr Cancer Drug Targets 4:97–104
16. Burns T, el-Deiry W (2003) Cell death signaling in maligancy. Cancer Treat Res 115:319–343
17. Kerr JFR, Wyllie AH, Currie AR (1972) Apoptosis: a basic biological phenomenon with wide-ranging implications in tissue kinetics. Br J Cancer 26:239–257
18. Philchenkov A (2004) Caspases: potential targets for regulating cell death. J Cell Mol Med 8:432–444
19. Lushnikov EF, Zagrebin VM (1987) Cellular apoptosis: its morphology, biological role and the mechanisms of its development. Arkh Patol 49:84–89
20. Wyllie AH, Kerr JF, Currie AR (1980) Cell death: the significance of apoptosis. Int Rev Cytol 68:251–306
21. Jin Z, El-Deiry WS (2005) Overview of cell death signaling pathways. Cancer Biol Ther 4:139–163
22. Schultz DR, Harrington WJ Jr (2003) Apoptosis: programmed cell death at a molecular level. Semin Arthritis Rheum 32:345–369

23. Askkenazi A, Dixit V (1999) Apoptosis control by death and decoy receptors. Curr Opin Cell Biol 11:255–260
24. Naismith JH, Sprang SR (1998) Modularity in the TNF-receptor family. Trends Biochem Sci 23:74–79
25. Schneider P, Tschopp J (2000) Apoptosis induced by death receptors. Pharm Acta Helv 74:281–286
26. Salvesen GS, Dixit VM (1999) Caspase activation: the induced-proximity model. Proc Natl Acad Sci U S A 96:10964–10967
27. Logue S, Martin S (2008) Caspase activation cascades in apoptosis. Biochem Soc Trans 36:1–9
28. Ferreira K, Clemens K, MacNelly S, Neubert K, Haber A, Bogyo M, Timmer J, Borner C (2012) Caspase-3 feeds back on caspase-8, Bid and XIAP in type I Fas signaling in primary mouse hepatocytes. Apoptosis 17:503–515
29. Cohen GM (1997) Caspases: the executioners of apoptosis. Biochem J 326(Pt 1):1–16
30. Safa AR (2012) c-FLIP, a master anti-apoptotic regulator. Exp Oncol 34:176–184
31. Silke J, Meier P (2013) Inhibitor of apoptosis (IAP) proteins-modulators of cell death and inflammation. Cold Spring Harb Perspect Biol 5(2):a008730
32. Tartaglia L, Ayres T, Wong G, Goeddel D (1993) A novel domain within the 55 kd TNF receptor signals cell death. Cell 74:845–853
33. Hsu H, Xiong J, Goeddel D (1995) The TNF receptor 1-associated protein TRADD signals cell death and NF-kappa B activation. Cell 81:495–504
34. Chan FK (2007) Three is better than one: pre-ligand receptor assembly in the regulation of TNF receptor signaling. Cytokine 37:101–107
35. Chan FK, Chun HJ, Zheng L, Siegel RM, Bui KL, Lenardo MJ (2000) A domain in TNF receptors that mediates ligand-independent receptor assembly and signaling. Science 288:2351–2354
36. Jiang Y, Woronicz J, Liu W, Goeddel D (1999) Prevention of constitutive TNF receptor 1 signaling by silencer of death domains. Science 283:543–546
37. Di Pietro R, Zauli G (2004) Emerging non-apoptotic functions of tumor necrosis factor-related apoptosis-inducing ligand (TRAIL)/Apo2L. J Cell Physiol 201:331–340
38. Nagata S (1999) FAS ligand-induced apoptosis. Annu Rev Genet 33:29–55
39. Sessler T, Healy S, Samali A, Szegezdi E (2013) Structural determinants of DISC function: new insights into death receptor-mediated apoptosis signalling. Pharmacol Ther 140:186–199
40. Yan Q, McDonald JM, Zhou T, Song Y (2013) Structural insight for the roles of fas death domain binding to FADD and oligomerization degree of the Fas-FADD complex in the death-inducing signaling complex formation: a computational study. Proteins 81:377–385
41. Wang L, Yang JK, Kabaleeswaran V, Rice AJ, Cruz AC, Park AY, Yin Q, Damko E, Jang SB, Raunser S, Robinson CV, Siegel RM, Walz T, Wu H (2010) The Fas-FADD death domain complex structure reveals the basis of DISC assembly and disease mutations. Nat Struct Mol Biol 17:1324–1329
42. Stennicke H, Salvesen G (2000) Caspases – controlling intracellular signals by protease zymogen activation. Biochim Biophys Acta 1477:299–306
43. Wang J, Chun HJ, Wong W, Spencer DM, Lenardo MJ (2001) Caspase-10 is an initiator caspase in death receptor signaling. Proc Natl Acad Sci U S A 98:13884–13888
44. Krueger A, Schmitz I, Baumann S, Krammer P, Kirchhoff S (2001) Cellular flice-inhibitory protein splice variants inhibit different steps of caspase-8 activation at the CD95 death inducing signaling complex. J Biol Chem 276:20633–20640
45. Jin Z, Mcdonald IE, Dicker D, El-Deiry W (2004) Deficient tumor necrosis factor-related apoptosis-inducing ligand (Trail) death receptor transport to the cell surface in human colon cancer cells selected for resistance to trail-induced apoptosis. J Biol Chem 279:35829–35839
46. Yeh W, Itie A, Elia A, Ng M, Shu H, Wakeham A, Mirtsos C, Suzuki N, Bonnard M, Goeddel D, Mak T (2000) Requirement for casper (C-Flip) in regulation of death receptor-induced apoptosis and embryonic development. Immunity 12:633–642

47. Chang D, Xing Z, Pan Y, Algeciras-Schimnich A, Barnhart B, Yaish-Ohad S, Peter M, Yang X (2002) c-FLIPL is a dual function regulator for caspase-8 activation and CD95-mediated apoptosis. EMBO J 21:3704–3714
48. Micheau O, Thome M, Schneider P, Holler N, Tschopp J, Nicholson DW, Briand C, Grutter M (2002) The long form of FLIP is an activator of caspase-8 at the Fas death-inducing signaling complex. J Biol Chem 277:45162–45171
49. Boatright K, Deis C, Denault J-B, Sutherlin D, Salvesen G (2004) Activation of caspases-8 and -10 by FLIPL. Biochem J 382:651–657
50. Chen G, Goeddel DV (2002) TNF-R1 signaling: a beautiful pathway. Science 296:1634–1635
51. Wang E, Marcotte R, Petroulakis E (1999) Signaling pathway for apoptosis: a racetrack for life or death. J Cell Biochem Suppl 32–33:95–102
52. Kelliher MA, Grimm S, Ishida Y, Kuo F, Stanger BZ, Leder P (1998) The death domain kinase RIP mediates the TNF-induced NF-kappaB signal. Immunity 8:297–303
53. Stanger B, Leder P, Lee T, Kim E, Seed B (1995) RIP: a novel protein containing a death domain that interacts with FAS/Apo-1 (CD95) in yeast and causes cell death. Cell 81:513–523
54. Pitti R, Marsters S, Ruppert S, Donahue C, Moore A, Ashkenaz IA (1996) Induction of apoptosis by Apo-2 ligand, a new member of the tumor necrosis factor cytokine family. J Biol Chem 271:12687–12690
55. Wiley SR, Schooley K, Smolak PJ, Din WS, Huang CP, Nicholl JK, Sutherland GR, Smith TD, Rauch C, Smith CA et al (1995) Identification and characterization of a new member of the TNF family that induces apoptosis. Immunity 3:673–682
56. Thomas WD, Hersey P (1998) TNF-related apoptosis-inducing ligand (TRAIL) induces apoptosis in Fas ligand-resistant melanoma cells and mediates CD4 T cell killing of target cells. J Immunol 161:2195–2200
57. Smyth MJ, Cretney E, Takeda K, Wiltrout RH, Sedger LM, Kayagaki N, Yagita H, Okumura K (2001) Tumor necrosis factor-related apoptosis-inducing ligand (TRAIL) contributes to interferon gamma-dependent natural killer cell protection from tumor metastasis. J Exp Med 193:661–670
58. Takeda K, Hayakawa Y, Smyth MJ, Kayagaki N, Yamaguchi N, Kakuta S, Iwakura Y, Yagita H, Okumura K (2001) Involvement of tumor necrosis factor-related apoptosis-inducing ligand in surveillance of tumor metastasis by liver natural killer cells. Nat Med 7:94–100
59. Cretney E, Takeda K, Yagita H, Glaccum M, Peschon JJ, Smyth MJ (2002) Increased susceptibility to tumor initiation and metastasis in TNF-related apoptosis-inducing ligand-deficient mice. J Immunol 168:1356–1361
60. Pan G, O'Rourke K, Chinnaiyan AM, Gentz R, Ebner R, Ni J, Dixit VM (1997) The receptor for the cytotoxic ligand TRAIL. Science 276:111–113
61. Walczak H, Degli-Esposti MA, Johnson RS, Smolak PJ, Waugh JY, Boiani N, Timour MS, Gerhart MJ, Schooley KA, Smith CA, Goodwin RG, Rauch CT (1997) TRAIL-R2: a novel apoptosis-mediating receptor for TRAIL. EMBO J 16:5386–5397
62. Wu GS, Burns TF, McDonald ER 3rd, Jiang W, Meng R, Krantz ID, Kao G, Gan DD, Zhou JY, Muschel R, Hamilton SR, Spinner NB, Markowitz S, Wu G, el-Deiry WS (1997) KILLER/DR5 is a DNA damage-inducible p53-regulated death receptor gene. Nat Genet 17:141–143
63. Sheridan JP, Marsters SA, Pitti RM, Gurney A, Skubatch M, Baldwin D, Ramakrishnan L, Gray CL, Baker K, Wood WI, Goddard AD, Godowski P, Ashkenazi A (1997) Control of TRAIL-induced apoptosis by a family of signaling and decoy receptors. Science 277:818–821
64. Pan G, Ni J, Wei YF, Yu G, Gentz R, Dixit VM (1997) An antagonist decoy receptor and a death domain-containing receptor for TRAIL. Science 277:815–818
65. Emery JG, McDonnell P, Burke MB, Deen KC, Lyn S, Silverman C, Dul E, Appelbaum ER, Eichman C, DiPrinzio R, Dodds RA, James IE, Rosenberg M, Lee JC, Young PR (1998) Osteoprotegerin is a receptor for the cytotoxic ligand TRAIL. J Biol Chem 273:14363–14367

66. Ashkenazi A (2002) Targeting death and decoy receptors of the tumour-necrosis factor superfamily. Nat Rev Cancer 2:420–430
67. Wagner KW, Punnoose EA, Januario T, Lawrence DA, Pitti RM, Lancaster K, Lee D, von Goetz M, Yee SF, Totpal K, Huw L, Katta V, Cavet G, Hymowitz SG, Amler L, Ashkenazi A (2007) Death-receptor O-glycosylation controls tumor-cell sensitivity to the proapoptotic ligand Apo2L/TRAIL. Nat Med 13:1070–1077
68. Deng Y, Lin Y, Wu X (2002) TRAIL-induced apoptosis requires Bax-dependent mitochondrial release of Smac/DIABLO. Genes Dev 16:33–45
69. Verhagen AM, Silke J, Ekert PG, Pakusch M, Kaufmann H, Connolly LM, Day CL, Tikoo A, Burke R, Wrobel C, Moritz RL, Simpson RJ, Vaux DL (2002) HtrA2 promotes cell death through its serine protease activity and its ability to antagonize inhibitor of apoptosis proteins. J Biol Chem 277:445–454
70. Cook AL, Frydenberg M, Haynes JM (2002) Protein kinase G activation of K(ATP) channels in human-cultured prostatic stromal cells. Cell Signal 14:1023–1029
71. Ashkenazi A, Pai RC, Fong S, Leung S, Lawrence DA, Marsters SA, Blackie C, Chang L, McMurtrey AE, Hebert A, DeForge L, Koumenis IL, Lewis D, Harris L, Bussiere J, Koeppen H, Shahrokh Z, Schwall RH (1999) Safety and antitumor activity of recombinant soluble Apo2 ligand. J Clin Invest 104:155–162
72. Zhang XD, Franco A, Myers K, Gray C, Nguyen T, Hersey P (1999) Relation of TNF-related apoptosis-inducing ligand (TRAIL) receptor and FLICE-inhibitory protein expression to TRAIL-induced apoptosis of melanoma. Cancer Res 59:2747–2753
73. Kim JH, Ajaz M, Lokshin A, Lee YJ (2003) Role of antiapoptotic proteins in tumor necrosis factor-related apoptosis-inducing ligand and cisplatin-augmented apoptosis. Clin Cancer Res 9:3134–3141
74. Li L, Thomas RM, Suzuki H, De Brabander JK, Wang X, Harran PG (2004) A small molecule Smac mimic potentiates TRAIL- and TNFalpha-mediated cell death. Science 305: 1471–1474
75. Shankar S, Srivastava RK (2004) Enhancement of therapeutic potential of TRAIL by cancer chemotherapy and irradiation: mechanisms and clinical implications. Drug Resist Updat 7:139–156
76. Kroemer G (2003) Mitochondrial control of apoptosis: an introduction. Biochem Biophys Res Commun 304:433–435
77. Fesik SW (2000) Insights into programmed cell death through structural biology. Cell 103:273–282
78. Hengartner MO (2000) The biochemistry of apoptosis. Nature 407:770–776
79. Zong WX, Lindsten T, Ross AJ, MacGregor GR, Thompson CB (2001) BH3-only proteins that bind pro-survival Bcl-2 family members fail to induce apoptosis in the absence of Bax and Bak. Genes Dev 15:1481–1486
80. Cheng EH, Wei MC, Weiler S, Flavell RA, Mak TW, Lindsten T, Korsmeyer SJ (2001) BCL-2, BCL-X(L) sequester BH3 domain-only molecules preventing BAX- and BAK-mediated mitochondrial apoptosis. Mol Cell 8:705–711
81. Schuler M, Green DR (2001) Mechanisms of p53-dependent apoptosis. Biochem Soc Trans 29:684–688
82. Chipuk JE, Green DR (2003) p53's believe it or not: lessons on transcription-independent death. J Clin Immunol 23:355–361
83. Datta SR, Dudek H, Tao X, Masters S, Fu H, Gotoh Y, Greenberg ME (1997) Akt phosphorylation of BAD couples survival signals to the cell-intrinsic death machinery. Cell 91:231–241
84. Li H, Zhu H, Xu CJ, Yuan J (1998) Cleavage of BID by caspase 8 mediates the mitochondrial damage in the Fas pathway of apoptosis. Cell 94:491–501
85. Luo X, Budihardjo I, Zou H, Slaughter C, Wang X (1998) Bid, a Bcl2 interacting protein, mediates cytochrome c release from mitochondria in response to activation of cell surface death receptors. Cell 94:481–490

86. Zamzami N, Marchetti P, Castedo M, Hirsch T, Susin SA, Masse B, Kroemer G (1996) Inhibitors of permeability transition interfere with the disruption of the mitochondrial transmembrane potential during apoptosis. FEBS Lett 384:53–57
87. Waring P, Beaver J (1996) Cyclosporin A rescues thymocytes from apoptosis induced by very low concentrations of thapsigargin: effects on mitochondrial function. Exp Cell Res 227:264–276
88. Scarlett JL, Murphy MP (1997) Release of apoptogenic proteins from the mitochondrial intermembrane space during the mitochondrial permeability transition. FEBS Lett 418:282–286
89. Ghiotto F, Fais F, Bruno S (2010) BH3-only proteins: the death-puppeteer's wires. Cytometry A 77:11–21
90. De Giorgi F, Lartigue L, Bauer MK, Schubert A, Grimm S, Hanson GT, Remington SJ, Youle RJ, Ichas F (2002) The permeability transition pore signals apoptosis by directing Bax translocation and multimerization. FASEB J 16:607–609
91. Tait SW, Green DR (2010) Mitochondria and cell death: outer membrane permeabilization and beyond. Nat Rev Mol Cell Biol 11:621–632
92. Eskes R, Desagher S, Antonsson B, Martinou JC (2000) Bid induces the oligomerization and insertion of Bax into the outer mitochondrial membrane. Mol Cell Biol 20:929–935
93. Wei MC, Lindsten T, Mootha VK, Weiler S, Gross A, Ashiya M, Thompson CB, Korsmeyer SJ (2000) tBID, a membrane-targeted death ligand, oligomerizes BAK to release cytochrome c. Genes Dev 14:2060–2071
94. Hsu YT, Wolter KG, Youle RJ (1997) Cytosol-to-membrane redistribution of Bax and Bcl-X(L) during apoptosis. Proc Natl Acad Sci U S A 94:3668–3672
95. Lovell JF, Billen LP, Bindner S, Shamas-Din A, Fradin C, Leber B, Andrews DW (2008) Membrane binding by tBid initiates an ordered series of events culminating in membrane permeabilization by Bax. Cell 135:1074–1084
96. George NM, Evans JJ, Luo X (2007) A three-helix homo-oligomerization domain containing BH3 and BH1 is responsible for the apoptotic activity of Bax. Genes Dev 21:1937–1948
97. Dewson G, Kratina T, Sim HW, Puthalakath H, Adams JM, Colman PM, Kluck RM (2008) To trigger apoptosis, Bak exposes its BH3 domain and homodimerizes via BH3: groove interactions. Mol Cell 30:369–380
98. Dewson G, Kratina T, Czabotar P, Day CL, Adams JM, Kluck RM (2009) Bak activation for apoptosis involves oligomerization of dimers via their alpha6 helices. Mol Cell 36:696–703
99. Saito M, Korsmeyer SJ, Schlesinger PH (2000) BAX-dependent transport of cytochrome c reconstituted in pure liposomes. Nat Cell Biol 2:553–555
100. Nechushtan A, Smith CL, Lamensdorf I, Yoon SH, Youle RJ (2001) Bax and Bak coalesce into novel mitochondria-associated clusters during apoptosis. J Cell Biol 153:1265–1276
101. Zhou L, Chang DC (2008) Dynamics and structure of the Bax-Bak complex responsible for releasing mitochondrial proteins during apoptosis. J Cell Sci 121:2186–2196
102. Goldstein JC, Waterhouse NJ, Juin P, Evan GI, Green DR (2000) The coordinate release of cytochrome c during apoptosis is rapid, complete and kinetically invariant. Nat Cell Biol 2:156–162
103. Muchmore SW, Sattler M, Liang H, Meadows RP, Harlan JE, Yoon HS, Nettesheim D, Chang BS, Thompson CB, Wong SL, Ng SL, Fesik SW (1996) X-ray and NMR structure of human Bcl-xL, an inhibitor of programmed cell death. Nature 381:335–341
104. Suzuki M, Youle RJ, Tjandra N (2000) Structure of Bax: coregulation of dimer formation and intracellular localization. Cell 103:645–654
105. Martinez-Caballero S, Dejean LM, Kinnally MS, Oh KJ, Mannella CA, Kinnally KW (2009) Assembly of the mitochondrial apoptosis-induced channel, MAC. J Biol Chem 284:12235–12245
106. Basanez G, Nechushtan A, Drozhinin O, Chanturiya A, Choe E, Tutt S, Wood KA, Hsu Y, Zimmerberg J, Youle RJ (1999) Bax, but not Bcl-xL, decreases the lifetime of planar phospholipid bilayer membranes at subnanomolar concentrations. Proc Natl Acad Sci U S A 96:5492–5497

107. Basanez G, Sharpe JC, Galanis J, Brandt TB, Hardwick JM, Zimmerberg J (2002) Bax-type apoptotic proteins porate pure lipid bilayers through a mechanism sensitive to intrinsic monolayer curvature. J Biol Chem 277:49360–49365
108. Hardwick JM, Polster BM (2002) Bax, along with lipid conspirators, allows cytochrome c to escape mitochondria. Mol Cell 10:963–965
109. Polster BM, Basanez G, Etxebarria A, Hardwick JM, Nicholls DG (2005) Calpain I induces cleavage and release of apoptosis-inducing factor from isolated mitochondria. J Biol Chem 280:6447–6454
110. Joza N, Susin SA, Daugas E, Stanford WL, Cho SK, Li CY, Sasaki T, Elia AJ, Cheng HY, Ravagnan L, Ferri KF, Zamzami N, Wakeham A, Hakem R, Yoshida H, Kong YY, Mak TW, Zuniga-Pflucker JC, Kroemer G, Penninger JM (2001) Essential role of the mitochondrial apoptosis-inducing factor in programmed cell death. Nature 410:549–554
111. Cipolat S, Rudka T, Hartmann D, Costa V, Serneels L, Craessaerts K, Metzger K, Frezza C, Annaert W, D'Adamio L, Derks C, Dejaegere T, Pellegrini L, D'Hooge R, Scorrano L, De Strooper B (2006) Mitochondrial rhomboid PARL regulates cytochrome c release during apoptosis via OPA1-dependent cristae remodeling. Cell 126:163–175
112. Frezza C, Cipolat S, Martins de Brito O, Micaroni M, Beznoussenko GV, Rudka T, Bartoli D, Polishuck RS, Danial NN, De Strooper B, Scorrano L (2006) OPA1 controls apoptotic cristae remodeling independently from mitochondrial fusion. Cell 126:177–189
113. Monian P, Jiang X (2012) Clearing the final hurdles to mitochondrial apoptosis: regulation post cytochrome C release. Exp Oncol 34:185–191
114. Yuan S, Akey CW (2013) Apoptosome structure, assembly, and procaspase activation. Structure 21:501–515
115. Szabo I, Zoratti M (1993) The mitochondrial permeability transition pore may comprise VDAC molecules. I. Binary structure and voltage dependence of the pore. FEBS Lett 330:201–205
116. Neer EJ, Schmidt CJ, Nambudripad R, Smith TF (1994) The ancient regulatory-protein family of WD-repeat proteins. Nature 371:297–300
117. Acehan D, Jiang X, Morgan DG, Heuser JE, Wang X, Akey CW (2002) Three-dimensional structure of the apoptosome: implications for assembly, procaspase-9 binding, and activation. Mol Cell 9:423–432
118. Zou H, Li Y, Liu X, Wang X (1999) An APAF-1.cytochrome c multimeric complex is a functional apoptosome that activates procaspase-9. J Biol Chem 274:11549–11556
119. Shi Y (2002) Apoptosome: the cellular engine for the activation of caspase-9. Structure 10:285–288
120. Reubold TF, Eschenburg S (2012) A molecular view on signal transduction by the apoptosome. Cell Signal 24:1420–1425
121. Riedl SJ, Salvesen GS (2007) The apoptosome: signalling platform of cell death. Nat Rev Mol Cell Biol 8:405–413
122. Yuan S, Yu X, Asara JM, Heuser JE, Ludtke SJ, Akey CW (2011) The holo-apoptosome: activation of procaspase-9 and interactions with caspase-3. Structure 19:1084–1096
123. Du C, Fang M, Li Y, Li L, Wang X (2000) Smac, a mitochondrial protein that promotes cytochrome c-dependent caspase activation by eliminating IAP inhibition. Cell 102:33–42
124. Suzuki Y, Imai Y, Nakayama H, Takahashi K, Takio K, Takahashi R (2001) A serine protease, HtrA2, is released from the mitochondria and interacts with XIAP, inducing cell death. Mol Cell 8:613–621
125. Hegde R, Srinivasula SM, Datta P, Madesh M, Wassell R, Zhang Z, Cheong N, Nejmeh J, Fernandes-Alnemri T, Hoshino S, Alnemri ES (2003) The polypeptide chain-releasing factor GSPT1/eRF3 is proteolytically processed into an IAP-binding protein. J Biol Chem 278:38699–38706
126. Xiao R, Gao Y, Shen Q, Li C, Chang W, Chai B (2013) Polypeptide chain release factor eRF3 is a novel molecular partner of survivin. Cell Biol Int 37(4):359–369
127. Li LY, Luo X, Wang X (2001) Endonuclease G is an apoptotic DNase when released from mitochondria. Nature 412:95–99

128. Jensen LE, Bultynck G, Luyten T, Amijee H, Bootman MD, Roderick HL (2013) Alzheimer's disease-associated peptide Abeta42 mobilizes ER Ca(2+) via InsP3R-dependent and -independent mechanisms. Front Mol Neurosci 6:36
129. Rao RV, Ellerby HM, Bredesen DE (2004) Coupling endoplasmic reticulum stress to the cell death program. Cell Death Differ 11:372–380
130. Wojcik C, Rowicka M, Kudlicki A, Nowis D, McConnell E, Kujawa M, DeMartino GN (2006) Valosin-containing protein (p97) is a regulator of endoplasmic reticulum stress and of the degradation of N-end rule and ubiquitin-fusion degradation pathway substrates in mammalian cells. Mol Biol Cell 17:4606–4618
131. Groenendyk J, Michalak M (2005) Endoplasmic reticulum quality control and apoptosis. Acta Biochim Pol 52:381–395
132. Grimm S (2012) The ER-mitochondria interface: the social network of cell death. Biochim Biophys Acta 1823:327–334
133. Szegezdi E, Fitzgerald U, Samali A (2003) Caspase-12 and ER-stress-mediated apoptosis: the story so far. Ann N Y Acad Sci 1010:186–194
134. Verkhratsky A, Toescu EC (2003) Endoplasmic reticulum Ca(2+) homeostasis and neuronal death. J Cell Mol Med 7:351–361
135. Taylor RC, Cullen SP, Martin SJ (2008) Apoptosis: controlled demolition at the cellular level. Nat Rev Mol Cell Biol 9:231–241
136. Scaffidi C, Fulda S, Srinivasan A, Friesen C, Li F, Tomaselli KJ, Debatin KM, Krammer PH, Peter ME (1998) Two CD95 (APO-1/Fas) signaling pathways. EMBO J 17:1675–1687
137. Bennett M, Macdonald K, Chan SW, Luzio JP, Simari R, Weissberg P (1998) Cell surface trafficking of Fas: a rapid mechanism of p53-mediated apoptosis. Science 282:290–293
138. Kasibhatla S, Brunner T, Genestier L, Echeverri F, Mahboubi A, Green DR (1998) DNA damaging agents induce expression of Fas ligand and subsequent apoptosis in T lymphocytes via the activation of NF-kappa B and AP-1. Mol Cell 1:543–551
139. Vandenabeele P, Galluzzi L, Vanden Berghe T, Kroemer G (2010) Molecular mechanisms of necroptosis: an ordered cellular explosion. Nat Rev Mol Cell Biol 11:700–714
140. Tait SW, Green DR (2008) Caspase-independent cell death: leaving the set without the final cut. Oncogene 27:6452–6461
141. Ellis HM, Horvitz HR (1986) Genetic control of programmed cell death in the nematode C. elegans. Cell 44:817–829
142. Huh JR, Foe I, Muro I, Chen CH, Seol JH, Yoo SJ, Guo M, Park JM, Hay BA (2007) The Drosophila inhibitor of apoptosis (IAP) DIAP2 is dispensable for cell survival, required for the innate immune response to gram-negative bacterial infection, and can be negatively regulated by the reaper/hid/grim family of IAP-binding apoptosis inducers. J Biol Chem 282:2056–2068
143. Haraguchi M, Torii S, Matsuzawa S, Xie Z, Kitada S, Krajewski S, Yoshida H, Mak TW, Reed JC (2000) Apoptotic protease activating factor 1 (Apaf-1)-independent cell death suppression by Bcl-2. J Exp Med 191:1709–1720
144. Lindsten T, Ross AJ, King A, Zong WX, Rathmell JC, Shiels HA, Ulrich E, Waymire KG, Mahar P, Frauwirth K, Chen Y, Wei M, Eng VM, Adelman DM, Simon MC, Ma A, Golden JA, Evan G, Korsmeyer SJ, MacGregor GR, Thompson CB (2000) The combined functions of proapoptotic Bcl-2 family members bak and bax are essential for normal development of multiple tissues. Mol Cell 6:1389–1399
145. Cecconi F, Alvarez-Bolado G, Meyer BI, Roth KA, Gruss P (1998) Apafl (CED-4 homolog) regulates programmed cell death in mammalian development. Cell 94:727–737
146. Yoshida H, Kong YY, Yoshida R, Elia AJ, Hakem A, Hakem R, Penninger JM, Mak TW (1998) Apafl is required for mitochondrial pathways of apoptosis and brain development. Cell 94:739–750
147. Hao Z, Duncan GS, Chang CC, Elia A, Fang M, Wakeham A, Okada H, Calzascia T, Jang Y, You-Ten A, Yeh WC, Ohashi P, Wang X, Mak TW (2005) Specific ablation of the apoptotic functions of cytochrome C reveals a differential requirement for cytochrome C and Apaf-1 in apoptosis. Cell 121:579–591

148. Chipuk JE, Green DR (2005) Do inducers of apoptosis trigger caspase-independent cell death? Nat Rev Mol Cell Biol 6:268–275
149. Johnson CE, Huang YY, Parrish AB, Smith MI, Vaughn AE, Zhang Q, Wright KM, Van Dyke T, Wechsler-Reya RJ, Kornbluth S, Deshmukh M (2007) Differential Apaf-1 levels allow cytochrome c to induce apoptosis in brain tumors but not in normal neural tissues. Proc Natl Acad Sci U S A 104:20820–20825
150. Sanchis D, Mayorga M, Ballester M, Comella JX (2003) Lack of Apaf-1 expression confers resistance to cytochrome c-driven apoptosis in cardiomyocytes. Cell Death Differ 10:977–986
151. Goldstein JC, Kluck RM, Green DR (2000) A single cell analysis of apoptosis. Ordering the apoptotic phenotype. Ann N Y Acad Sci 926:132–141
152. Cauwels A, Janssen B, Waeytens A, Cuvelier C, Brouckaert P (2003) Caspase inhibition causes hyperacute tumor necrosis factor-induced shock via oxidative stress and phospholipase A2. Nat Immunol 4:387–393
153. Holler N, Zaru R, Micheau O, Thome M, Attinger A, Valitutti S, Bodmer JL, Schneider P, Seed B, Tschopp J (2000) Fas triggers an alternative, caspase-8-independent cell death pathway using the kinase RIP as effector molecule. Nat Immunol 1:489–495
154. Temkin V, Huang Q, Liu H, Osada H, Pope RM (2006) Inhibition of ADP/ATP exchange in receptor-interacting protein-mediated necrosis. Mol Cell Biol 26:2215–2225
155. Ricci JE, Munoz-Pinedo C, Fitzgerald P, Bailly-Maitre B, Perkins GA, Yadava N, Scheffler IE, Ellisman MH, Green DR (2004) Disruption of mitochondrial function during apoptosis is mediated by caspase cleavage of the p75 subunit of complex I of the electron transport chain. Cell 117:773–786
156. Colell A, Ricci JE, Tait S, Milasta S, Maurer U, Bouchier-Hayes L, Fitzgerald P, Guio-Carrion A, Waterhouse NJ, Li CW, Mari B, Barbry P, Newmeyer DD, Beere HM, Green DR (2007) GAPDH and autophagy preserve survival after apoptotic cytochrome c release in the absence of caspase activation. Cell 129:983–997
157. Gasser SM, Daum G, Schatz G (1982) Import of proteins into mitochondria. Energy-dependent uptake of precursors by isolated mitochondria. J Biol Chem 257:13034–13041
158. Ewen CL, Kane KP, Bleackley RC (2012) A quarter century of granzymes. Cell Death Differ 19:28–35
159. Orrenius S, Zhivotovsky B, Nicotera P (2003) Regulation of cell death: the calcium-apoptosis link. Nat Rev Mol Cell Biol 4:552–565
160. Odake S, Kam CM, Narasimhan L, Poe M, Blake JT, Krahenbuhl O, Tschopp J, Powers JC (1991) Human and murine cytotoxic T lymphocyte serine proteases: subsite mapping with peptide thioester substrates and inhibition of enzyme activity and cytolysis by isocoumarins. Biochemistry 30:2217–2227
161. Adrain C, Murphy BM, Martin SJ (2005) Molecular ordering of the caspase activation cascade initiated by the cytotoxic T lymphocyte/natural killer (CTL/NK) protease granzyme B. J Biol Chem 280:4663–4673
162. Andrade F, Roy S, Nicholson D, Thornberry N, Rosen A, Casciola-Rosen L (1998) Granzyme B directly and efficiently cleaves several downstream caspase substrates: implications for CTL-induced apoptosis. Immunity 8:451–460
163. Darmon AJ, Nicholson DW, Bleackley RC (1995) Activation of the apoptotic protease CPP32 by cytotoxic T-cell-derived granzyme B. Nature 377:446–448
164. Martin SJ, Amarante-Mendes GP, Shi L, Chuang TH, Casiano CA, O'Brien GA, Fitzgerald P, Tan EM, Bokoch GM, Greenberg AH, Green DR (1996) The cytotoxic cell protease granzyme B initiates apoptosis in a cell-free system by proteolytic processing and activation of the ICE/CED-3 family protease, CPP32, via a novel two-step mechanism. EMBO J 15:2407–2416
165. Medema JP, Toes RE, Scaffidi C, Zheng TS, Flavell RA, Melief CJ, Peter ME, Offringa R, Krammer PH (1997) Cleavage of FLICE (caspase-8) by granzyme B during cytotoxic T lymphocyte-induced apoptosis. Eur J Immunol 27:3492–3498

166. Van de Craen M, Van den Brande I, Declercq W, Irmler M, Beyaert R, Tschopp J, Fiers W, Vandenabeele P (1997) Cleavage of caspase family members by granzyme B: a comparative study in vitro. Eur J Immunol 27:1296–1299
167. Atkinson EA, Barry M, Darmon AJ, Shostak I, Turner PC, Moyer RW, Bleackley RC (1998) Cytotoxic T lymphocyte-assisted suicide. Caspase 3 activation is primarily the result of the direct action of granzyme B. J Biol Chem 273:21261–21266
168. Darmon AJ, Ley TJ, Nicholson DW, Bleackley RC (1996) Cleavage of CPP32 by granzyme B represents a critical role for granzyme B in the induction of target cell DNA fragmentation. J Biol Chem 271:21709–21712
169. Sorimachi H, Ishiura S, Suzuki K (1997) Structure and physiological function of calpains. Biochem J 328(Pt 3):721–732
170. Ray SK, Hogan EL, Banik NL (2003) Calpain in the pathophysiology of spinal cord injury: neuroprotection with calpain inhibitors. Brain Res Brain Res Rev 42:169–185
171. Saido TC, Sorimachi H, Suzuki K (1994) Calpain: new perspectives in molecular diversity and physiological-pathological involvement. FASEB J 8:814–822
172. Stys PK, Jiang Q (2002) Calpain-dependent neurofilament breakdown in anoxic and ischemic rat central axons. Neurosci Lett 328:150–154
173. Santella L, Carafoli E (1997) Calcium signaling in the cell nucleus. FASEB J 11:1091–1109
174. Elmore S (2007) Apoptosis: a review of programmed cell death. Toxicol Pathol 35:495–516
175. Vande Walle L, Lamkanfi M, Vandenabeele P (2008) The mitochondrial serine protease HtrA2/Omi: an overview. Cell Death Differ 15:453–460
176. Sheikh MS, Fornace AJ Jr (2000) Death and decoy receptors and p53-mediated apoptosis. Leukemia 14:1509–1513
177. Goyal L (2001) Cell death inhibition: keeping caspases in check. Cell 104:805–808
178. White MK, McCubrey JA (2001) Suppression of apoptosis: role in cell growth and neoplasia. Leukemia 15:1011–1021
179. Vila M, Przedborski S (2003) Targeting programmed cell death in neurodegenerative diseases. Nat Rev Neurosci 4:365–375
180. Lev N, Melamed E, Offen D (2003) Apoptosis and Parkinson's disease. Prog Neuropsychopharmacol Biol Psychiatry 27:245–250
181. Jellinger KA (2001) Cell death mechanisms in neurodegeneration. J Cell Mol Med 5:1–17
182. Ameisen JC (1992) The programmed cell death theory of AIDS pathogenesis: implications, testable predictions, and confrontation with experimental findings. Immunodefic Rev 3:237–246
183. Copeland KF, Heeney JL (1996) T helper cell activation and human retroviral pathogenesis. Microbiol Rev 60:722–742
184. Bush JA, Li G (2002) Cancer chemoresistance: the relationship between p53 and multidrug transporters. Int J Cancer 98:323–330
185. Nielsen LL, Maneval DC (1998) P53 tumor suppressor gene therapy for cancer. Cancer Gene Ther 5:52–63
186. Zhan M, Yu D, Lang A, Li L, Pollock RE (2001) Wild type p53 sensitizes soft tissue sarcoma cells to doxorubicin by down-regulating multidrug resistance-1 expression. Cancer 92:1556–1566
187. Maccioni RB, Munoz JP, Barbeito L (2001) The molecular bases of Alzheimer's disease and other neurodegenerative disorders. Arch Med Res 32:367–381
188. Butterfield DA, Bader Lange ML, Sultana R (2010) Involvements of the lipid peroxidation product, HNE, in the pathogenesis and progression of Alzheimer's disease. Biochim Biophys Acta 1801:924–929
189. Battaglia F, Trinchese F, Liu S, Walter S, Nixon RA, Arancio O (2003) Calpain inhibitors, a treatment for Alzheimer's disease: position paper. J Mol Neurosci 20:357–362
190. Boland B, Campbell V (2003) beta-Amyloid (1-40)-induced apoptosis of cultured cortical neurones involves calpain-mediated cleavage of poly-ADP-ribose polymerase. Neurobiol Aging 24:179–186

191. Fischer U, Schulze-Osthoff K (2005) New approaches and therapeutics targeting apoptosis in disease. Pharmacol Rev 57:187–215
192. Kroemer G, Galluzzi L, Brenner C (2007) Mitochondrial membrane permeabilization in cell death. Physiol Rev 87:99–163
193. Galluzzi L, Vitale I, Kepp O, Seror C, Hangen E, Perfettini JL, Modjtahedi N, Kroemer G (2008) Methods to dissect mitochondrial membrane permeabilization in the course of apoptosis. Methods Enzymol 442:355–374
194. Hisatomi T, Ishibashi T, Miller JW, Kroemer G (2009) Pharmacological inhibition of mitochondrial membrane permeabilization for neuroprotection. Exp Neurol 218:347–352
195. Siman R, Card JP, Davis LG (1990) Proteolytic processing of beta-amyloid precursor by calpain I. J Neurosci 10:2400–2411
196. Samantaray S, Ray SK, Banik NL (2008) Calpain as a potential therapeutic target in Parkinson's disease. CNS Neurol Disord Drug Targets 7:305–312
197. Camins A, Crespo-Biel N, Junyent F, Verdaguer E, Canudas AM, Pallas M (2009) Calpains as a target for therapy of neurodegenerative diseases: putative role of lithium. Curr Drug Metab 10:433–447
198. Scholzke MN, Potrovita I, Subramaniam S, Prinz S, Schwaninger M (2003) Glutamate activates NF-kappaB through calpain in neurons. Eur J Neurosci 18:3305–3310
199. Lopes JP, Oliveira CR, Agostinho P (2010) Neurodegeneration in an Abeta-induced model of Alzheimer's disease: the role of Cdk5. Aging Cell 9:64–77
200. Rami A (2003) Ischemic neuronal death in the rat hippocampus: the calpain-calpastatin-caspase hypothesis. Neurobiol Dis 13:75–88
201. Wilcock GK (2003) Memantine for the treatment of dementia. Lancet Neurol 2:503–505
202. Molinuevo JL, Llado A, Rami L (2005) Memantine: targeting glutamate excitotoxicity in Alzheimer's disease and other dementias. Am J Alzheimers Dis Other Demen 20:77–85
203. Goni-Oliver P, Avila J, Hernandez F (2009) Memantine inhibits calpain-mediated truncation of GSK-3 induced by NMDA: implications in Alzheimer's disease. J Alzheimers Dis 18:843–848
204. Xie X, Zhao X, Liu Y, Zhang J, Matusik RJ, Slawin KM, Spencer DM (2001) Adenovirus-mediated tissue-targeted expression of a caspase-9-based artificial death switch for the treatment of prostate cancer. Cancer Res 61:6795–6804
205. Komata T, Kondo Y, Kanzawa T, Hirohata S, Koga S, Sumiyoshi H, Srinivasula SM, Barna BP, Germano IM, Takakura M, Inoue M, Alnemri ES, Shay JW, Kyo S, Kondo S (2001) Treatment of malignant glioma cells with the transfer of constitutively active caspase-6 using the human telomerase catalytic subunit (human telomerase reverse transcriptase) gene promoter. Cancer Res 61:5796–5802
206. Jia LT, Zhang LH, Yu CJ, Zhao J, Xu YM, Gui JH, Jin M, Ji ZL, Wen WH, Wang CJ, Chen SY, Yang AG (2003) Specific tumoricidal activity of a secreted proapoptotic protein consisting of HER2 antibody and constitutively active caspase-3. Cancer Res 63:3257–3262
207. Xu YM, Wang LF, Jia LT, Qiu XC, Zhao J, Yu CJ, Zhang R, Zhu F, Wang CJ, Jin BQ, Chen SY, Yang AG (2004) A caspase-6 and anti-human epidermal growth factor receptor-2 (HER2) antibody chimeric molecule suppresses the growth of HER2-overexpressing tumors. J Immunol 173:61–67
208. Buckley CD, Pilling D, Henriquez NV, Parsonage G, Threlfall K, Scheel-Toellner D, Simmons DL, Akbar AN, Lord JM, Salmon M (1999) RGD peptides induce apoptosis by direct caspase-3 activation. Nature 397:534–539
209. Skorko-Glonek J, Zurawa-Janicka D, Koper T, Jarzab M, Figaj D, Glaza P, Lipinska B (2013) HtrA protease family as therapeutic targets. Curr Pharm Des 19:977–1009

Chapter 2
Caspases – Key Players in Apoptosis

Christine E. Cade and A. Clay Clark

Abstract Caspases are the terminal proteases involved in apoptosis, as well as being involved in inflammation. The apoptotic caspases can be classified as either initiator or effector caspases based on both their position in the caspase cascade and their activation mechanism. Initiator caspases require dimerization to be activated, and cleavage of a loop called the intersubunit linker stabilizes the active enzyme. Effector caspases, on the other hand, are found as dimers in the cell and cleavage of the intersubunit linker is the key step in their activation.

The name caspase is short for cysteinyl aspartate-specific protease. As their name suggests, these enzymes hydrolyze peptide bonds after certain aspartate residues using a catalytic cysteine (with the aid of an active-site histidine residue). Caspases can be inhibited by endogenous inhibitors such as XIAP, by synthetic inhibitors which target either the active site or an allosteric site, or by post-translational modification. Further research is needed to find novel activators and inhibitors of caspases to treat diseases which involve misregulation of apoptosis.

Keywords Caspase • Apoptosis • Protease • Intersubunit linker • Prodomain • Dimerization • Allostery

---

Funding: This work was supported by a grant from the NIH (National Institutes of Health) [grant number GM065970 (to A.C.C.)].

C.E. Cade
Department of Molecular and Structural Biochemistry, North Carolina State University, 128 Polk Hall, Raleigh, NC 27695-7622, USA

A.C. Clark (✉)
Department of Molecular and Structural Biochemistry and Center for Comparative Medicine and Translational Research, North Carolina State University, 128 Polk Hall, Raleigh, NC 27695-7622, USA

Present affiliation: Professor and Chair, Department of Biology, University of Texas at Arlington, Arlington, TX 76019, USA
e-mail: clay_clark@ncsu.edu

K. Bose (ed.), *Proteases in Apoptosis: Pathways, Protocols and Translational Advances*, DOI 10.1007/978-3-319-19497-4_2

Abbreviations

ASC	Apoptosis-associated speck-like protein containing a CARD
CARD	Caspase activation and recruitment domain
Caspase	Cysteinal aspartate-specific protease
DAMPs	Danger-associated molecular patterns
DD	Death domain
DISC	Death inducing signaling complex
FADD	Fas-associated death domain
FasL	Fas ligand
FLICE	FADD-like interleukin 1β-converting enzyme
FLIP	FLICE-like inhibitory protein
$FLIP_L$	Long splice variant of FLIP which forms a heterodimer with caspase-8
$FLIP_S$	Short splice variant of FLIP which blocks caspase-8 from binding death receptor
ICE	Interleukin 1β-converting enzyme
IL-1β	Interleukin 1β
IL-18	Interleukin 18
PS	Phosphatidylserine
Smac	Second mitochondrial activator of caspases
TNFR	Tumor necrosis factor receptor
XIAP	X-linked inhibitor of apoptosis protein

Caspases

Caspases (**c**ysteinal**asp**artate-specific prote**ases**) [1] are enzymes which utilize a catalytic cysteine to cleave their peptide substrates after specific aspartate residues. The first caspase was discovered in 1992 and because of its function was named interleukin-1-β converting enzyme (ICE) [2, 3] but was later renamed to caspase-1. In 1993, Ced-3 from *C. elegans* was found to be homologous to ICE [4] and the corresponding human protein CPP32 (later named caspase-3) was found in 1994 [5]. The official caspase nomenclature was decided on in 1996 to alleviate the confusion that went along with discovery of ten different caspases, some with multiple names [1].

1 Structure

Caspases are expressed as proenzymes (zymogens) called procaspases, which then become activated to the mature caspase form. Procaspase structure can be divided into three domains: an N-terminal prodomain, a large subunit, and a small subunit. The first step in maturation is dimerization. Then, proteolytic processing removes the prodomain and cleaves a loop called the intersubunit linker between the large and small subunits.

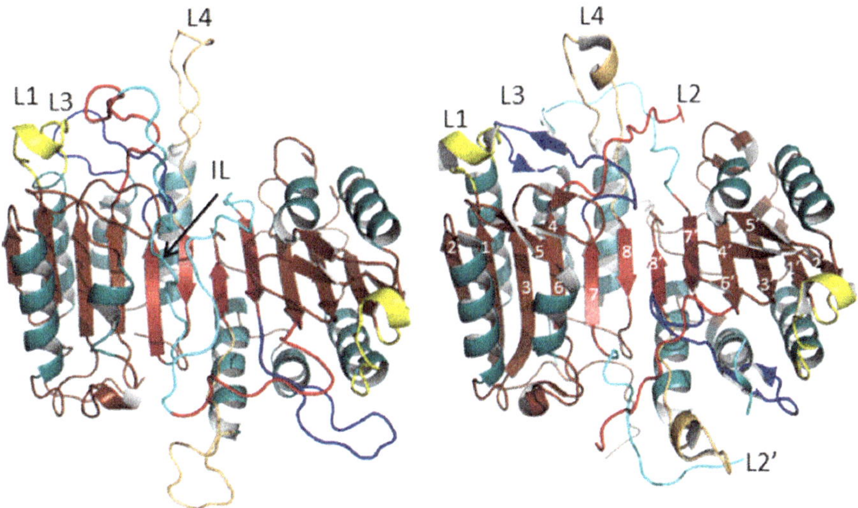

Fig. 2.1 Procaspase-3 model and crystal structure of caspase-3. Active site loop coloring: *yellow* = L1, *red* = L2, *cyan* = L2′, *blue* = L3, *tan* = L4

The secondary structure of mature caspases consists of six core β-strands in a slightly twisted sheet in each monomer, with two main helices on one face (the "front") of the protein and three helices on the other face (the "back") of the protein (Fig. 2.1). The first four core β-strands and helices 1–3 form the large subunit, whereas the last two core β-strands and helices 4–5 form the small subunit.

The dimer interface consists of the final β-strand from each monomer, side-by-side in an antiparallel manner. The two monomers are related through a C2 axis of symmetry such that one monomer is "upside-down" compared to the other monomer.

Five loops are important for the formation of the active site. Once the intersubunit linker is cleaved, the two halves of the cleaved linker are called L2 and L2′. Active site loops L1, L2, L3, and L4 come from one monomer, and loop L2′ comes from the other. The catalytic cysteine is part of loop L2, and the catalytic histidine is part of a loop extending from the C terminal end of β3.

2 Classification

Caspases are divided into two main categories based on their function: apoptotic caspases and inflammatory caspases. The apoptotic caspases are further divided into two categories based on time of entry into the apoptotic cascade: initiator caspases and effector caspases.

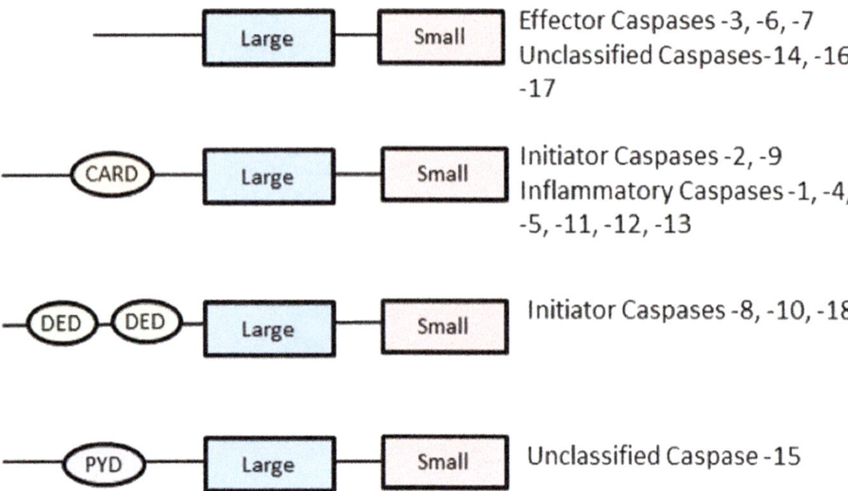

Fig. 2.2 Domain arrangement of mammalian caspases

2.1 *Apoptotic Caspases*

2.1.1 Initiator Caspases

Initiator caspases are stable monomers in the cell until they are activated by dimerization. Once dimerized, initiator caspases have sufficient activity to autoprocess, cleaving their prodomain and intersubunit linker. An induced proximity model for dimerization was first invoked for caspases-8 and -10 but now seems to be generalizable to initiator caspases as a whole. This model says that activation complexes increase the local concentration of the initiator caspases, enabling them to dimerize [6]. The prodomains of initiator caspases contain either a CARD (caspase activation and recruitment domain) or DED (death effector domain), which allow initiator caspases to bind to activation complexes (Fig. 2.2).

The initiator caspases-2 and -9 are involved in the intrinsic pathway, which is activated by mitochondrial damage, cytotoxic stress, chemotherapeutic drugs or certain developmental cues [7]. Activation of caspase-2 leads to release of cytochrome *c* from the mitochondria, which then binds to Apaf-1 and forms the heptameric apoptosome. The apoptosome binds procaspase-9 to dimerize and therefore activate it. Once active, caspase-9 activates downstream effector caspases.

The initiator caspases-8 and -10 are activated by the extrinsic pathway: in order to eliminate excess cells created during development or remove cells with tumorigenic properties, a molecule binds to a death receptor at the membrane which is part of the tumor necrosis factor receptor (TNFR) superfamily [8, 9]. One such ligand/receptor pair is FasL (Fas ligand) and CD95(APO-1/Fas). The cytosolic death domains (DD) of the receptor recruit an adaptor molecule such as FADD (Fas-associated death domain), allowing the complex to bind initiator procaspases-8 or -10 to forma death-inducing signaling complex (DISC). Once the procaspases are part of the DISC, they are able to dimerize and therefore become active. The active caspase-8 or -10 then activates downstream effector caspases such as caspase-3.

2.1.2 Effector Caspases

The effector caspases-3, -6, and -7, are found as inactive dimers in the cell. They are activated once an initiator caspase cleaves their intersubunit linkers. Because they do not require death scaffolds for dimer formation [10, 11], their prodomains are short and lack the CARD and DED domains typical of initiator caspases. Their prodomains are, however, likely to be involved in targeting within the cell [12–15].

2.2 Inflammatory Caspases

Similarly to the initiator caspases, the inflammatory caspases-1, -4, -5, -11, -12, and -13 are activated by dimerization. Their prodomains contain a CARD which allows them to bind to activation complexes. Similarly to apoptosome formation, a multiprotein complex called the inflammasome consists of a NOD-like receptor such as NLRP1, an adaptor protein such as ASC (apoptosis-associated speck-like protein containing a CARD), and the inflammatory procaspase, particularly procaspase-1 [16]. In some cases, the procaspase can also be recruited to CARD domains in the receptor directly, without the aid of an adaptor molecule [17].

Once the inflammatory caspases become active, they are activators of cytokines through cleavage of their preforms. In monocytes and macrophages, caspase-1 activates interleukin-1β (IL-1β) [3] and interleukin-18 (IL-18). These cytokines mediate innate immunity and inflammation [18].

The mouse caspase-11 is a homolog of human caspase-4 [19]. In humans, caspase-12 is generally truncated due to a premature stop codon, but in some people of African descent, a read-through mutation causes expression of the full-length protein, causing increased risk of sepsis due to decreased inflammatory and immune response to endotoxins [20]. Caspase-13 is a bovine ortholog of human caspase-4 [21].

2.3 Other or Unclassified Caspases

Caspase-14 expression is restricted to epidermal keratinocytes and is involved in differentiation [22]. Like the effector caspases, it has a short prodomain with no adaptor regions. Several caspases are not yet classified: 15, 16, and 17 [23]. Caspase-15 is expressed in several mammalian species including pigs, dogs, and cattle [24]. It contains a pyrin-like region in its prodomain similar to that found in zebrafish caspases caspy and caspy2 [25]. Caspase-16 is found in marsupials and placental mammals and contains a short prodomain with no adaptor regions [23]. Caspase-17 is found in vertebrates except for marsupials and placental mammals and also does not contain adaptor regions in its prodomain. Caspase-18 is found in opossums and chickens and, like caspases-8 and -10, contains two DED regions in its prodomain, so it is likely also an initiator apoptotic caspase [23].

3 Mechanisms

3.1 Activation

Activation of caspases generally requires two events: they must be a dimer and the intersubunit linker must be cleaved. Removal of the prodomain is not necessary for activation; in fact, the prodomain may serve to stabilize the active enzyme [26].

After dimerization, cleavage of the intersubunit linker occurs first, followed by cleavage of the prodomain. Prior to cleavage, the intersubunit linker from one monomer occupies the dimer interface. Upon cleavage of the intersubunit linker, the C-terminal portion of the linker, L2′, vacates the central cavity and rotates about 180 degrees toward the active site, forming contacts with L2, L3, and L4 from the opposite monomer. These loop bundle contacts stabilize the active site. The movement of L2′ out of the dimer interface allows L3 to slide in towards the interface and form the substrate binding pocket. Rotation of a key arginine on L2 from a solvent-exposed position into the interface allows its neighboring residue, the catalytic cysteine, to assume its proper position for catalysis.

For effector caspases, equilibrium favors the inactive dimer. For initiator caspases, however, dimerization is the main challenge to be overcome for activation. Addition of kosmotropes such as sodium citrate causes caspase-8 to dimerize and become activated [27]. At least partly because the initiator caspases have longer intersubunit linkers than effector caspases, cleavage of the intersubunit linker is not necessary for activation, but rather, stabilizes the active conformation.

Effector caspase mutants, particularly procaspase-3 V266E, can also be activated without cleavage of the intersubunit linker [28]. This mutant is even more effective at inducing apoptosis than the wild-type (WT) enzyme [29]. The enhancement of activation caused by the mutation is predicted to occur because the mutation keeps the intersubunit linker from binding to the dimer interface. In general, when the intersubunit linker is in the dimer interface, the protein is inactive, whereas when it is out of the interface it can become active.

The conformational ensemble of effector procaspases includes both active and inactive conformers. Although the inactive ensemble is favored, binding of allosteric activators could shift the equilibrium to the active ensemble. On the other hand, binding of allosteric inhibitors to the active caspase could inactivate it. Manipulating the position of the intersubunit linker could lead to allosteric activation or inhibition. A drug which binds at the dimer interface and holds the intersubunit linker in place could inactivate the enzyme. Conversely, a drug which binds at the dimer interface and keeps the intersubunit linker from binding could activate the procaspase. In fact, a small molecule has been suggested to activate procaspase-3 by this mechanism [30].

Additionally, Wells and coworkers have found a small molecule termed 1541 which forms nanofibrils that act as a scaffold for (pro)caspase-3 binding and increase activation of the procaspase [31]. They suggest that the procaspase is activated through induced proximity, similar to the activation of initiator caspases. *In vitro*, amyloid-β (residues 1–40) fibrils were also shown to activate procaspase-3. The activation of caspases by fibrils may play a role in neurodegenerative diseases [32].

3.2 *Catalysis*

Proteases all have some mechanistic features in common. The trigonal planar peptide bond of the substrate is forced into a tetrahedral intermediate [33]. As this tetrahedral intermediate forms, a nucleophile attacks the carbonyl carbon of the peptide bond. Then, the amino nitrogen of the leaving group is protonated.

Caspases contain a catalytic dyad consisting of a cysteine and a histidine [33]. Based on the catalytic mechanism accepted for cysteine proteases, the mechanism for caspases has been thought to be as follows (See Fig. 2.3a): First, the catalytic histidine abstracts a proton from the catalytic cysteine. The catalytic cysteine acts

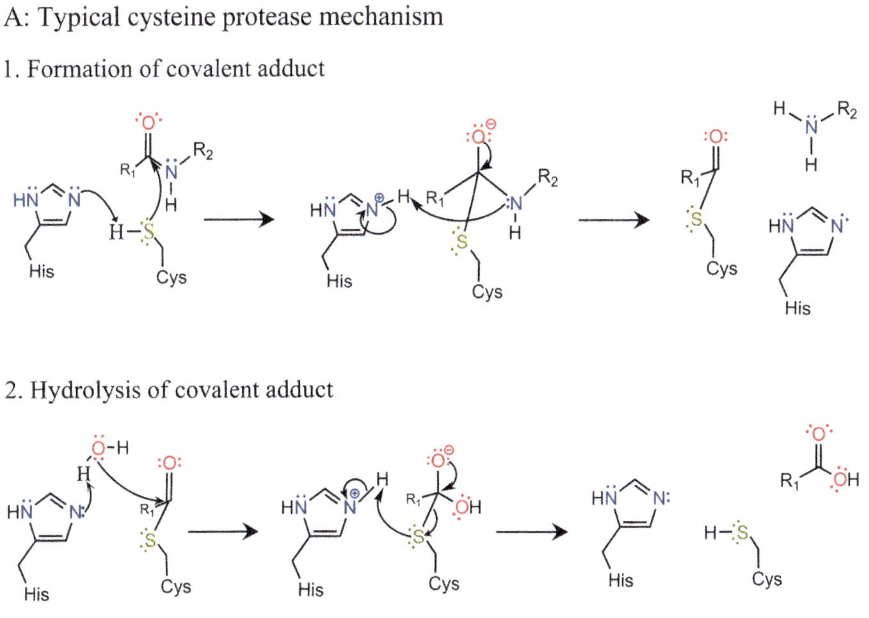

A: Typical cysteine protease mechanism

1. Formation of covalent adduct

2. Hydrolysis of covalent adduct

Fig. 2.3 Two proposed mechanisms of caspase catalysis (Adapted from Miscione et al. [34])

B. Simulated caspase mechanism

1. Formation of covalent adduct

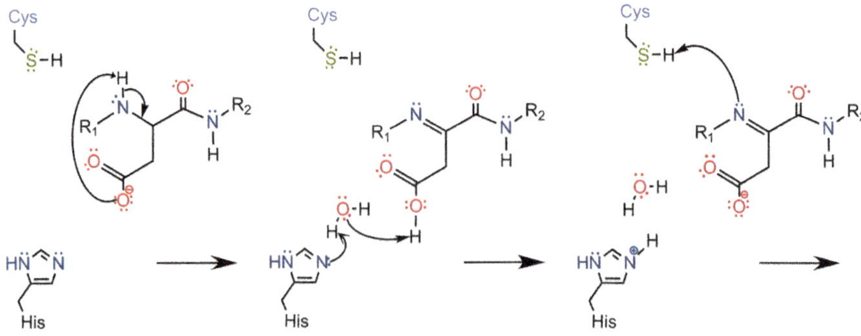

2. Hydrolysis of covalent adduct

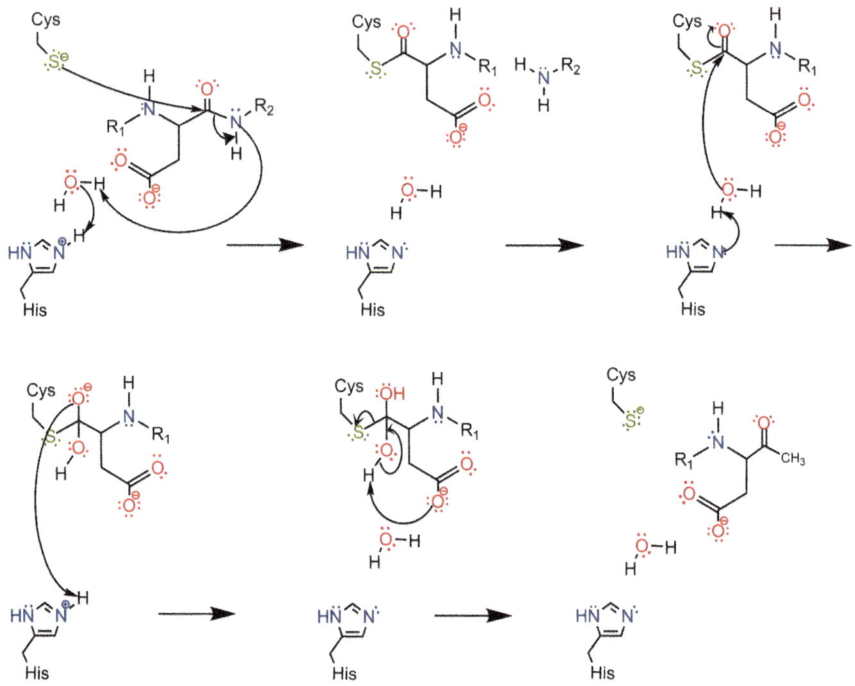

Fig. 2.3 (continued)

as the nucleophile to form a covalent tetrahedral intermediate with the peptide substrate. Once the cysteine has bound, the histidine donates the proton to the amino moiety of the peptide leaving group. The peptide bond is cleaved, with the N-terminal part of the peptide remaining covalently attached to the cysteine while the C-terminal part of the peptide leaves. Finally, hydrolysis frees the N-terminal part of the peptide and re-protonates the catalytic histidine.

An oxyanion hole, a pocket in the enzyme that hydrogen bonds to the carbonyl oxygen of the substrate, is also thought to be key for catalysis [33]. It is formed by the backbone nitrogens of a conserved glycine (238 in caspase-1) and the catalytic cysteine (285 in caspase-1). The oxyanion hole is thought to be important for polarizing and stabilizing the scissile carbonyl group [34].

However, there are some problems with the proposed mechanism. The 6–7 Å distance between the two catalytic residues is larger than found in most proteases, and makes direct hydrogen transfer unlikely [33]. Molecular dynamics simulations have shown that the catalytic residues cannot exist as a charged pair prior to catalysis [35]. Therefore, the deprotonation of the cysteine likely occurs during catalysis. Also, the histidine residue is not in an optimal location for protonating the amino leaving group [36].

A DFT study of the first part of the catalytic process (Fig. 2.3b, part 1) has been carried out for caspase-7 [34]. Miscione and coworkers found that first, a proton is transferred from the backbone nitrogen of the P1 aspartate to the carboxylate group of the P1 aspartate. In the second step, a proton is transferred from the aspartate to a water molecule, and from that water to the catalytic histidine. In the third step, a proton is transferred from the catalytic cysteine to the backbone nitrogen of the P1 aspartate. The overall result of these first three steps is the protonation of the catalytic histidine and the deprotonation of the catalytic cysteine. In a fourth step, the catalytic cysteine nucleophile attacks the carbonyl carbon of the substrate to form a tetrahedral intermediate, the peptide bond is cleaved, and a proton is transferred from the catalytic histidine to a second water, which transfers a proton to the amino nitrogen of the leaving group.

A QM/MM simulation focused on the hydrolysis of the covalent adduct (Fig. 2.3b, part 2) [37]. In the reaction scheme proposed by Sulpizi and coworkers, the catalytic histidine deprotonates a water molecule, which attacks the scissile carbonyl carbon (as in the original proposed mechanism). Then the proton from the catalytic histidine is abstracted by the now negatively-charged carbonyl oxygen, such that a diol is formed. A second water molecule interacts with the catalytic histidine and one of the diolhydroxy groups. Finally, a proton is transferred from that diol hydroxyl group to the P1 aspartate residue, causing cleavage of the covalent adduct. If this is true, it could more cogently explain the specificity for a P1 aspartate residue.

4 Functions

4.1 Apoptosis

The activation of caspases commits the cell to apoptosis. The main hallmarks of apoptosis include rounding of cells and retraction from neighbors, membrane blebbing to form vesicles called apoptotic bodies, nuclear fragmentation, chromatin condensation, hydrolysis of genomic DNA to approximately 200 bp fragments, and translocation of phosphatidylserine (PS) to the external surface of cells as an "eat me" signal to phagocytes. The apoptotic caspases are necessary for conferring all of these phenotypes.

In addition to the systematic dismantling of the cell, caspases are also involved in producing "find-me" signals to cause chemotaxis of phagocytes to apoptotic cells [38–40]. The recruitment of phagocytes keeps cells from releasing their contents into extracellular space and activating an immune response which could be harmful to the tissue.

When the number of apoptotic cells is too great for consumption by phagocytes, secondary necrosis can occur. When this happens, the cell releases its contents into extracellular space. However, immune cells are somehow able to recognize the cells undergoing apoptosis (and secondary necrosis) differently from necrotic cells. This is likely due to the actions of caspases. Caspases keep danger-associated molecular patterns (DAMPs) and alarmins from being activated [41]. This can be thought of as a "tolerate me" signal.

Caspases are also involved in turning off transcription and translation [42]. This keeps any infecting viral particles from replicating using the host's machinery. They also fragment the Golgi, ER, and mitochondria [43, 44].

4.2 *Inflammatory Response*

In contrast to the actions of apoptotic caspases, which systematically dismantle the cell to avoid an immune response, the actions of inflammatory caspases lead to cell lysis and activation of the immune response in a process called pyroptosis [45]. In order to activate an immune response, caspases cleave cytokine IL-1β and IL-18 to produce the mature form [46].

In addition to activation of cytokines, procaspase-1 is also able to activate the pro-inflammatory transcription factor NF-κB [47]. Rather than using its catalytic activity, the CARD domain of procaspase-1 binds to a CARD domain in the kinase RIP2, which is involved in NF-κB activation.

4.3 *Other Functions*

Caspase expression is kept below a certain threshold required for apoptosis by IAPs (inhibitor of apoptosis proteins). At these subthresholdlevels they are able to play roles that are neither apoptotic nor inflammatory. Caspase-3 activity was found to be important for differentiation of erythroblasts, [48] skeletal muscle, [49] bone marrow stromal stem cells, [50] and neural stem cells [51].

Caspase-3 has several other non-apoptotic functions in nerve cells. In addition to neural cell differentiation, caspase-3 has also been implicated in neuronal migration and plasticity, [52] axon pruning, and synapse elimination [53].

Caspases have been shown to play a role in cell migration and invasion under certain circumstances [54]. They can also induce neighboring cells to proliferate to replace dying cells in a process called apoptosis-induced proliferation [55]. These roles for caspases have implications for cancer: moderate activation of caspases could, in fact, cause cancer to progress rather than regress [54, 55].

In addition to its apoptotic function, caspase-8 has an anti-apoptotic function when it forms a heterodimer with $FLIP_L$ (a protein similar to caspase-8 but lacking a catalytic site) [56]. This protein complex is able to activate the NF-κB signaling pathway leading to proliferation [57]. In another pro-survival capacity, the caspase-8/$FLIP_L$ complex is also able to inhibit RIPK3-dependent necrosis [56].

5 Substrates and Inhibitors

5.1 Synthetic Substrates and Substrate Specificity

Caspase substrate specificity is determined by a series of 4–5 substrate residues which bind to the active site of the caspase. These residues are named P1-P4 or P5, with P1 always being an aspartate residue (Fig. 2.4). The P4 residue is especially important in determining specificity for a given caspase [58].

Because of this 4–5 residue contribution to specificity, substrates used for measuring caspase activity typically have a tetrapeptide preceded by an acetyl group (Ac) on the N terminus and followed by a fluorophore on the C terminus: for example, Ac-DEVD-AFC. When the peptide is cleaved by the caspase, the fluorophore is released and activity can be determined by fluorescence. Some typical fluorophores include AMC (7-amino-4-methylcoumarin) and AFC (7-amino-4-trifluoromethylcoumarin). Addition of p-nitroanilide (pNA) instead of a fluorophore to the C-terminus allows caspase activity to be determined colorimetrically.

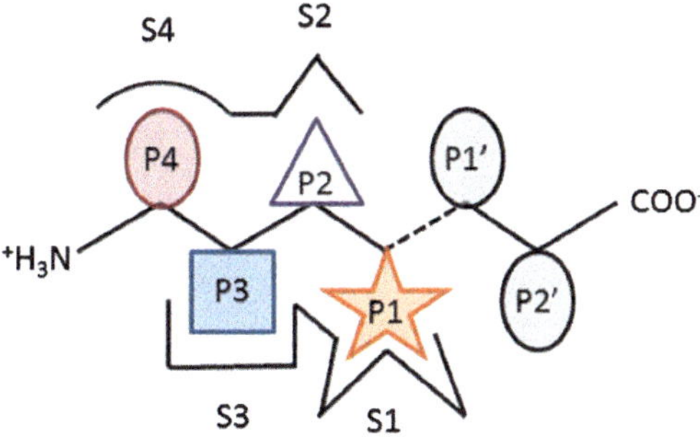

Fig. 2.4 Caspase substrate binding site (Adapted from Stennicke and Salvesen [33])

A positional scanning combinatorial library approach has been used with these synthetic substrates to determine the substrate specificity for most of the mammalian caspases [58, 59]. Caspases-3 and -7 share the same substrate specificity: DEVD. The optimal sequence for caspase-1 is WEHD, and the optimal sequence for both caspase-4 and caspase-5 is (W/L) EHD. The optimal sequence for caspase-2 is DEHD, for caspase-6 is VEHD, for caspase-9 is LEHD, for caspase-8 is LETD, and for caspase-10 is LE(Nle)D (Nle = norleucine).

The P1-P4 residues fit into the S1-S4 pockets in the active site of the caspase. The S1 pocket, consisting of R179, R341, and Q283 (caspase-1 numbering), is nearly 100 % conserved; its basicity and its size make it ideally suited for binding an aspartate residue [60].

The S2 pocket of caspases-3 and -7 is formed by aromatic residues and accommodates small aliphatic amino acids [61]. A substitution of a valine or alanine in place of a tyrosine opens up the S2 subsite to larger residues in the initiator and inflammatory caspases.

The S3 pocket consists of main-chain interactions with R341 (caspase-1 numbering) [61]. In caspases-8, and -9, nearby basic residues enhance the binding of glutamic acid residues to the S3 subsite [27, 62, 63].

The S4 subsite of inflammatory caspases is long, shallow, and hydrophobic, accommodating bulky aromatic side chains such as a tryptophan [59]. On the other hand, in apoptotic caspases, a tryptophan (214 in caspase-3) reduces the size of the subsite, causing a preference for an aspartate or a small aliphatic residue in the S4 pocket [60]. An asparagine in caspases-2 and -3 or a glutamine in caspase-7 enhances interaction with a P4 aspartate [60].

Caspase-2 requires a P5 residue to occupy a S5 subsite [60]. The reason for this specificity may be that binding of a small hydrophobic residue to this subsite may enhance the burial of a P4 aspartate [64].

5.1.1 Endogenous Substrates

To date more than 700 substrates of caspases have been catalogued [65]. A searchable database can be found at http://bioinf.gen.tcd.ie/casbah/. Caspase substrates are involved in conferring an apoptotic phenotype to cells. They are also involved in producing "find-me" and "tolerate-me" signals during apoptosis.

5.1.1.1 Substrates Involved in the Apoptotic Phenotype

The following are some of the substrates of caspases which are involved in producing the apoptotic phenotype. The rounding of cells is likely in part due to caspase cleavage of components of actin microfilaments and microtubular proteins [42]. Retraction of cells from their neighbors likely facilitates phagocytosis and is caused in part by caspase cleavage of components of focal adhesion sites, components of cell-cell adherens junctions, cadherins, and desmosome-associated proteins [42]. Caspase cleavage of Rho effector ROCK1 which regulates movement of the actin cytoskeleton is a factor in blebbing and nuclear fragmentation [42]. Nuclear fragmentation also involves caspase cleavage of lamins A, B, and C [66]. Chromatin condensation is caused by caspase cleavage of Mst1 kinase [67]. Hydrolyisis of genomic DNA to small fragments is carried out by caspase-activated DNAse (CAD/DFF) [68]. Translocation of PS to the external surface of the cell is also caspase-dependent, but not fully understood [69].

5.1.1.2 Substrates Involved in Other Aspects of Apoptosis

Caspases are also involved in producing "find-me" signals to cause chemotaxis of phagocytes to apoptotic cells. Caspase-3 cleaves calcium-independent phospholipase A2, causing phosphatidylcholine in the membrane to become hydrolyzed to produce lysophatidylcholine (LPC) [38]. The C-terminal fragment of endothelial monocyte-activating polypeptide II (EMAPII) is produced by caspase-dependent proteolysis and acts as a "find-me" signal to attract monocytes [39]. Caspase-dependent cleavage of the membrane channel pannexin-1 causes release of modest amounts of ATP, which may also act as a "find-me" signal [40].

Caspases also function to keep danger-associated molecular patterns (DAMPs) and alarmins from being activated. This function can be thought of as a "tolerate-me" signal, and is important for avoiding autoimmunity [41]. As mentioned above, caspase activation leads to hydrolysis of genomic DNA (which acts as a DAMP) into short fragments [68]. Additionally, the alarmin IL-33 is inactivated by caspase-3/-7-dependent proteolysis [70].

5.1.2 Synthetic Inhibitors

5.1.2.1 Active Site Inhibitors

Active-site inhibitors bind in the place of substrate and are therefore competitive inhibitors. These inhibitors can be peptidic or nonpeptidic and can bind reversibly or irreversibly.

Peptidic inhibitors can have as few as one amino acid (for example, Boc-Asp-FMK), but typically have four (for example, Ac-DEVD-FMK) [71]. Peptides linked to leaving groups such as halomethylketones [for example, chloromethylketone (CMK) and fluoromethylketone (FMK)], acylomethylketones, and (phosphinyloxy) methyl ketones bind irreversibly, whereas peptides linked to non-leaving groups such as aldehyde (CHO) bind reversibly. The electrophilic carbonyl of the aldehyde or ketone binds to the catalytic cysteine, inhibiting it.

Several different peptidomimetics have been designed as inhibitors for caspases. These include urazolopyrazine-based β-strand peptidomimeticsdesigned as inhibitors for caspase-3 and caspase-8, [72] hydantoin-based peptidomimetics as inhibitors of caspase-3 [73], dipeptidylaspartylfluormethylketones with unnatural amino acids [74], 1-(2-acylhydrazinocarbonyl)-cycloalkyl carboxamides, [75] 8,5-fused bicyclic compounds, [76] and peptidomimetics containing a caprolactam ring [77].

Non-peptide inhibitors have also been discovered. These include isatins, [78, 79] indole aspartyl ketones, fuchsone derivatives, and pyrrolo[3,4-c]quinolone-1,3-diones [80].

5.1.2.2 Allosteric Inhibitors

Caspases-3 and -7 were found to contain an allosteric site at the dimer interface [81]. The drugs FICA and DICA form disulfide bonds with cysteines in the dimer interface of those caspases and inactivate the protein. The structural changes brought about by binding of these drugs involves massive loop rearrangements to a structure very similar to that of the proenzyme.

Mutation of valine 266 to a histidine at the dimer interface of caspase-3 also caused allosteric inactivation of the protein [28]; however, the structural changes brought about by the mutation were much more subtle than those that occurred upon binding of FICA or DICA [82]. Instead of conversion to a structure like that of the proenzyme, inactivation may be caused by a series of steric clashes, disordering of loop L1, and/or destabilization of helix 3.

A drug called compound 34 was found to bind to cysteines near the dimer interface of caspase-1 [83]. Similarly to the binding of FICA and DICA, the inactive structure was like that of the proenzyme.

Another set of allosteric inhibitors was found to inhibit caspases-3, -7, -8, and -9 [84]. A crystal structure with caspase-7 indicates that one and likely all of these compounds binds to the dimer interface. One of the compounds, Comp-A, inhibits dimerization of caspase-8; however, caspase-7 remained a dimer upon binding of the drug. As with FICA and DICA inhibition, the inhibited form was similar to that of the zymogen. However, these new compounds are reversible inhibitors, unlike FICA and DICA.

One of the urazolering peptomimetic inhibitors which bind at the active site was also found to bind near the dimer interface of caspase-8 [72]. Some of the interacting residues of caspase-8 are Tyr334, Thr337, Glu396, and Phe399.

Caspase-2 was allosterically inhibited through binding of a designed ankyrin repeat protein (DARPin) [85]. Binding causes the caspase to be fixed in an inactive conformation different from that of the proenzyme.

A novel allosteric site was found on caspase-6 [86]. Phage display produced a peptide pep419 which binds near helix 2 and causes tetramerization and therefore inactivation of caspase-6. Interestingly, it was found that at pH 8, the zymogen of caspase-6 is a tetramer in solution, whereas at pH 5.5, the zymogen is a dimer, but can be induced to form a tetramer through the binding of pep419 or a related peptide pep420. The pH changes in the cell brought about by apoptosis could potentially lead to dissociation of caspase-6 tetramers to the dimeric form, leading to activation of the protein.

5.1.3 Endogenous Inhibitors

Both viral and endogenous inhibitors can block caspase activity by competing for binding to activation complexes. Viral inhibitors target caspase activity of their host cells in order to counter an immune response. Several γ-herpesviruses and molluscipoxvirus use v-FLIPs to block caspase access to the DISC. Similarly, endogenous $FLIP_s$ blocks procaspase-8 recruitment to DISC, [87] and ICEBERG blocks caspase-1 recruitment to form the inflammasome [88].

Most protease inhibitors bind to the protease and block substrate access [60]. Suicide inhibitors are cleaved and cause a conformational change to occur in either the inhibitor or the protease. Although it is typically a serine protease inhibitor, the serpinCrmA is also able to inhibit caspases-1, [89] -8, [90] and -9, [91] likely by forming a covalent attachment with the caspase and undergoing a conformational change upon cleavage of the scissile P1–P1′ bond to place the caspase on the "bottom" of the inhibitor [92]. Similarly, the baculovirus protein p35 becomes covalently attached to the catalytic cysteine, the scissile bond is cleaved, but the protein is not liberated because it blocks the hydrolytic water from gaining access to the active site [93, 94].

Anothercategory of inhibitors are IAPs (inhibitor of apoptosis proteins). They were first discovered using baculovirus lacking a functional *p35* gene [95]. They contain a 70–80 residue Zn^{2+} binding module named BIR. The most well-studied is X-linked inhibitor of apoptosis protein (XIAP) [96].

XIAP targets caspases in two different ways. A linker to the BIR1 domain and The BIR2 domain of XIAP target effector caspases-3 and -7 [97]. Residues in the active site, particularly in loop L1 make critical contacts with the inhibitor. Loop L2′ also makes contacts with XIAP. The necessity of ordered active site loops and cleaved intersubunit linker to form L2′ mean that XIAP only binds the active caspase rather than the inactive procaspase.

Unlike XIAP binding to the effector caspases, BIR3 and RING of XIAP target initiator caspase-9 [97, 98]. Also, instead of binding to the active site, it binds to the dimer interface of the monomer and blocks dimer formation. Loop L2′ of caspase-9 binds BIR3 in a similar manner to how loop L2′ of caspase-3 binds BIR2. The pocket where loop L2′ of caspase-9 binds BIR3 is called the Smac (second mitochondrial activator of caspases) pocket because Smac can also bind there to derepress caspase activation.

Caspase activity is also controlled endogenously through the use of post-translational modifications. The RING domain of XIAP acts as an E3 ubiquitin ligase toubiquitylate effector caspases-3 and -7, leading to proteasomal degradation [99, 100]. Sumoylation of procaspase-2 [101] and caspase-8 [102] likely leads to localization of the protein in the nucleus.

Phosphorylation is a third post-translational modification which affects caspase activity. p38-MAPK phosphorylates S150 of caspase-3, inhibiting it [103]. Phosphorylation by PKC-δ at an as yet unknown site, on the other hand, enhances caspase-3 activity [104]. PAK2 phosphorylates caspase-7 at three sites, decreasing its activity [105]. For caspase-9, ERK phosphorylates T125, [106] c-Abl phosphorylates Y153, [107] and Akt phosphorylates S196, [108] leading to decreased activity of the protein.

References

1. Alnemri ES, Livingston DJ, Nicholson DW, Salvesen G, Thornberry NA, Wong WW, Yuan J (1996) Human ICE/CED-3 protease nomenclature. Cell 87:171
2. Cerretti DP, Kozlosky CJ, Mosley B, Nelson N, Van Ness K, Greenstreet TA, March CJ, Kronheim SR, Druck T, Cannizzaro LA (1992) Molecular cloning of the interleukin-1 beta converting enzyme. Science 256:97–100
3. Thornberry NA, Bull HG, Calaycay JR, Chapman KT, Howard AD, Kostura MJ, Miller DK, Molineaux SM, Weidner JR, Aunins J (1992) A novel heterodimeric cysteine protease is required for interleukin-1 beta processing in monocytes. Nature 356:768–774
4. Miura M, Zhu H, Rotello R, Hartwieg EA, Yuan J (1993) Induction of apoptosis in fibroblasts by IL-1 beta-converting enzyme, a mammalian homolog of the C. elegans cell death gene ced-3. Cell 75:653–660
5. Fernandes-Alnemri T, Litwack G, Alnemri ES (1994) CPP32, a novel human apoptotic protein with homology to Caenorhabditis elegans cell death protein Ced-3 and mammalian interleukin-1 beta-converting enzyme. J Biol Chem 269:30761–30764
6. Muzio M, Stockwell BR, Stennicke HR, Salvesen GS, Dixit VM (1998) An induced proximity model for caspase-8 activation. J Biol Chem 273:2926–2930
7. Lowe SW, Lin AW (2000) Apoptosis in cancer. Carcinogenesis 21:485–495
8. Ashkenazi A, Dixit VM (1998) Death receptors: signaling and modulation. Science 281:1305–1308
9. Smith CA, Farrah T, Goodwin RG (1994) The TNF receptor superfamily of cellular and viral proteins: activation, costimulation, and death. Cell 76:959–962
10. Milam SL, Clark AC (2009) Folding and assembly kinetics of procaspase-3. Protein Sci 18:2500–2517
11. Pop C, Chen YR, Smith B, Bose K, Bobay B, Tripathy A, Franzen S, Clark AC (2001) Removal of the pro-domain does not affect the conformation of the procaspase-3 dimer. Biochemistry 40:14224–14235
12. Mao PL, Jiang Y, Wee BY, Porter AG (1998) Activation of caspase-1 in the nucleus requires nuclear translocation of pro-caspase-1 mediated by its prodomain. J Biol Chem 273:23621–23624
13. Baliga BC, Colussi PA, Read SH, Dias MM, Jans DA, Kumar S (2003) Role of prodomain in importin-mediated nuclear localization and activation of caspase-2. J Biol Chem 278:4899–4905
14. Colussi PA, Harvey NL, Kumar S (1998) Prodomain-dependent nuclear localization of the caspase-2 (Nedd2) precursor. A novel function for a caspase prodomain. J Biol Chem 273:24535–24542
15. Yaoita Y (2002) Inhibition of nuclear transport of caspase-7 by its prodomain. Biochem Biophys Res Commun 291:79–84
16. Martinon F, Burns K, Tschopp J (2002) The inflammasome: a molecular platform triggering activation of inflammatory caspases and processing of proIL-beta. Mol Cell 10:417–426
17. Faustin B, Lartigue L, Bruey JM, Luciano F, Sergienko E, Bailly-Maitre B, Volkmann N, Hanein D, Rouiller I, Reed JC (2007) Reconstituted NALP1 inflammasome reveals two-step mechanism of caspase-1 activation. Mol Cell 25:713–724
18. Newton K, Dixit VM (2012) Signaling in innate immunity and inflammation. Cold Spring Harbor Perspect Biol 4:a006049, 006019 pp

19. Wang S, Miura M, Jung YK, Zhu H, Li E, Yuan J (1998) Murine caspase-11, an ICE-interacting protease, is essential for the activation of ICE. Cell 92:501–509
20. Saleh M, Vaillancourt JP, Graham RK, Huyck M, Srinivasula SM, Alnemri ES, Steinberg MH, Nolan V, Baldwin CT, Hotchkiss RS, Buchman TG, Zehnbauer BA, Hayden MR, Farrer LA, Roy S, Nicholson DW (2004) Differential modulation of endotoxin responsiveness by human caspase-12 polymorphisms. Nature 429:75–79
21. Koenig U, Eckhart L, Tschachler E (2001) Evidence that caspase-13 is not a human but a bovine gene. Biochem Biophys Res Commun 285:1150–1154
22. Rendl M, Ban J, Mrass P, Mayer C, Lengauer B, Eckhart L, Declerq W, Tschachler E (2002) Caspase-14 expression by epidermal keratinocytes is regulated by retinoids in a differentiation-associated manner. J Invest Dermatol 119:1150–1155
23. Eckhart L, Ballaun C, Hermann M, VandeBerg JL, Sipos W, Uthman A, Fischer H, Tschachler E (2008) Identification of novel mammalian caspases reveals an important role of gene loss in shaping the human caspase repertoire. Mol Biol Evol 25:831–841
24. Eckhart L, Ballaun C, Uthman A, Kittel C, Stichenwirth M, Buchberger M, Fischer H, Sipos W, Tschachler E (2005) Identification and characterization of a novel mammalian caspase with proapoptotic activity. J Biol Chem 280:35077–35080
25. Masumoto J, Zhou W, Chen FF, Su F, Kuwada JY, Hidaka E, Katsuyama T, Sagara J, Taniguchi S, Ngo-Hazelett P, Postlethwait JH, Núñez G, Inohara N (2003) Caspy, a zebrafish caspase, activated by ASC oligomerization is required for pharyngeal arch development. J Biol Chem 278:4268–4276
26. Feeney B, Clark AC (2005) Reassembly of active caspase-3 is facilitated by the propeptide. J Biol Chem 280:39772–39785
27. Boatright KM, Renatus M, Scott FL, Sperandio S, Shin H, Pedersen IM, Ricci JE, Edris WA, Sutherlin DP, Green DR, Salvesen GS (2003) A unified model for apical caspase activation. Mol Cell 11:529–541
28. Pop C, Feeney B, Tripathy A, Clark AC (2003) Mutations in the procaspase-3 dimer interface affect the activity of the zymogen. Biochemistry 42:12311–12320
29. Walters J, Pop C, Scott FL, Drag M, Swartz P, Mattos C, Salvesen GS, Clark AC (2009) A constitutively active and uninhibitable caspase-3 zymogen efficiently induces apoptosis. Biochem J 424:335–345
30. Schipper JL, MacKenzie SH, Sharma A, Clark AC (2011) A bifunctional allosteric site in the dimer interface of procaspase-3. Biophys Chem 159:100–109
31. Zorn JA, Wolan DW, Agard NJ, Wells JA (2012) Fibrils colocalize caspase-3 with procaspase-3 to foster maturation. J Biol Chem 287:33781–33795
32. Wellington CL, Hayden MR (2000) Caspases and neurodegeneration: on the cutting edge of new therapeutic approaches. Clin Genet 57:1–10
33. Stennicke HR, Salvesen GS (1999) Catalytic properties of the caspases. Cell Death Differ 6:1054–1059
34. Miscione GP, Calvaresi M, Bottoni A (2010) Computational evidence for the catalytic mechanism of caspase-7. A DFT investigation. J Phys Chem B 114:4637–4645
35. Sulpizi M, Rothlisberger U, Carloni P (2003) Molecular dynamics studies of caspase-3. Biophys J 84:2207–2215
36. Brady KD, Giegel DA, Grinnell C, Lunney E, Talanian RV, Wong W, Walker N (1999) A catalytic mechanism for caspase-1 and for bimodal inhibition of caspase-1 by activated aspartic ketones. Bioorg Med Chem 7:621–631
37. Sulpizi M, Laio A, VandeVondele J, Cattaneo A, Rothlisberger U, Carloni P (2003) Reaction mechanism of caspases: insights from QM/MM Car-Parrinello simulations. Proteins 52:212–224
38. Lauber K, Bohn E, Kröber SM, Xiao YJ, Blumenthal SG, Lindemann RK, Marini P, Wiedig C, Zobywalski A, Baksh S, Xu Y, Autenrieth IB, Schulze-Osthoff K, Belka C, Stuhler G, Wesselborg S (2003) Apoptotic cells induce migration of phagocytes via caspase-3-mediated release of a lipid attraction signal. Cell 113:717–730

39. Knies UE, Behrensdorf HA, Mitchell CA, Deutsch U, Risau W, Drexler HC, Clauss M (1998) Regulation of endothelial monocyte-activating polypeptide II release by apoptosis. Proc Natl Acad Sci U S A 95:12322–12327
40. Chekeni FB, Elliott MR, Sandilos JK, Walk SF, Kinchen JM, Lazarowski ER, Armstrong AJ, Penuela S, Laird DW, Salvesen GS, Isakson BE, Bayliss DA, Ravichandran KS (2010) Pannexin 1 channels mediate 'find-me' signal release and membrane permeability during apoptosis. Nature 467:863–867
41. Martin SJ, Henry CM, Cullen SP (2012) A perspective on mammalian caspases as positive and negative regulators of inflammation. Mol Cell 46:387–397
42. Taylor RC, Cullen SP, Martin SJ (2008) Apoptosis: controlled demolition at the cellular level. Nat Rev Mol Cell Biol 9:231–241
43. Frank S, Gaume B, Bergmann-Leitner ES, Leitner WW, Robert EG, Catez F, Smith CL, Youle RJ (2001) The role of dynamin-related protein 1, a mediator of mitochondrial fission, in apoptosis. Dev Cell 1:515–525
44. Lane JD, Lucocq J, Pryde J, Barr FA, Woodman PG, Allan VJ, Lowe M (2002) Caspase-mediated cleavage of the stacking protein GRASP65 is required for Golgi fragmentation during apoptosis. J Cell Biol 156:495–509
45. Lamkanfi M (2011) Emerging inflammasome effector mechanisms. Nat Rev Immunol 11:213–220
46. Miao EA, Rajan JV, Aderem A (2011) Caspase-1-induced pyroptotic cell death. Immunol Rev 243:206–214
47. Lamkanfi M, Kalai M, Saelens X, Declercq W, Vandenabeele P (2004) Caspase-1 activates nuclear factor of the kappa-enhancer in B cells independently of its enzymatic activity. J Biol Chem 279:24785–24793
48. Zermati Y, Garrido C, Amsellem S, Fishelson S, Bouscary D, Valensi F, Varet B, Solary E, Hermine O (2001) Caspase activation is required for terminal erythroid differentiation. J Exp Med 193:247–254
49. Fernando P, Kelly JF, Balazsi K, Slack RS, Megeney LA (2002) Caspase 3 activity is required for skeletal muscle differentiation. Proc Natl Acad Sci U S A 99:11025–11030
50. Miura M, Chen XD, Allen MR, Bi Y, Gronthos S, Seo BM, Lakhani S, Flavell RA, Feng XH, Robey PG, Young M, Shi S (2004) A crucial role of caspase-3 in osteogenic differentiation of bone marrow stromal stem cells. J Clin Invest 114:1704–1713
51. Fernando P, Brunette S, Megeney LA (2005) Neural stem cell differentiation is dependent upon endogenous caspase 3 activity. FASEB J 19:1671–1673
52. Gulyaeva NV (2003) Non-apoptotic functions of caspase-3 in nervous tissue. Biochemistry (Mosc) 68:1171–1180
53. Hyman BT, Yuan J (2012) Apoptotic and non-apoptotic roles of caspases in neuronal physiology and pathophysiology. Nat Rev Neurosci 13:395–406
54. Rudrapatna VA, Bangi E, Cagan RL (2013) Caspase signalling in the absence of apoptosis drives Jnk-dependent invasion. EMBO Rep 14:172–177
55. Ryoo HD, Bergmann A (2012) The role of apoptosis-induced proliferation for regeneration and cancer. Cold Spring Harb Perspect Biol 4:a008797
56. Oberst A, Dillon CP, Weinlich R, McCormick LL, Fitzgerald P, Pop C, Hakem R, Salvesen GS, Green DR (2011) Catalytic activity of the caspase-8-FLIP(L) complex inhibits RIPK3-dependent necrosis. Nature 471:363–367
57. Kataoka T, Tschopp J (2004) N-terminal fragment of c-FLIP(L) processed by caspase 8 specifically interacts with TRAF2 and induces activation of the NF-kappaB signaling pathway. Mol Cell Biol 24:2627–2636
58. Thornberry NA, Chapman KT, Nicholson DW (2000) Determination of caspase specificities using a peptide combinatorial library. Methods Enzymol 322:100–110
59. Thornberry NA, Rano TA, Peterson EP, Rasper DM, Timkey T, Garcia-Calvo M, Houtzager VM, Nordstrom PA, Roy S, Vaillancourt JP, Chapman KT, Nicholson DW (1997) A combinatorial approach defines specificities of members of the caspase family and granzyme B. Functional relationships established for key mediators of apoptosis. J Biol Chem 272:17907–17911

60. Fuentes-Prior P, Salvesen GS (2004) The protein structures that shape caspase activity, specificity, activation and inhibition. Biochem J 384:201–232
61. Chéreau D, Kodandapani L, Tomaselli KJ, Spada AP, Wu JC (2003) Structural and functional analysis of caspase active sites. Biochemistry 42:4151–4160
62. Blanchard H, Kodandapani L, Mittl PR, Marco SD, Krebs JF, Wu JC, Tomaselli KJ, Grütter MG (1999) The three-dimensional structure of caspase-8: an initiator enzyme in apoptosis. Structure 7:1125–1133
63. Renatus M, Stennicke HR, Scott FL, Liddington RC, Salvesen GS (2001) Dimer formation drives the activation of the cell death protease caspase 9. Proc Natl Acad Sci U S A 98:14250–14255
64. Schweizer A, Briand C, Grutter MG (2003) Crystal structure of caspase-2, apical initiator of the intrinsic apoptotic pathway. J Biol Chem 278:42441–42447
65. Lüthi AU, Martin SJ (2007) The CASBAH: a searchable database of caspase substrates. Cell Death Differ 14:641–650
66. Rao L, Perez D, White E (1996) Lamin proteolysis facilitates nuclear events during apoptosis. J Cell Biol 135:1441–1455
67. Ura S, Masuyama N, Graves JD, Gotoh Y (2001) Caspase cleavage of MST1 promotes nuclear translocation and chromatin condensation. Proc Natl Acad Sci U S A 98:10148–10153
68. Enari M, Sakahira H, Yokoyama H, Okawa K, Iwamatsu A, Nagata S (1998) A caspase-activated DNase that degrades DNA during apoptosis, and its inhibitor ICAD. Nature 391:43–50
69. Martin SJ, Reutelingsperger CP, McGahon AJ, Rader JA, van Schie RC, LaFace DM, Green DR (1995) Early redistribution of plasma membrane phosphatidylserine is a general feature of apoptosis regardless of the initiating stimulus: inhibition by overexpression of Bcl-2 and Abl. J Exp Med 182:1545–1556
70. Lüthi AU, Cullen SP, McNeela EA, Duriez PJ, Afonina IS, Sheridan C, Brumatti G, Taylor RC, Kersse K, Vandenabeele P, Lavelle EC, Martin SJ (2009) Suppression of interleukin-33 bioactivity through proteolysis by apoptotic caspases. Immunity 31:84–98
71. Callus BA, Vaux DL (2007) Caspase inhibitors: viral, cellular and chemical. Cell Death Differ 14:73–78
72. Wang Z, Watt W, Brooks NA, Harris MS, Urban J, Boatman D, McMillan M, Kahn M, Heinrikson RL, Finzel BC, Wittwer AJ, Blinn J, Kamtekar S, Tomasselli AG (2010) Kinetic and structural characterization of caspase-3 and caspase-8 inhibition by a novel class of irreversible inhibitors. Biochim Biophys Acta 1804:1817–1831
73. Vázquez J, García-Jareño A, Mondragón L, Rubio-Martinez J, Pérez-Payá E, Albericio F (2008) Conformationally restricted hydantoin-based peptidomimetics as inhibitors of caspase-3 with basic groups allowed at the S3 enzyme subsite. ChemMedChem 3:979–985
74. Wang Y, Jia S, Tseng B, Drewe J, Cai SX (2007) Dipeptidyl aspartyl fluoromethylketones as potent caspase inhibitors: peptidomimetic replacement of the P(2) amino acid by 2-aminoaryl acids and other non-natural amino acids. Bioorg Med Chem Lett 17:6178–6182
75. Soper DL, Sheville J, O'Neil SV, Wang Y, Laufersweiler MC, Oppong KA, Wos JA, Ellis CD, Fancher AN, Lu W, Suchanek MK, Wang RL, De B, Demuth TP (2006) Synthesis and evaluation of novel 1-(2-acylhydrazinocarbonyl)-cycloalkyl carboxamides as interleukin-1beta converting enzyme (ICE) inhibitors. Bioorg Med Chem Lett 16:4233–4236
76. Soper DL, Sheville JX, O'Neil SV, Wang Y, Laufersweiler MC, Oppong KA, Wos JA, Ellis CD, Baize MW, Chen JJ, Fancher AN, Lu W, Suchanek MK, Wang RL, Schwecke WP, Cruze CA, Buchalova M, Belkin M, Wireko F, Ritter A, De B, Wang D, Demuth TP (2006) Synthesis and evaluation of novel 8,5-fused bicyclic peptidomimetic compounds as interleukin-1beta converting enzyme (ICE) inhibitors. Bioorg Med Chem 14:7880–7892
77. Wang Y, O'Neil SV, Wos JA, Oppong KA, Laufersweiler MC, Soper DL, Ellis CD, Baize MW, Fancher AN, Lu W, Suchanek MK, Wang RL, Schwecke WP, Cruze CA, Buchalova M, Belkin M, De B, Demuth TP (2007) Synthesis and evaluation of unsaturated caprolactams as interleukin-1beta converting enzyme (ICE) inhibitors. Bioorg Med Chem 15:1311–1322

78. Lee D, Long SA, Adams JL, Chan G, Vaidya KS, Francis TA, Kikly K, Winkler JD, Sung CM, Debouck C, Richardson S, Levy MA, DeWolf WE, Keller PM, Tomaszek T, Head MS, Ryan MD, Haltiwanger RC, Liang PH, Janson CA, McDevitt PJ, Johanson K, Concha NO, Chan W, Abdel-Meguid SS, Badger AM, Lark MW, Nadeau DP, Suva LJ, Gowen M, Nuttall ME (2000) Potent and selective nonpeptide inhibitors of caspases 3 and 7 inhibit apoptosis and maintain cell functionality. J Biol Chem 275:16007–16014
79. Lee D, Long SA, Murray JH, Adams JL, Nuttall ME, Nadeau DP, Kikly K, Winkler JD, Sung CM, Ryan MD, Levy MA, Keller PM, DeWolf WE (2001) Potent and selective nonpeptide inhibitors of caspases 3 and 7. J Med Chem 44:2015–2026
80. Kravchenko DV, Kysil VM, Tkachenko SE, Maliarchouk S, Okun IM, Ivachtchenko AV (2005) Pyrrolo[3,4-c]quinoline-1,3-diones as potent caspase-3 inhibitors. Synthesis and SAR of 2-substituted 4-methyl-8-(morpholine-4-sulfonyl)-pyrrolo[3,4-c]quinoline-1,3-diones. Eur J Med Chem 40:1377–1383
81. Hardy JA, Lam J, Nguyen JT, O'Brien T, Wells JA (2004) Discovery of an allosteric site in the caspases. Proc Natl Acad Sci U S A 101:12461–12466
82. Walters J, Schipper JL, Swartz P, Mattos C, Clark AC (2012) Allosteric modulation of caspase 3 through mutagenesis. Biosci Rep 32:401–411
83. Datta D, Scheer JM, Romanowski MJ, Wells JA (2008) An allosteric circuit in caspase-1. J Mol Biol 381:1157–1167
84. Feldman T, Kabaleeswaran V, Jang SB, Antczak C, Djaballah H, Wu H, Jiang X (2012) A class of allosteric caspase inhibitors identified by high-throughput screening. Mol Cell 47:585–595
85. Schweizer A, Roschitzki-Voser H, Amstutz P, Briand C, Gulotti-Georgieva M, Prenosil E, Binz HK, Capitani G, Baici A, Plückthun A, Grütter MG (2007) Inhibition of caspase-2 by a designed ankyrin repeat protein: specificity, structure, and inhibition mechanism. Structure 15:625–636
86. Stanger K, Steffek M, Zhou L, Pozniak CD, Quan C, Franke Y, Tom J, Tam C, Elliott JM, Lewcock JW, Zhang Y, Murray J, Hannoush RN (2012) Allosteric peptides bind a caspase zymogen and mediate caspase tetramerization. Nat Chem Biol 8:655–660
87. Irmler M, Thome M, Hahne M, Schneider P, Hofmann K, Steiner V, Bodmer JL, Schröter M, Burns K, Mattmann C, Rimoldi D, French LE, Tschopp J (1997) Inhibition of death receptor signals by cellular FLIP. Nature 388:190–195
88. Humke EW, Shriver SK, Starovasnik MA, Fairbrother WJ, Dixit VM (2000) ICEBERG: a novel inhibitor of interleukin-1beta generation. Cell 103:99–111
89. Komiyama T, Ray CA, Pickup DJ, Howard AD, Thornberry NA, Peterson EP, Salvesen G (1994) Inhibition of interleukin-1 beta converting enzyme by the cowpox virus serpin CrmA. An example of cross-class inhibition. J Biol Chem 269:19331–19337
90. Zhou Q, Snipas S, Orth K, Muzio M, Dixit VM, Salvesen GS (1997) Target protease specificity of the viral serpin CrmA. Analysis of five caspases. J Biol Chem 272:7797–7800
91. Ryan CA, Stennicke HR, Nava VE, Burch JB, Hardwick JM, Salvesen GS (2002) Inhibitor specificity of recombinant and endogenous caspase-9. Biochem J 366:595–601
92. Gettins PG (2002) Serpin structure, mechanism, and function. Chem Rev 102:4751–4804
93. Xu G, Cirilli M, Huang Y, Rich RL, Myszka DG, Wu H (2001) Covalent inhibition revealed by the crystal structure of the caspase-8/p35 complex. Nature 410:494–497
94. Xu G, Rich RL, Steegborn C, Min T, Huang Y, Myszka DG, Wu H (2003) Mutational analyses of the p35-caspase interaction. A bowstring kinetic model of caspase inhibition by p35. J Biol Chem 278:5455–5461
95. Crook NE, Clem RJ, Miller LK (1993) An apoptosis-inhibiting baculovirus gene with a zinc finger-like motif. J Virol 67:2168–2174
96. Rehm M, Huber HJ, Dussmann H, Prehn JH (2006) Systems analysis of effector caspase activation and its control by X-linked inhibitor of apoptosis protein. EMBO J 25:4338–4349
97. Takahashi R, Deveraux Q, Tamm I, Welsh K, Assa-Munt N, Salvesen GS, Reed JC (1998) A single BIR domain of XIAP sufficient for inhibiting caspases. J Biol Chem 273:7787–7790

98. Deveraux QL, Leo E, Stennicke HR, Welsh K, Salvesen GS, Reed JC (1999) Cleavage of human inhibitor of apoptosis protein XIAP results in fragments with distinct specificities for caspases. EMBO J 18:5242–5251
99. Vucic D, Dixit VM, Wertz IE (2011) Ubiquitylation in apoptosis: a post-translational modification at the edge of life and death. Nat Rev Mol Cell Biol 12:439–452
100. Suzuki Y, Nakabayashi Y, Takahashi R (2001) Ubiquitin-protein ligase activity of X-linked inhibitor of apoptosis protein promotes proteasomal degradation of caspase-3 and enhances its anti-apoptotic effect in Fas-induced cell death. Proc Natl Acad Sci U S A 98:8662–8667
101. Shirakura H, Hayashi N, Ogino S, Tsuruma K, Uehara T, Nomura Y (2005) Caspase recruitment domain of procaspase-2 could be a target for SUMO-1 modification through Ubc9. Biochem Biophys Res Commun 331:1007–1015
102. Besnault-Mascard L, Leprince C, Auffredou MT, Meunier B, Bourgeade MF, Camonis J, Lorenzo HK, Vazquez A (2005) Caspase-8 sumoylation is associated with nuclear localization. Oncogene 24:3268–3273
103. Alvarado-Kristensson M, Melander F, Leandersson K, Rönnstrand L, Wernstedt C, Andersson T (2004) p38-MAPK signals survival by phosphorylation of caspase-8 and caspase-3 in human neutrophils. J Exp Med 199:449–458
104. Voss OH, Kim S, Wewers MD, Doseff AI (2005) Regulation of monocyte apoptosis by the protein kinase Cdelta-dependent phosphorylation of caspase-3. J Biol Chem 280:17371–17379
105. Li X, Wen W, Liu K, Zhu F, Malakhova M, Peng C, Li T, Kim HG, Ma W, Cho YY, Bode AM, Dong Z (2011) Phosphorylation of caspase-7 by p21-activated protein kinase (PAK) 2 inhibits chemotherapeutic drug-induced apoptosis of breast cancer cell lines. J Biol Chem 286:22291–22299
106. Allan LA, Morrice N, Brady S, Magee G, Pathak S, Clarke PR (2003) Inhibition of caspase-9 through phosphorylation at Thr 125 by ERK MAPK. Nat Cell Biol 5:647–654
107. Raina D, Pandey P, Ahmad R, Bharti A, Ren J, Kharbanda S, Weichselbaum R, Kufe D (2005) c-Abl tyrosine kinase regulates caspase-9 autocleavage in the apoptotic response to DNA damage. J Biol Chem 280:11147–11151
108. Cardone MH, Roy N, Stennicke HR, Salvesen GS, Franke TF, Stanbridge E, Frisch S, Reed JC (1998) Regulation of cell death protease caspase-9 by phosphorylation. Science 282:1318–1321

Chapter 3
Calpains and Granzymes: Non-caspase Proteases in Cell Death

Raja Reddy Kuppili and Kakoli Bose

Abstract Proteolysis is the fundamental requirement of the process known as apoptosis or programmed cell death. Despite caspases being the primary molecules for apoptosis, other non-caspase proteases including calpains, granzymes, cathepsins, and the HtrA family of proteins also play pivotal roles in mediating and promoting cell death. This chapter is an attempt to discuss the different aspects of two of these non-caspase proteases: calpains and granzymes.

Calpains, a family of Ca^{2+} activated cysteine proteases are localized to the cytosol and mitochondria. On the other hand, granzymes (Grs), a set of serine proteases that are present in cytotoxic T and natural killer cells are capable of eliminating virally infected and malignant cells. An increase in intracellular Ca^{2+} level is thought to trigger calpain activation, whereas, the granzymes work in concert with the FasL-Fas signaling system. Once activated, calpains degrade cell membrane, cytoplasmic as well as nuclear substrates thus leading to the breakdown of cellular architecture and finally apoptosis. Grs execute their functions via caspase-dependent apoptotic pathway by either cleaving death antagonists or directly processing caspase-3. However, Grs also contribute to antiviral immunity by triggering pro-inflammatory response and thus resemble caspases in their ability to influence both apoptosis and inflammation. Since calpains and granzymes play pivotal roles in mitochondria related disorders and cancer respectively, delineating their specific functions in this complex cellular biochemical cascade might assist in identifying rate-limiting steps for targeted therapy.

Keywords Apoptosis • Non-caspases • Calpains • Granzymes • Serine proteases • Cysteine proteases

---

R.R. Kuppili • K. Bose (✉)
Integrated Biophysics and Structural Biology Lab, Advanced Centre for Treatment, Research and Education in Cancer (ACTREC), Tata Memorial Centre, Navi Mumbai, Maharashtra 410210, India
e-mail: kbose@actrec.gov.in

K. Bose (ed.), *Proteases in Apoptosis: Pathways, Protocols and Translational Advances*, DOI 10.1007/978-3-319-19497-4_3

1 Calpains

Calpains are calcium-regulated non-lysosomal papain-like thiol proteases that catalyze limited proteolysis of their cellular substrates. These substrates or targets are generally the ones involved in cytoskeletal remodeling and signal transduction. The first reports on calpain came from two different groups in the 1960s who noted the presence of a calcium-activated proteolytic activity in soluble extracts from rat brain [1] and skeletal muscle [2]. In the year 1976, the enzyme was purified to homogeneity from skeletal muscle by Dayton and co-workers [3] and was called CANP (calcium-activated neutral protease), where 'neutral' refers to requirement of optimal pH for its activity. The calcium-dependent activity and intracellular localization along with the limited proteolysis of its substrates highlighted its role as a regulatory, rather than a digestive protease. The word 'calpain' was originally used by Murachi et al. in 1980 to recognize it as a hybrid of two well-known proteins at the time: the calcium-regulated signalling protein called calmodulin and the cysteine protease papain [4]. The calpain family has two abundant isoforms found in nearly all tissues and cell types in mammals and are known as the classical or conventional calpains. They were originally referred to as μ-and m-calpain to reflect the differences observed from the [Ca^{2+}] required for half-maximal activation. These were the first two calpains to be characterized and hence designated as classical calpains [5, 6]. The μ-and m-calpains were renamed as calpain 1 and calpain 2 respectively since more comprehensive studies reported that the difference in activating [Ca^{2+}] for these two molecules was not significant enough to be categorized in that manner.

1.1 Classification

The calpain (CAPN) family is well conserved from fungi to humans. The activity of calpains was initially found to be attributable to two main isoforms, denoted as μ-calpain and m-calpain. Each of these is a heterodimer composed of an 80 kDa or 80 K large catalytic domain and a common 30 kDa regulatory subunit or 30 K.

There are 15 human calpain genes identified for members of the 80 K catalytic subunit family denoted by *CAPNn* (n = 1–3 and 5–16) apart from two genes for 30 K regulatory subunits *CAPNS1* and *CAPNS2*, and one gene for endogenous inhibitor calpastatin denoted by *CAST*. The predicted product for the gene *CAPN16*, encoding calpain-16 includes only the first half of the protease core. These genes and their products that make up the calpain superfamily are summarized in Fig. 3.1.

Calpains are often broadly classified as conventional/typical calpains and atypical calpains based on their domain composition. Calpains 1, 2, 3, 8, 9, 11, 12, 13 and 14 that are composed of four domains resembling structural organization originally found in classical calpains are called typical calpains. Among these, calpains 1, 2 and 9 associate with 30 K to form the heterodimer, while calpains 3, 8, 11, 12, 13

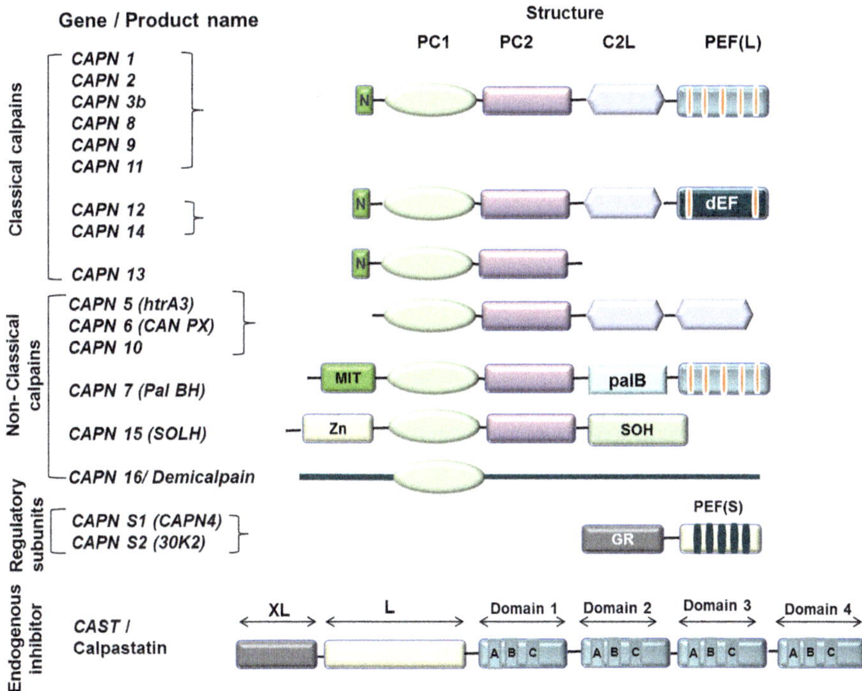

Fig. 3.1 Domain architecture of the calpain family members: Typical calpains (80 K) are composed of four domains viz. Protease core domain 1 (*PC1*), Protease core domain 2 (*PC2*), C2 like domain (*C2*), penta E-helix-F-helix motif PEF (*L*). In atypical calpains, certain domains of typical calpains are deleted or replaced. The small subunit of calpain (30 K) is composed of Glycine rich (GR) and PEF (S) domains. The only endogenous calpain inhibitor is calpastatin which is composed of five domains viz. *XL*, *L* and four domain repeats which are subdivided into *A*, *B* and *C* and are actually involved in the inhibition

and 14 do not interact with 30 K despite the existence of calmodulin-like domain that is important for this association. The atypical calpains include CAPN 5, 6, 7, 8b, 10a, 15 and 16. The term 'atypical' denotes the absence or deletion of certain domains found in classical calpains [3, 7]. Their inability to form dimer with 30 K is due to the lack of domain IV that is required for this function.

Another basis of categorizing calpains is their tissue specificity. Based on their expression profile, calpains 1, 2, 5, 7, 10, 13, and 15 are referred to as ubiquitous calpains, while calpains 3, 6, 8, 9, 11, and 12 are tissue-specific. For example, CAPN 3 is skeletal muscle specific, CAPN 8 is gastrointestinal tract specific [8–10], CAPN6 is expressed in placenta and embryonic striated muscle [11], CAPN11 expression is predominant in the testis [11, 12], and CAPN12 is primarily found in the hair follicles [13]. Another unique tissue specific member found in this family is the calpain oligomer, G-calpain, which is associated with the gastrointestinal tract [10]. This enzyme appears to be a heterodimer of CAPN8 and CAPN9 and is also referred to as calpain-8/9 [8–10]. The significance of this tissue specific differential expression profile of calpains, is manifested by the different effects they have. The ubiquitously expressing calpains are thus bound to have a global and sometimes lethal effect while others that are tissue specific are associated with abnormalities in specific cells or tissues.

1.2 Nomenclature and Structural Assembly

The nomenclature for calpains has undergone several modifications over the course of time. According to the more recent and widely accepted nomenclature, the μ- and m-calpains are known as calpain 1 and 2 respectively. The classical calpain has a heterodimeric structure composed of catalytic and regulatory subunits, which are composed of four and two domains, respectively. The catalytic subunit comprises an N terminal anchor helix, PC1 and PC2 domains that make up the protease core, C2L or C2-like domain followed by the PEF (penta E-helix-loop-F-helix motif) domain of the large (L) subunit or PEF (L). Similarly, the regulatory subunit is composed of GR (glycine-rich) domain and the PEF domain of the small subunit or PEF (S) (Fig. 3.1). The above domain denotations are based on the current nomenclature. So, the protease core domains PC1 and PC2 were originally known as I and II or IIa and IIb. Domain III is now termed as C2L, and the PEF hand domains IV and VI are now widely known as PEF (L) and PEF (S) domains, respectively. Furthermore, the GR domain was earlier referred to as Domain V.

Calpain activity was first identified from tissue extracts wherein activities corresponding to the classical calpains 1 and 2 were detected [10]. Further improvisation and *in vitro* purifications led to isolation of calpains in homogeneity. With structural and functional genomics gaining impetus, numerous calpain structures and their domain combinations have been solved. The first structure solved for a calpain family member was the PEF domain of CAPNS1 in 1997 [14, 15]. Although this provided only a partial picture, functionally more relevant structures for the apo-forms of rat and human calpain-2 (CAPN2/S1) were soon published in 1999 and 2000 respectively [16, 17]. Later, three dimensional (3D) structures of the protease (CysPc) domains of CAPN1, 2, and 9 in the presence of Ca^{2+} were elucidated, revealing the possible molecular mechanism of calpain activation [18–20]. Nonetheless, the structure providing the most comprehensive information was that of the calcium bound active form of rat CAPN2/S1 [m-calpain] co-crystallized with calpastatin fragment. Calpastatin being the only known endogenous inhibitor for calpains [19, 21], this structure not only provided snapshots of structural details of the interaction but also furnished important insights into the calpain activation mechanism.

The crystal structures also provide crucial information on the nature of interaction involved in heterodimer formation in classical calpains. The 80 K subunit of the heterodimer begins with a 19 residue N-terminal anchor region. This region is involved in interaction with the PEF (S) and hence anchors the large subunit to the

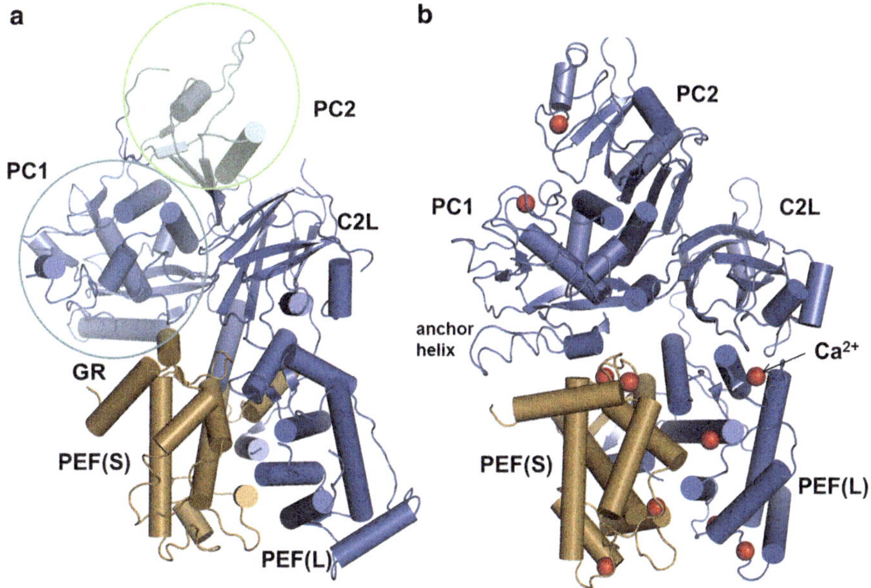

Fig. 3.2 Cartoon representation of m-calpain structure: (**a**) Crystal structure of m-calpain without Ca^{2+} (PDB ID: 1KFU) (**b**) Crystal structure of m-calpain with Ca^{2+} (PDB ID: 1DFO). The large (L) and small (S) subunits are shown in *blue* and *light brown* respectively. The *red spheres* are the calcium molecules bound to calcium-binding sites in the respective domains. The figures are generated using PyMOL (DeLano Scientific, USA)

small subunit. This anchor-helix is autolysed when activated by Ca^{2+}, resulting in calpain to function at a lower Ca^{2+} concentration with different substrate specificity and rate of subunit dissociation. Therefore, autolysis of this anchor helix was believed to be one of the most important regulatory steps in calpain activation.

PC1 is the first catalytic core domain of the 80 K and contains the site where proteolysis of substrates is undertaken (Fig. 3.2). The active site Cys present in PC1 interacts with both the substrate and the inhibitory regions of calpastatin as shown in Fig. 3.3 [10, 16]. PC2 is the second protease core domain, bearing the other two residues of the triad and forms a cleft along with PC1 to make up the active protease (CysPc) domain. This cleft formation is triggered only upon binding of a single Ca^{2+} to each of the Ca^{2+}-binding site in both core domains (CBS-1 and −2). Prior to activation, the inactive conformation is stabilized by PC2-C2L interaction on one side and the PC1-PEF (S) interaction on the other [17, 22, 23].

With no apparent sequence homology to other proteins, the exact function of C2L domain is unknown. However, the higher order conformation of this β-sandwich structure resembles C2 domains found in proteins including phospholipase C, protein kinase C and synaptotagmin. It has also been proposed to bind Ca^{2+} and phospholipids which can be attributed to the presence of clusters of acidic residues on its surface [24–26]. Mutational studies further provided evidence for C2L to act

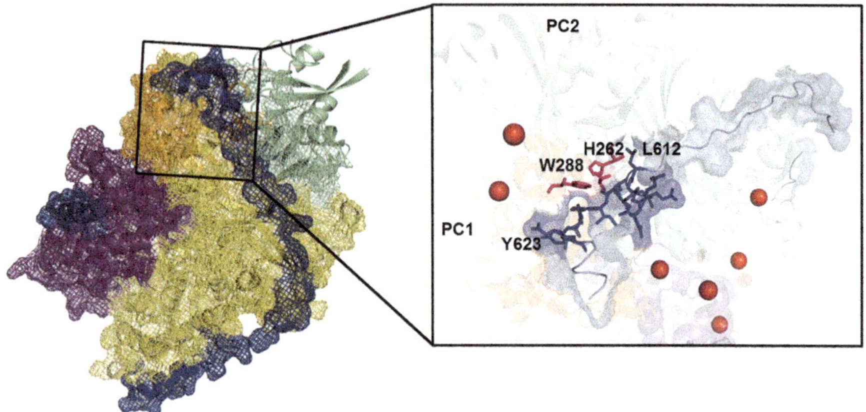

Fig. 3.3 m-Calpain in complex with calpastatin: The mesh representation of m-calpain with bound capastatin and Ca^{2+} (PDB ID 3BOW). Large Subunit (L) – *yellow*, small subunit (S) – *purple*, calpastatin – *blue* and the critical residues are shown as *sticks*

as an electrostatic switch for maintenance of both the catalytic core in an inactive form as well as subsequent stabilization of the active enzyme [27, 28]. In the structure, C2L is followed by PEF (L) at the C-terminal end of the large subunit. PEF (L) domain is approximately 170 amino acids, predominantly α-helical, bearing structural similarity to calmodulin with five Ca^{2+} binding sites. Of these, four EF hands are involved in binding Ca^{2+} while the last assists in making contacts between the two subunits.

The GR domain, which is the N-terminal region of the regulatory subunit, might act as a membrane anchor. This small subunit domain, with five proline and 40 glycine residues including two contiguous stretches of 11 and 20 glycines, is one of the most hydrophobic regions of the protease. Due to its high conformational plasticity, this domain is not seen in the Ca^{2+}- bound crystal structure. The other domain present in the regulatory subunit is PEF (S) that resides at the extreme C-terminal end of the small subunit [29]. Both the PEF (L) and PEF (S) domains have a very similar sequence and structure with approximately 45 % sequence identity and an RMSD (root mean square deviation) of approximately 1.5 Å for Cα atoms. PEF domain is a unique structural feature of calpains as this domain is typically involved in either heterodimerization, usually with the small subunit, CAPNS1, or homodimerization as in the case of calpain-3 [30, 31].

Additionally, studies with domain combinations reveal their significance and indispensability in the overall protease functions. For example, although it was observed that the calpain-1 core forms a functional protease with properties similar to the whole enzyme including substrate specificity, inhibitor sensitivity and calcium requirements, its turnover number is greatly reduced to merely ~5–10 % of the wild type protease. Studies also demonstrated that although 30 K acts as a chaperone to fold 80 K into its correct conformation, its role for the 80 K protease activity is dispensable [30, 31]. A common understanding that could be derived from these structures is that, in contrast to most of the other allosterically regulated enzymes where activation relieves steric hindrance at a preformed active site, classical calpains require a conformational change to realign the active site residues (Cys, His, and Asn). This realignment is mandatory for it to be catalytically competent as discussed elaborately in the next section.

1.3 Activation Mechanism

The activation mechanism of calpains has been deduced based on studies on calpains 1 and 2. Calpain exists as an inactive enzyme in the cytosol and translocates to membranes in response to increased cellular level of Ca^{2+}. Calpain is thus activated in the presence of Ca^{2+} and phospholipids at the membrane [20].

After their release from the cell-membrane, activated calpain hydrolyzes substrate proteins at the membrane or in the cytosol [17, 23]. The original concept of calpain activation was sequence- based, postulating that the C-terminal PEF domains act as calmodulin-like regulators. According to this, the binding of Ca^{2+} to PEF domains, induces a conformational change in the calpain dimer resulting in autoproteolytic cleavage of the N terminal anchor region, thereby activating the enzyme [32]. This theory remained largely unsubstantiated and with comprehensive studies has been replaced by a more plausible model described in Fig. 3.4. This more recent and widely accepted model suggests the following events to be responsible for calpain activation.

The most important and primary factor responsible for calpain activation is the presence of calcium. In the absence of Ca^{2+}, the active site is misaligned as revealed by the apo structure of calpain 2. The two protease core domains PC1 and PC2 are

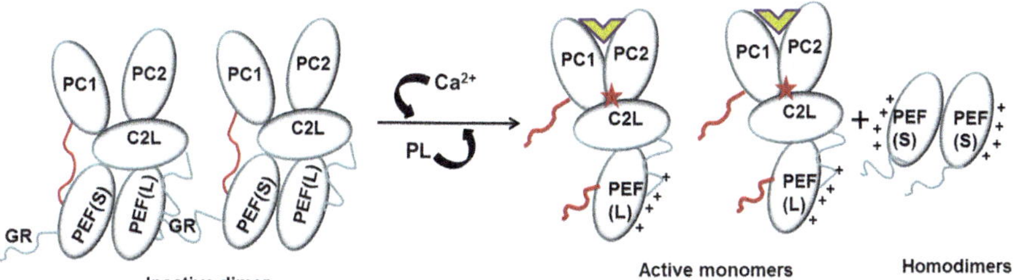

Fig. 3.4 Model representing activation mechanism of calpains: Binding of Ca^{2+} and phospholipids (PL) to m-calpain induces conformational changes, which bring PC1 and PC2 closer together to form a cleft shown as 'V' and a functional catalytic site (represented as a *red star*). This leads to dissociation of 30 K from 80 K thus resulting in PEF(S) homodimer formation. There are at least three different calcium-binding sites in m-calpain, two calmodulin-like EF-hand structures in domains PEF (L) and PEF (S), an acidic loop in C2L, and two non–EF-hand calcium-binding sites in PC1 and PC2 (Adapted from Suzuki et al. 2004 [32])

separated by structural constraints imposed by domain interaction thus rendering the protease inactive due to absence of functional cleft formation. As a consequence, the catalytic triad residues Cys 115, His 272 and Asn 296 are at distances incapable of forming a functional active site. The distance between the cysteine and histidine side chains in this inactive conformation is more than 8.5 Å which is too far for the deprotonation of Cys 115 and subsequent nucleophilic attack on the substrate. The complex mechanism of activation thus involves multiple factors in concert to bring these residues within a distance of ~3.5 Å thus rendering it functional.

The protease domain, which is the epicenter of proteolysis, has two calcium-binding sites: one in each of the sub-domains PC1 and PC2. Upon Ca^{2+} binding, these act co-operatively to convert the core into a functional cysteine protease. Each site comprises two flexible loops that supply backbone carbonyl and side-chain carboxy groups for co-ordinating the Ca^{2+} ions. In the presence of an environment of high [Ca^{2+}], this co-operativity causes a huge conformational change in the core that reorients the two domains. Reorientation enables the core to form a functional active site cleft by a rotation of the two protease core domains relative to each other. A characteristic feature of the cleft so formed is its extremely buried nature which explains the role of calpains in limited proteolysis and preference for substrate molecules with slightly disordered or open conformation. The relative inter-domain movement also brings the catalytic triad into an orientation that is conducive for peptide bond cleavage. The key functional re-orientation here is the repositioning of Cys 115 in PC1 close enough to His 272 of PC2 for deprotonation of the former to occur. Since the core is supported in its active conformation by the other domains, it has been hypothesized that these domains constrain the core in the absence of Ca^{2+} and prevent unregulated activation. The drastically reduced enzymatic activity of the core in the absence of these domains explains the significance of these inter-domain contacts.

Unlike the original belief that the PEF domains are primary regulators of calpain activity, studies suggest that it has an additional role in regulating constitutive activation of calpain. One constraint that is freed upon activation is the anchor helix, which bridges the N-terminus of the large subunit to the small subunit. It has been postulated that the co-ordination of Ca^{2+} by PEF(S) immediately opposite the anchor helix causes a charge repulsion of the basic residues in the helix. Consistent with this idea, the anchor helix is both displaced and unstructured in the calpastatin-inhibited holo structure of calpain-2. The other inter-domain interactions include those shared by domains PC2 and C2L with the linker region between PEF (L) and C2L. This interaction is pre-dominantly defined by extensive series of electrostatic contacts between these regions. It has also been suggested that subtle conformational changes derived from Ca^{2+} binding to the PEF domains might be transduced through the linker to the contacts between PC2 and C2L. For example, binding of Ca^{2+} to the third EF-hand brings about shift in the relative positions of C2L and PEF (L) thus resulting in an increase in the contact area between the two domains by rendering the linker region more flexible. Simultaneously, the subunits dissociate from each other releasing two PEF (S) domains. The role of post-translational modifications in regulating calpain activity also cannot be denied.

One such observation was phosphorylation at Ser 369 in C2L by protein kinase A which restricts domain movement and freezes calpain-1 in an inactive state [25]. Further studies along this line would evaluate the physiological importance of phosphorylation and other post-translational modifications in the regulation of calpain activity.

1.4 Catalytic Mechanism

Calpains (EC 3.4.22.17; Clan CA, family C02) constitute a distinct group of intracellular cysteine proteases found in almost all eukaryotes and a few bacteria. This family of proteins executes its physiological functions via its proteolytic activity. Although most of the calpain family members have been fully characterized at the protein or enzyme level, the protease activity has not yet been identified for calpains 7, 10, and 15. Calpains have a characteristic catalytic triad of cysteine, asparagine, and histidine, which is conserved throughout the entire family. An exception in this group is calpain 6 where the active site Cys has been replaced with lysine, making it inactive. Thus the role of this calpain lies in its interaction with other proteins and microtubule stabilization rather than as a protease. The general mode of catalysis by cysteine proteases has been exemplified for caspases in Chap. 2 of this volume.

1.5 Functions of Calpains

In vivo and *in vitro* studies have led to the identification of several substrate proteins for calpains and the potential pathways they are involved in. Despite these findings, the physiological function of calpain is still not fully understood. The cellular relevance of calpain has been studied using gene knockout models of the same. One such study was the homozygous disruption of the murine *CAPN4* (30 K) that eliminated both μ- and m-calpain activities. This loss in activity had no significant effect on the survival and proliferation of cultured embryonic stem cells [24, 33]. Although, $CAPN4^{+/-}$ mice possessed calpain activity and were phenotypically normal, $CAPN4^{-/-}$ embryos died at 11.5 embryonic day with apparent defects in cardiovascular system and erythropoiesis [22, 34]. These studies thus prove the indispensability of the calpain system in the cell. Similar work has led to various important findings that define the physiological significance of the calpains.

1.5.1 Role in Apoptosis

It was in 1993 that the possible involvement of calpain in apoptosis was first suggested based on the findings in different cell types [35]. Several studies since then have documented calpains to be important in ischemia/reperfusion injury in the heart, mitochondrial permeability transition, and necrotic/ apoptotic cell death. It has also been implicated in myocardial infarction resulting in mitochondrial damage and apoptosis of cardiomyocytes among others [36]. Neurode-generative disorders such as cerebral ischemia, Alzheimer's, Parkinson's and Huntington's disease, amyotrophic lateral sclerosis and multiple sclerosis, where the neurons in the central nervous system die, are some of the pathogenic manifestations of calpain activation.

Calpains 1 and 2 have been shown to promote apoptosis and necrosis in renal cells, by cleaving cytoskeletal proteins, thereby increasing plasma membrane permeability followed by cleavage of caspases [37]. Another example of the physiological effect of calpains includes decreased mitochondrial respiration due to cleavage of the electron transport chain proteins by calpain 10. Thus excessive activation or inhibition of calpain 10 has been shown to induce mitochondrial dysfunction and apoptosis [38]. In cardiomyocytes, calpain 1 activates caspase 3 and poly-ADP ribose polymerase during tumour necrosis factor-α-induced apoptosis [38]. In addition, it also cleaves apoptosis-inducing factor (AIF) after Ca^{2+} overload [36].

The first direct 'cause-effect relationship' among calpains was established for a condition known as calpainopathy, or Limb girdle muscular dystrophy type II A abbreviated as LGMD2A. This is an autosomal recessive disorder affecting the proximal muscles leading to impairment in movement, and the gene for calpain 3/p94 (CAPN3) was found to be primarily responsible for this condition [39]. Binding of the skeletal muscle-specific protease CAPN3 to titin/connectin at the N2 and M-lines [30] protects titin and other skeletal muscle-associated proteins from degradation. This modulates signal transduction around skeletal muscle and is an important event for normal muscle cell development [30, 31]. It has been identified by point mutation analyses that the functional rather than structural defect of calpain 3 is responsible for this disease [40, 41].

Contrary to the above observations, calpain has been found to have pro-survival role in neuronal cells. In apoptotic response, calpains inhibit p53-dependent cell death by hydrolyzing it and hence lowering its *in vivo* level [42, 43]. In addition, calpain can promote survival through activation of NF-κB, a family of related transcription factors. This is executed by cleavage of NF-κB inhibitor IκBα, which has been reported to be mediated by either calapin-1,-2 or –3 [44]. Calpain-mediated IκBα cleavage can occur in response to tumour necrosis factor (TNF), and activation of the epidermal growth factor receptor (EGFR) family member ERBB2 in breast cancer [45]. In another mechanism, calpain-1 was shown to regulate receptor activation of NF-κB ligand (RANKL)-supported osteoclastogenesis by activating NF-κB [46].

Despite the identification of several substrates of calpains both *in vitro* and *in vivo*, more studies are required to elucidate its precise role in calcium-regulated cellular functions. Although calpastatin is a unique *in vivo* regulator of calpain, other cellular proteins interacting with it such as Gas2 [47] appears to act like calpastatin thus regulating calpain activity [48]. Identification and characterization of inhibitors specific for each calpain homolog are essential, however present literature is limited by lack of such inhibitors that are truly calpain-specific [49, 50].

1.5.2 Other Functions

Various calpain substrate proteins are associated with carcinogenesis through transcription factors such as ras, c-fos, c-jun, p53, pp60src, and merlin [34]. In gastric cancer, digestive tract-specific nCL-4 (*CAPN9*) is downregulated which suggests role of calpain 9 as a tumor suppressor that degrades oncogene products important for carcinogenesis in digestive tracts [47]. During the mitotic clonal expansion phase of pre-adipocyte differentiation, calpain degrades cyclin-dependent kinase inhibitor p27 [47, 51, 52]. p27 degradation releases the inhibition on the cyclin D-CDK4 complex and phosphorylates retinoblastoma gene product (Rb). This in turn releases the Rb sequestration of transcription factors necessary for differentiation and subsequent activation [39]. Although reports indicate important roles of calpain in cell cycle regulation and differentiation, the mechanism and the species involved at an early stage of differentiation are yet to be conclusively delineated. For example, calpain-mediated turnover of p27 might not be critical since it appears to be permissive in p27-deficient mice [47, 51, 52].

Calpain 10 is a ubiquitously expressing atypical calpain [23, 53–55] with highest expression in the heart, followed by brain, liver, kidney, and pancreas in humans [7]. Some of the pathological conditions that CAPN10 has been found to be associated with is the elevated levels of free fatty acids and insulin resistance [53]. Studies have shown that the activation of protein kinase C by free fatty acids results in hyper-phosphorylation of insulin receptors which leads to reduction in the kinase activity of the receptors thus enhancing resistance to insulin [54]. Therefore, down-regulation of protein kinase C activity appears to be an important factor for maintaining proper phosphorylation levels of insulin receptors [56]. Since protein kinase C is a well-known substrate of calpain, lower calpain levels lead to upregulation of its activity. This in turn leads to reduction in insulin signaling consequently bringing about insulin resistance [56]. The lower calpain mRNA level has been attributed to polymorphism within intron 3 of *CAPN10* [24]. Further studies using calpain inhibitors suggest that calpain species other than calpain 10 also participate in insulin secretion which leads to type 2 diabetes susceptibility [57]. The molecular and physiological mechanism explaining the association of calpain with type 2-diabetes requires further clarification due to the different levels they are involved in. Calpain also play important roles in membrane fusion [55] and hydrolyzes various proteins that participate in cellular signaling such as kinases, receptors, and transcription factors [38, 58]. In addition, calpain is also important for differentiation of pre-adipocytes into adipocytes [51]. In a nutshell, it can be commented that the activity of calpain is tightly regulated by Ca^{2+}, both temporally and spatially. Deregulation of this protease activity causes excessive degradation or accumulation of co-existing cellular proteins resulting in severe cellular damage and pathological conditions [25].

1.6 Substrates and Inhibitors

1.6.1 Endogenous Substrates

Calpains cause limited proteolysis of their substrates, mainly within inter-domain unstructured regions. As a consequence, it hydrolyzes substrate proteins in a limited manner, producing large fragments with their domains intact. Two exceptions to this are, casein and myelin basic protein that are proteolyzed exhaustively by calpains with casein being used as the most common substrate for *in vitro* calpain assays. Calpain is regarded as a modulator protease, since hydrolysis by calpain results in modulation of the properties of the substrate rather than its degradation [1–5]. This was exemplified by the treatment of protein kinase C with calpain that produces an intact kinase domain independent of the effectors such as diacylglycerol and calcium [1–5].

A large number of proteins have been reported to be degraded by calpains, which include Bax, calcineurin, caspases, calmodulin-protein kinase, G protein, IκB, p53 and protein kinase C [59–61]. The substrate list also include cytoskeletal proteins such as α-fodrin, neurofilaments, membrane proteins such as ion channels, growth factor receptors, adhesion molecules as well as enzymes and protein constituents of myelin.

The nuclear substrates of calpains include the nucleoskeletal proteins lamin A and B. Interestingly, calpastatin, the well-known calpain inhibitor can also act as its substrate. One such unique observation reported by Barnoy et al. shows the degradation of calpastatin by calpain 3 with an increase in the calpain/calpastatin ratio [62]. Identifying more *in vivo* calpain substrates will open avenues to study the respective pathways in a different light altogether.

1.6.2 Synthetic Substrates and Specificity

Synthetic oligopeptides designed in conjunction with fluorescent probes are also used as *in vitro* substrates for calpains (Table 3.1). A common limitation of these substrates is that they are not calpain-specific. For example, SLY-MCA is a good substrate for cathepsin- 1 like protease [63], SLLVY-MCA is also cleaved by chymotrypsin and proteasomes [64], and BocLM-CMCA is cleaved by fiber cell globulizing aminopeptidase [65] as shown in Table 3.1. Since the short oligopeptides are generally poor substrates for calpains, some longer peptide substrates were developed using calpain substrate sequences to improve specificity and efficacy (Table 3.2). These substrates, however, are also proteolyzed by other proteases and are not calpain specific.

Calpain specificity has been a matter of great contention over the years with overwhelming suggestions that it recognizes the overall 3D structure, rather than the primary sequence of their substrates [66, 67]. Nevertheless, some sequence preferences have been extracted by comparing the amino acid sequences around the

Table 3.1 Calpain inhibitors and their chemical names

Calpain inhibitors (general name)	Chemical name
Calpain inhibitors I	N-acetyl-Leu-Leu-norleucinal
Calpain inhibitors II	N-acetyl-Leu-Leu-methioninal
Calpain inhibitor VI	4-Fluorophenyl sulfonyl-Val-Leu-CHO
Calpain inhibitor XI	Z-Leu-Abu-CONH(CH2)3-morpholine
EGTA	Ethyleneglycol-bis (b- aminoethyl ether) N, N, N′, N′-tetraacetic acid
E64	N-(N-(L-3-trans-carboxyoxirane-2-carbonyl)-L-leucyl)agmatin
E64c	N-(N-(L-3-trans-carboxyoxirane-2-carbonyl)-L-leucyl) isoamylamine
E64d	N-(N-(L-3-trans-ethoxycarbonyloxyoxirane-2-carbonyl)-L-leucyl
AK295	Benzyloxycarbonyl-Leu-aminobutyric acid- CONH (CH2)3-morpholine
AK275	Benzyloxycarbonyl-Leu-aminobutyric acid-CONH-CH2CH3
CX275	Active isomer of the diastereomeric mixture of AK275
ZLLal	Benzyloxycarbonyl-Leu-Leu-leucinal
ZLLLal	Benzyloxycarbonyl-Leu-leucinal
ZLLL-MCA	ZLLL-4-methylcoumaryl-7-amide
LLVY-MCA	Succinyl-LLVY-MCA
PD150606	3-(4-Iodophenyl)-2-mercapto-(Z)-2-propenoic acid
Calpastatin peptide	(Ac-)DPMSSTYIEELGKREVTIPPKTRELLA(-NH 2)

Table 3.2 Calpain synthetic substrates and their cleavage sites

Calpain synthetic substrates	Chemical name/cleavage site
SLY-MCA	Succinyl 4-methylcoumaryl-7-amide (7-amino-4-methylcoumarin)
SLLVYMCA	SLLVY/ methylcoumaryl-7-amide (7-amino-4-methylcoumarin)
BocLM-CMCA	t-butoxycarbonyl-7-amino-4-chloromethylcoumarin
KEVYGMMKK	K(- ε -N-5(6)-FAM)-EVY-/-GMM-K- ε -N-4,4-Dabcyl
TPLKSPPPSPR	Dabcyl-TPLK-/-SPPPSP-R-5-EDANS
TPLKSPPPSPRE-R7	Dabcyl-TPLKSPPPSPR-E (-5-EDANS)-RRRRRRR-NH
EPLFAERK	EDANS-EPLF-/-AER-K- ε -N-4,4-DABCYL

proteolytic sites in calpain substrates. Studies using various small peptide substrates revealed that the P3, P2, P1, and P1′ positions of the calpain proteolytic site were preferentially associated with F/W/L/V, L/V, R/K, and R/K/L residues of the substrates respectively [64, 68, 69].

Other reports on comprehensive analyses of the calpain cleavage sites [70, 71] have identified a position-specific scoring matrix (PSSM) for amino acids (aa) around the site [72]. The most recently extended transformed PSSM version assists in discriminating favored and disfavored cleavage sites when considered alongside the Amino Acid (AA) index [73]. The calpain substrate PSSM has been examined to determine whether a specific AA index correlated with the aa scores. These findings suggest that P5′, P7′ and P9′ should prefer hydrophilic aa and that P4′ is likely to be unstructured. These predictions corroborate very well with the 3D structure as the aa closest to P5′, P7′, and P9′ in CAPN2 are the hydrophilic residues Q290, E251 and K161 respectively, whereas the substrate bends at P4′ proving it to be possibly unstructured.

These preferences could further be explained by the 3D structure of active CAPN2/S1 providing important information regarding substrate binding to calpains. However, no clear relationship between the PSSM and calpain subsites at other positions could be established making it difficult to deduce a general rule for characterizing the interface between calpain and different substrate sequences [69]. To further explore the substrate specificities of calpains, examination of the role of calpain domains other than the protease domains offers a reasonable approach. For example, the C2L domain adjacent to CysPc may be crucial for substrate recognition and specificity because C2L bears close contacts with calpastatin in the active CAPN2/S1 structure (Fig. 3.3). The interface between calpain and calpastatin provides useful information for discussing the role of different domains in specificity and subsequent events.

The calpastatin reactive site contains the consensus sequence Gxx[E/D]xTIPPxYR where G613 of calpastatin forces the next four aas in the sequence to loop-out from the calpain active site (Fig. 3.3). This looping-out prevents the calpastatin from being cleaved. However, the sequence IPPEYRHLL spanning residues 618–627 binds tightly to subsites within the PC1 and PC2 domains of CAPN2 (Fig. 3.3). G613 fits into the S1 subsite, while the sequence N-terminal to G613 associates with subsites, which extend into the C2L domain. Residues close to the bound calpastatin are highly conserved in the classical calpains as 20 out of 24 are found to be conserved in CAPN1 and CAPN2. This strongly suggests that CAPN1/S1 and CAPN2/S1 have very similar substrate specificities, which is believed to be shared among other classical calpains. As mentioned above, at least 20 aas of the bound calpastatin fragment are close to the surface of the calpain molecule. These 20 aas of calpastatin have high affinity for the corresponding calpain subsites and exert strong and specific inhibitory activity by stabilizing the E614 – D617 loop (Fig. 3.3). This further emphasizes the need to carry out studies on domains other than the core protease to understand context-dependent specificity.

The manual integration of all the above observations seem cumbersome and an *in silico* approach that simulates the proteolytic events elicited by calpains would provide very close understanding on how calpains assess the 3D structure and local sequences of substrate proteins and select the appropriate sites for proteolysis. This, in concert with the machine learning approach might be the way ahead in defining the cleavage sites for calpains as suggested by recent work [70, 74]. From all the studies to understand substrate specificity it can be concluded that calpain substrate specificity is context-dependent with a combination of factors governing it.

1.7 Inhibitors

1.7.1 Endogenous Inhibitors

The only endogenous calpain inhibitor, calpastatin, exists in the cytosol. It is specific and very effective regulator of calpains and archetypically is a protein of approximately 70 kDa [75]. A single gene (*CAST*) encodes calpastatin and it has multiple promoters that generate four distinct isoforms with variation in the N-terminal region (Fig. 3.1). Type I and type II calpastatin both contain the 'L' domain but have differing N-terminal sequences. Type III calpastatin, where the XL region is absent, is the product of a promoter that is associated with the untranslated exon 1u. Type IV calpastatin is a testis-specific isoform that is generated from a promoter between exons 14 and 15 and lacks the L domain and the inhibitory domain I [76, 77]. These promoters can be differentially regulated in a tissue-specific manner in response to agonists. The cellular consequence of multiple calpastatin splice variants has not been fully elucidated; however, the absence of some exons has been shown to promote the formation of intracellular storage aggregates.

Calpastatin has four repetitive inhibitory domains of ∼140 residues, having three conserved regions A, B, and C (Fig. 3.1). A and C interact with PEF (L) and PEF (S) respectively, in a Ca^{2+}-dependent manner while B shows inhibitory activity by itself, probably by binding at the active site [22]. Presence of the two calmodulin-like domains, PEF (L) and PEF (S) is necessary for effective inhibition by calpastatin. Thus, calpastatin inhibits only dimeric calpains bearing 30 K. It also directly binds to the Ca^{2+}-binding domains of both the large and small subunits of calpain [25, 55] as illustrated in Fig. 3.3. The details of calpastatin specificity and consensus sequences based on the calpain-calpastatin co-crystal structure (Fig. 3.3) are discussed in Sect. 1.6. However, calpastatin, with high molecular weight and membrane impermeability has limited use as a pharmacological molecule. Nevertheless, with a better understanding of the inhibitory regions of calpastatin, peptides ranging between 20 and 40 mers have been generated. These peptides correspond to reactive sites of calpastatin and are used as specific inhibitors of calpains (Table 3.2).

1.7.2 Synthetic Inhibitors

A variety of calpain inhibitors have been synthesized till date due to the limitations associated with calpastatin. The history for calpain inhibitors dates back to 1980 when Sugita and colleagues used several low molecular derivatives of E64 (Table 3.1) to prevent muscle degradation in patients with muscular dystrophy [26]. E64 was first isolated from the culture medium of *Aspergillus japonicus* as a papain inhibitor and its typical derivatives include E64c and E64d [78–80]. Although these inhibitors along with leupeptin efficiently inhibit both μ- and m-calpains, they are not very specific as they also inhibit other cysteine proteinases including cathepsins, papain, proteasomes as well as matrix metalloproteinase-2 [81]. It is noteworthy that E-64 and leupeptin do not suppress the autolysis of p94, the muscle specific calpain at all [82].

Among calpain inhibitors, the frequently used and commercially available ones include calpain inhibitors I and II [83, 84]. The major drawback associated with these is their non-specific inhibition of proteasome and other cysteine proteinases [85]. The more specific calpain inhibitors, calpain inhibitors VI and XI (Table 3.1) have been shown to protect retinal [86] and cortical neurons [87] respectively against ischemia- induced damage. Reports have shown that both inhibitors have a protective effect on the apoptosis of motor neurons in spinal cord slices [88]. Further evidences also demonstrate that both the calpain inhibitors VI and XI were able to block calpain activity in mouse retinal photoreceptors [87]. Studies on spinal cord slice culture has also shown that calpain inhibitor VI could inhibit the protease activity in motor neurons [88]. Although EGTA (Table 3.1) is not a calpain inhibitor, it mimics the effects of calpain inhibitors by chelating extracellular Ca^{2+} and inhibits calpain activity in motor neurons of spinal cord slices [88] as well as cultures of rat oligodendrocytes [89].

In the context of differential inhibition of calpain and proteasome, the studies using di- and tri-leucyl aldehydes shown by Tsubuki and colleagues are noteworthy [90]. Their work with the inhibitors ZLLLal and ZLLal demonstrated strong inhibition of calpains with an inhibitory constant, K_i of 1 μM. The inhibition of proteasome by other inhibitors ZLLL-MCA and succinyl-LLVY-MCA bears a K_i of 1 μM and 0.1 μM respectively, while that for ZLLal is above 100 μM which shows its minimal non-specific action on proteasome. These synthetic inhibitors are potentially useful for identifying the functions of calpain and determining its specificity.

Other group of calpain-specific and effective inhibitors included ketoamide inhibitor molecules developed by Powers and coworkers [49, 91]. Prominent among these were AK295, AK275, and CX275 (Table 3.1) which were more effective and calpain-specific than the above discussed inhibitors. The inhibitor constant value, K_i of AK295 is approximately 30 nM for μ- and m-calpains but almost 1,000 times higher for cathepsin B [91] suggesting specificity of its action. They also screened derivatives of peptidyl alpha-keto compounds to improve the specificity and K_i value and found that AK275 with a K_i of 15 nM for m-calpain and 19 nM for μ-calpain were the best inhibitors among 100 other molecules tested [91, 92].

In another study, a novel inhibitor PD150606 was developed by Wang and colleagues [93] that has a distinct inhibitory mechanism compared to the active site inhibitors. This is a cell permeable, non-competitive and very specific non-peptide inhibitor. K_i value of about 0.3 μM for μ- and m-calpains, but greater than 100 μM for cathepsin B and papain, for this inhibitor is a testimony to its high specificity for calpains relative to other proteinases [94]. It brings about its action by binding to the Ca^{2+}-binding PEF domain of calpain. The distinct features of PD150606 make it an ideal inhibitor to be used in combination with other active site-directed inhibitors such as AK295. Such combinatorial treatment producing a very specific inhibition of calpain is essential for *in vivo* studies toward understanding the physiological roles of calpain.

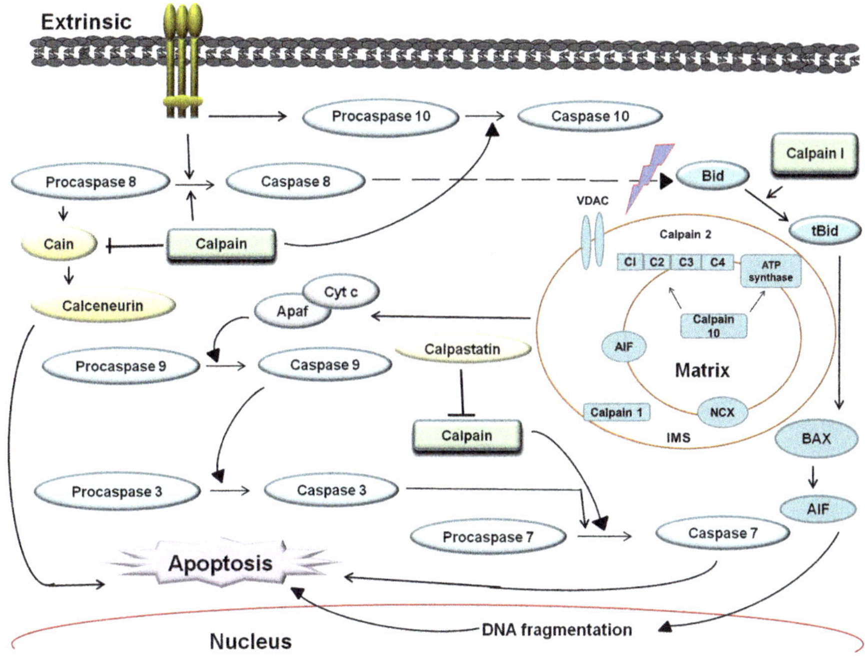

Fig. 3.5 Calpains in the apoptotic pathway: Schematic representation of the effect of calpains on the mitochondria as well as other cellular substrates. Calpain 10 cleaves electron transport chain proteins, calpain 1 degrades BH3-interacting domain death agonist Bid, AIF, and NCX, while calpain 2 cleaves voltage-dependent anion channel (VDAC). The effect of calpains on different substrates ultimately culminates in apoptosis

1.8 Future Perspective

Majority of calpain research performed so far has been focused on its pro-apoptotic role post Ca^{2+} overload in the mitochondria. Considering its potential significance, the cleavage of caspase 3 and/or AIF as represented in Fig. 3.5 [26] has evoked particular interest among researchers [95–97]. Apart from this, it is well established that calpain 10 regulates electron transport chain proteins and thus must be tightly regulated for mitochondrial and renal cell function. However, the underlying mechanism has not yet been elucidated [98–100] and there remains limited findings on the physiological functions of mitochondrial calpains except for calpain 10 which needs to be addressed in future.

Reports have delineated calpain I activation to be directly involved in triggering AIF release after ischemic neuronal injury [101]. AIF release and its nuclear translocation constitute an important caspase-independent cell death pathway in cerebral ischemia [95–97]. Hence novel therapeutic interventions aiming at inhibiting calpain I activity and preventing AIF release may offer favorable neuroprotective effects without interfering with the physiological functions of the latter within mitochondria. Despite several years of rigorous research on the role of calpains in pathophysiology, many discrepancies do exist raising pertinent questions.

Another major challenge is the conservation of calpain active site throughout the entire family which restricts the design of calpain isoform-specific inhibitors. Additionally, many popular calpain inhibitors also inhibit cathepsins, proteasome, and Lon protease (a mitochondrial matrix protease). Thus specificity of inhibitors is an important challenge concerning the field of calpain research. With better understanding of the activation mechanism and three-dimensional structure of different calpains, the objective of developing calpain-specific inhibitors seems more plausible. Establishing the specific roles of different isoforms using knockdown models will ensure better utility of calpain as a target with the best possible efficacy of the inhibitors.

Apart from these, a new perspective to the calpain family members that has surfaced lately is its evolutionary studies. Some calpains lack the protease activity which includes calpain 6, Drosophila CALPC, few *C. elegans* and all *Trypanosoma* homologs. With literature providing novel roles of some calpain homologs apart from its protease activity, it seems a field with more discoveries awaiting revelation.

2 Granzymes

One of the primary mechanisms of immune response by higher organisms to eliminate viruses and transformed cells is mediated by Cytotoxic T lymphocyte and natural killer cells. The introduction of family of proteases known as granzymes proved the conventional belief wrong that perforins were the only mediators of target cell destruction. Major ground breaking discoveries in the field of granzymes have been made by Jurg Tschopp group who first introduced this class of enzymes in 1980s [102]. 'Granule enzymes' or 'granzymes' are serine proteases, that make up about 90 % of the mass of cytolytic granules, encapsulated in specialized cellular compartments, the secretory lysosomes. These lysosomes are present in both the gatekeepers, cytotoxic T lymphocytes (CTLs) and natural killer (NK) cells [103]. CTLs and NK cells initiate cell death via two major mechanisms, the granule-mediated and death ligand-mediated cytotoxicity [104–106]. Both of these mechanisms require cell-to-cell contact and cause activation of the executioner caspases [107]. The death receptor pathway requires binding of ligand (e.g. Fas ligand) to receptors expressed on the target cells [108, 109]. In contrast, the granule exocytosis pathway requires the pore-forming protein perforin (pfp), and granzymes to cleave and activate effector molecules within the target cell [107]. Once inside the target cell that is virus-infected or transformed, granzymes cleave specific proteins and trigger apoptosis. Granzyme-mediated proteolysis of cellular substrates thus compensates for the timely death of infected or transformed cells in the absence of caspase activity due to mutations or its inhibition by viral proteins.

Structurally, this family is closely related to chymotrypsin, with a conserved characteristic catalytic triad comprising histidine, aspartic acid and serine [110].

They are also genetically linked to other serine proteases belonging to mast cells and monocytes further emphasizing their role as protective enzymes [111]. Although granzymes probably occur in other species with complex immune systems, but granzyme sequences have only been reported in mammals till date. This section tries to encompass the research findings on this family of proteases.

2.1 *Classification*

A total of ten granzymes (Grs) viz. A, B, C, D, E, F, G, K, M and N have been identified in the mouse, while only five are known in human such as Grs A, B, H, M and K (Fig. 3.6) [112]. Among these, Gr H appears to be specifically present in human whereas no human equivalents of mouse granzymes C-G are known [113]. All granzyme genes have a similar organization with their transcripts made from five exons encoded by three gene clusters (Fig. 3.6). The first exon encodes the leader sequence, while exons 2, 3 and 5 encode individual amino acids of the catalytic triad. Most of the granzyme genes encode only one transcript with the exception of granzyme H, which arises from alternative splicing of two mRNAs. Gr H gene appears to have arisen as a 'hybrid' made up of the first three exons and

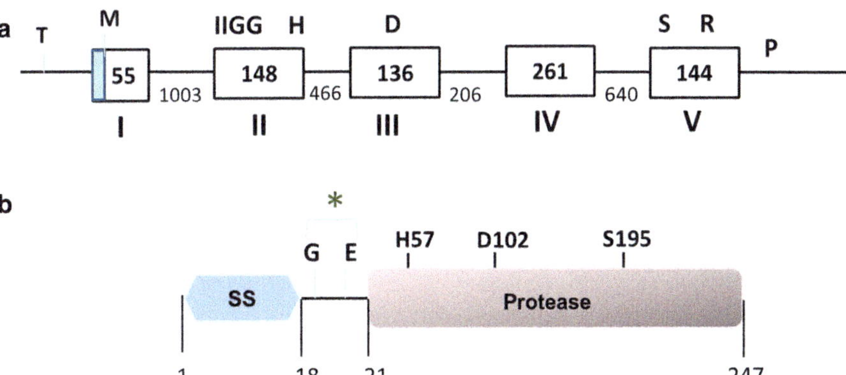

Fig. 3.6 Gene and domain organization for granzyme superfamily: Organization of the human granzyme B gene of 3.2 kb is a prototype of granzyme family. (**a**) A TATA box consensus sequence (T) is present approximately 30 bp from the initiation site for transcription. The five exons are indicated by Roman numerals. Exon I codes for the leader sequence, exon II contains the start of the mature protein and encodes the first of the essential catalytic triad residues at the active site, histidine (H), exon III encodes the aspartic acid (D), and exon V the most essential catalytic residue, serine (S). The side chain of Arg226 (R) forms the base of the substrate pocket and forms a salt bridge with aspartic acid at the P1 position of the substrate. The polyadenylation signal is shown as P and the 5′ untranslated sequence is shaded. The lengths are shown in base-pairs. (**b**) General granzyme domain organization represented by granzyme B displays the presence of an N terminal signal sequence followed by a dipeptide and finally the all-important serine protease domain

intervening introns of granzyme B and the remainder of another serine protease gene. In humans, granzyme genes from cytotoxic T lymphocyte map to three chromosomal loci with each subfamily having broad substrate specificity. HFSP (for human granzyme A) and TRYP2 map to chromosome 5q11-q12, a second cluster of serine protease genes on 14q11-q12 encode human granzyme B and H, whereas the genes encoding granzyme M are located along with the genes encoding azurocidin (AZU), neutrophil elastase (NE) and proteinase-3 (PR3) on chromosome 19p13.3. The corresponding loci in mice are chromosome 13 for granzyme A and other tryptases, 14D for granzymes B-F and 10C for met-ases [114]. These clusters have their own uniqueness, for example, azurocidin (AZU) and proteinase 3 (PR3) in mouse and human granzyme M gene, each have intron 1 located between residues 7 and 6 of the leader sequence, indicating a close evolutionary relationship [111]. Similarly in humans, the genes encoding granzymes B and H as well as cathepsin G are very closely linked, mapping to within 50 kb of each other. Overall, the granzyme subfamilies have trypsin-, chymotrypsin- and elastase-like specificities, and their genes are grouped under the 'tryptase', 'chymase', and 'metase' loci respectively [113].

2.2 Structural Assembly

Granzymes are synthesized as zymogens that are processed at the time of packaging into cytolytic granules. The processing involves cleavage of the leader peptide leaving two amino acids attached at the mature amino terminus. This dipeptide is then clipped off by dipeptidyl peptidase I (DPPI), a peptidase constitutively expressed in lysosomes. On cleavage of the amino-terminal dipeptide, the granzymes become enzymatically active. Since their optimum activity is at pH 7.5, they are maximally active following release from the secretory granules into the cytoplasm.

The catalytic activity of granzymes like other members of the serine protease family depends on the serine residue of the conserved active site triad [115]. Other common features of this group of proteases include an oxyanion hole to stabilize transition states of the enzyme-substrate complex and a substrate-binding pocket, the shape of which determines specificity of the protease. The crystal structures of several granzymes such as Grs A, B, C, H and M that have been solved provide insights into the differential preference and recognition of substrates by each of these granzymes, which is discussed in details in Sect. 2.5 [116, 117]. Here we discuss the crystallographic structure of human granzyme B (PDB ID: 1FQ3) as a representative of this family to understand the overall structural assembly among granzymes [115].

The secondary and tertiary structure of granzyme B generally resembles that of trypsin-like serine proteases. Like other serine proteases, it comprises two six-stranded L-barrels that are connected by three trans-domain segments. The catalytic residues His 57, Asp 102 and Ser 195, are located at the junction of the two L-barrels with the active site cleft perpendicular to this junction (Fig. 3.7). Other

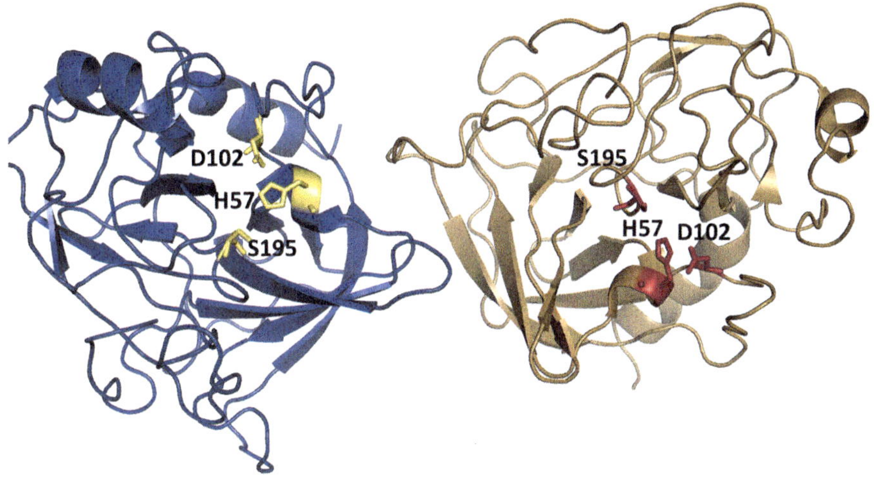

Fig. 3.7 Structure of granzyme: Crystal structure of granzyme B (PDB ID: 1FQ3) represented as a cartoon with the residues of the catalytic triad as sticks

regular secondary structure elements include a helical loop between Ala 56 and Cys 58, a helix involving residues between Asp 165 – Leu 172, and a long C-terminal helix. The structurally important proline pair Pro 224 and Pro 225 is in the cis conformation in granzyme B similar to that in cathepsin G, rat mast cell protease II, and human chymase. The proline pair is part of a shortened segment, Xaa 221-Xaa 224- Pro 225 that is common for this subfamily of serine proteases. In other serine proteases, for example, the longer segments of human leukocyte elastase, bovine chymotrypsinogen and bovine L-trypsin, Pro 225 is in the typical trans conformation when residue 224 is not a proline. A positively charged arginine side chain is oriented into the S1 subsite to interact optimally with the negatively charged P1 aspartic acid of the inhibitor, in the cis proline conformation.

It has been reported that granzyme B has a capacious substrate cleft capable of accommodating as many as eight substrate residues. The key contact in the S1 subsite is between Arg 226 from Gr B and the P1 residue at amino-terminal of the cleaved bond of the substrate. Despite their distinct catalytic mechanisms granzymes share a novel and stringent specificity for aspartic acid in the P1 position [118] with the optimal tetrapeptide recognition motif spanning P4-P1 and the conserved residues being IEPD [119]. The unique and stringent requirement for aspartic acid in the P1 position of substrates and inhibitors of granzyme B has been confirmed by the geometry and the chemical nature of this subsite. Among the S1 residues, 13 arginines and 19 lysines of granzyme B are solvent-exposed except for Arg 226 thus producing positively charged surface patches distant from the active site. Though Grs display superficial similarity to chymotrypsin, there are profound differences in the S1 subsite that account for the distinct macromolecular specificities and different biological functions of these enzymes.

The N-terminal amino acids of mature granzyme B lie in the interior of the molecule, whereas its ammonium group forms an internal salt bridge with the side chain carboxylate of Asp 194. This salt bridge formation is linked to the creation of oxyanion hole and the functional S1 pocket in the mature protease. Of this, the 'oxyanion hole' is presumed to stabilize the reaction intermediates formed during catalysis, and is made by the amide nitrogen atoms of Gly 193, Asp 194 and Ser 195. Two additional aspartic acids, Asp 102, a member of the catalytic triad, and Asp 194, salt-linked to Ile 16, are not solvent-exposed [115]. Apart from providing all this crucial information, the structure presents template for the design of highly specific inhibitors of human granzyme B.

2.3 Catalytic Mechanism

Granzyme catalysis is typical of a serine protease where the side chains of the three catalytic triad residues His, Ser and Asp work in concert. The residues of the catalytic triad are conserved in all of the serine proteases, and follow similar catalytic mechanism. The mechanism of catalysis in serine proteases is based on the chymotrypsin model. The hydroxyl group of Ser and the imidazole ring of His residues are so close to each other that they can form a hydrogen bond. The Asp residue of the triad too is very close to His but lies on the opposite side of Ser. Although the serine residue plays primary role in catalysis, the contribution of the other two catalytic triad residues towards stabilization of the intermediates is crucial.

The first step in the activation mechanism is substrate binding where the side-chain of the amino acid residue immediately before the scissile peptide bond in substrate can bind to the recognition site on the enzyme. This is followed by a nucleophilic attack where serine 195 acts as a nucleophile facilitated by His 57, which abstracts a proton from Ser 195 thus enhancing the reactivity of the hydroxyl group of serine residue. The result of this nucleophilic attack is a covalent bond between the Ser 195 side-chain oxygen and the substrate. The negative charge formed on the carbonyl oxygen of the peptide is stabilized by hydrogen bonds from two protease backbone amide protons thus forming a catalytically poised oxyanion hole. This is followed by proton donation by His 57 to the substrate amide nitrogen thus allowing release of the C-terminal part of the substrate as a free peptide. The final step is an attack by water on the ester bond between the peptide and the Ser 195 oxygen of the protease. This forms the second product peptide with a normal carboxyl group, and regenerates the serine hydroxyl. The second peptide then dissociates from the enzyme to allow another catalytic cycle to begin.

2.4 Functions of Granzymes

The physiological roles of the different cellular granzymes have been determined and established over time by knocking out these genes individually and in combination [120]. The functional implications or the phenotypic manifestation of these knockouts provide insights into the specific roles or lethality of these genes apart from their dispensability.

2.4.1 Role in Apoptosis

In the context of granzymes, target cell death is rapid and efficient that is initiated via two main pathways involving either the ligation of death receptors or through the granule-exocytosis pathway. The granule-exocytosis pathway has attained prominence over the past decade and consequently, a mechanism for granule-dependent killing has been well established. This is the major mechanism via which CTLs eliminate tumor and virus-infected cells [121]. In this pathway, CTLs release granules containing the pore-forming protein perforin and pro-apoptotic granzymes into the immunological synapse (Fig. 3.9). Pore formation by perforin facilitates entry of granzymes into the target cell where they induce apoptosis by promoting the activation of a family of death-inducing proteases called caspases (Fig. 3.8). Endocytosis is another alternate mode wherein Gr B leaves the endosomal compartment to access the cytosol of the target cell in a perforin-dependent manner. The key

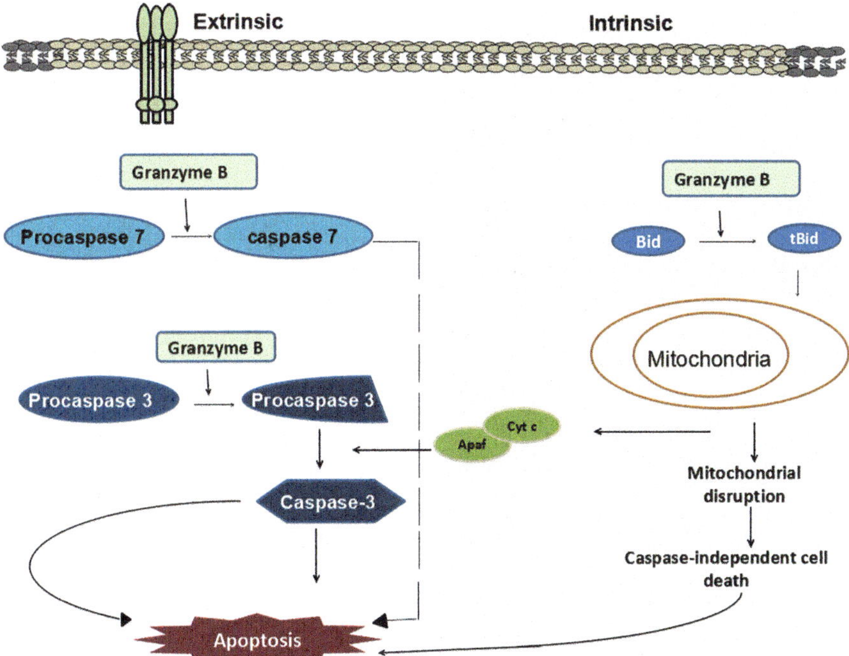

Fig. 3.8 Granzymes in apoptosis: The entry of the granzymes in the target cell is assisted by the perforins that punch holes in the target cell membrane. On entering the target cell, the granzymes cleave cytoplasmic and nuclear substrates that culminate into apoptosis

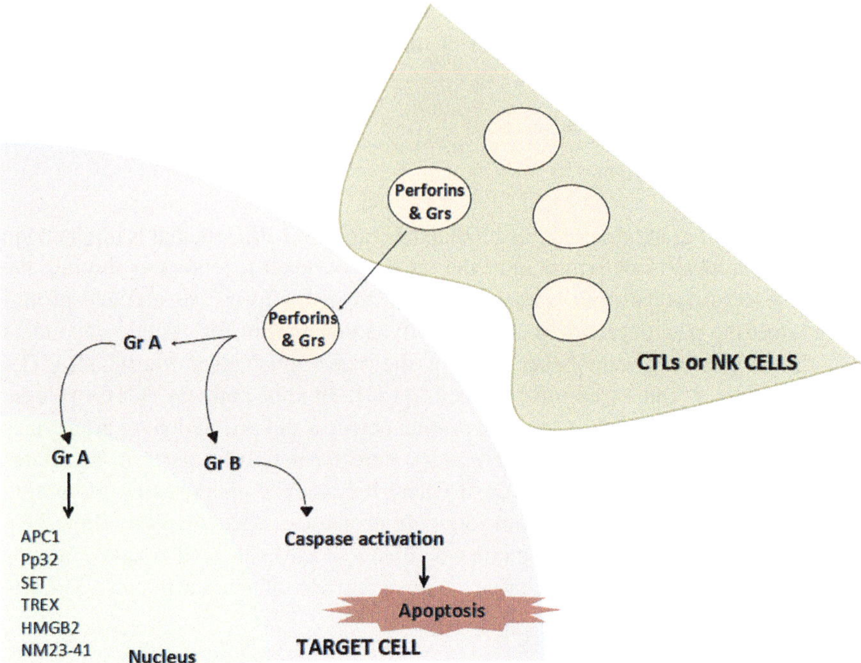

Fig. 3.9 Nuclear substrates of GrA: On entry into the target cell, GrA enters the nucleus to cleave several substrates ultimately leading to cell death

cytosolic substrate for granzyme B is the pro-apoptotic Bcl-2 family member Bid. Truncated Bid (tBid) then induces mitochondrial disruption which leads to release of other pro-apoptotic mediators such as cytochrome c and DIABLO/Smac. This subsequently induces caspase activation as illustrated in Fig. 3.8 [122, 123].

Similarly, when a CD8 T cell or NK cell is activated by its antigen receptor, the lytic granules move to cluster around the microtubule organizing center which is then followed by their alignment along the immunological synapse [124, 125]. The granule membrane fuses with the cell plasma membrane of the killer cells thereby releasing its contents, including perforin and granzymes into the synapse. In CTL, granule fusion appears to localize to a distinct secretory region of the central cluster (c-SMAC) that is separate from the signaling domain containing the T cell receptor and associated kinases [123, 124]. However, the process of granule fusion at the synapse may differ among killer cells, with cytotoxicity and granule fusion occuring even in the absence of a stable synapse [124, 126]. The mode of entry of granzymes into the target cell cytosol mediated by perforin is largely unclear.

The original model which proposed that the entry of granzyme is through perforin-mediated plasma membrane pores, no longer holds true [127]. A revised model highlights the role of perforin in making microscopic holes in the plasma membrane creating a calcium influx. This triggers a cellular plasma membrane response and rapid endocytosis of granzymes and other cell surface bound molecules. Granzymes are likely to dissociate from serglycin, a secretory granule core protein before they enter target cells [128]. They bind to the target cell membrane by electrostatic interactions owing to their positive pI values (~9–11), and negatively charged cell surface [129]. They are also held on the surface by specific receptors, one of them being the cation-independent mannose-6-phosphate receptor [130]. Though these receptors are shown to be involved, specific receptors are not necessary for binding to the cell surface owing to their dispensable roles in entry as well as cytotoxicity. The entry is in fact dynamin-dependent [130], and results in the formation of giant endosomes containing both granzymes and perforin (Fig. 3.9). The granzymes escape through perforin pores in the endosome within few minutes and are ultimately delivered into the cytosol.

Although perforin is the major molecule responsible for granzyme delivery, under some circumstances other molecules may serve this function as well. For example, bacterial and viral endosomolysins can substitute for perforin *in vitro* and therefore are now widely used for intracellular delivery and may potentially play a similar role *in vivo*. The heat shock protein hsp70, which is known to act as a chaperone for some peptides across cell membranes are also capable of carrying granzymes into the cells. Hsp70 is found on the surface of some stressed or tumor cells and thus may be assisting the removal of these cells from the body [131].

Once delivered in the cell of interest, the primary role of granzymes is to cleave specific substrates. Although many key granzyme targets including Bid and inhibitor of caspase activated DNase (ICAD) are cytosolic, Grs are also generally localized in membrane-bound cellular compartments, including the nucleus and mitochondrion. GrA and GrB translocate to the nucleus [132], where proteolytic cleavage of key substrates is important to induce programmed cell death. The nuclear substrates for GrA include SET, Ape1, lamins, histones, Ku70, and PARP1 whereas those for GrB are lamin B and PARP1 where nuclear translocation of the granzymes may be mediated by importin-α [132]. It was recently found that GrA also traffics into the mitochondrial matrix, which is necessary for it to initiate mitochondrial damage [133]. Mutations in the perforin gene have been found to be associated with profound immunodeficiency and familial hemophagocytic lymphohistocytosis syndrome [134–136].

The mode of action of Gr B in cleaving some caspases directly has been well established (Fig. 3.8) [137]. On the other hand, substrates for Granzymes C/H and K, the so called orphan enzymes have still not been identified [138]. Some studies show Granzyme-C-induced cell death is independent of caspase activation and the main feature of this pathway is thought to be rapid mitochondrial swelling and loss of mitochondrial membrane potential. Similarly Gr A, and Gr C induce single-stranded DNA nicks, but the DNase responsible for this effect is yet to be identified. Gr K on the other hand has been found to be involved in ROS production hinting at the possibility that its mechanism of cell death activation is similar to that of granzyme A [139].

In case of Gr M, cell death occurs rapidly and in a caspase- and mitochondria-independent fashion [140]. Cells treated with granzyme M display large cytoplasmic vacuoles, which may be indicative of autophagy, and display rapid plasma membrane permeabilization via an unknown mechanism. These claims have however been disproved based on apoptotic assays and identification of ICAD and PARP as potential substrates for granzyme M [141]. Moreover, granzyme M was also suggested to regulate granzyme B activity by cleavage of the endogenous inhibitor PI-9 [141]. Though the actual mechanism undertaken by this granzyme remains enigmatic, its use of alternative cleavage sites to cause cell death is almost certain.

In a more recent study, it was also shown that all human granzymes directly cleaved heterogeneous nuclear ribonucleoprotein K (hnRNP K), a DNA/RNA-binding protein. The study was performed with living tumor cells during lymphokine-activated killer cell-mediated attack. The cleavage of hnRNP K under physiological conditions when purified granzymes were delivered, led to the designation of hnRNP K as the first known pan-granzyme substrate [142]. Cleavage of hnRNP K was shown to be more efficient in the presence of RNA and occurred in two apparent proteolysis-sensitive amino acid regions, thereby dissecting the functional DNA/RNA-binding hnRNP K domains. HnRNP K is essential for tumor cell viability, since its knockdown results in spontaneous tumor cell apoptosis with caspase activation and reactive oxygen species production. This apoptosis was more pronounced at low tumor cell density where hnRNP K knockdown also triggered a caspase-independent apoptotic pathway. This suggests that hnRNP K promotes tumor cell survival in the absence of cell-cell contact. Silencing of hnRNP K protein expression rendered tumor cells more susceptible to cellular cytotoxicity. It was thus conclusively proven by studies that hnRNP K is indispensable for tumor cell viability. This study thus further suggested that targeting of hnRNP K by granzymes contributes to or reinforces the cell death mechanisms by which cytotoxic lymphocytes eliminate tumor cells. There are also other caspase-independent pathways to cell death, and although they are poorly characterized, an important step in this pathway appears to be cytoskeletal disruption.

Based on the destructive nature of the granules, the obvious question that arises is that how the killer cell, on the other side of the synapse are not injured by its own granules? Studies in these lines found an important protective mechanism against auto-destruction provided by the irreversible granzyme inhibitors called serpins expressed in the cytoplasm of killer cells [143]. Although serpins that inactivate GrB have been shown in killer cells, no such serpins are known to inactivate GrA. Another protective mechanism occurs via externalization of a cytotoxic granule membrane protein (cathepsin B) that is capable of proteolytically inactivating perforin during granule fusion [144]. Cathepsin B is thought to protect the killer cell membrane from any perforin that is redirected to the CTL side of the synapse. Cathepsin B–knockout mice however show that killer cells survive encounters with target cells. This suggests the likelihood of existence of other mechanisms to protect killer cells from their own agents of destruction.

2.4.2 Other Roles

The functions of granzymes A and B in inducing target-cell apoptosis have been investigated extensively *in vitro*, and they are better understood than the role of perforin at the molecular level. Severe immunodeficiency is observed in perforin deficient mice infected with lymphocytic choriomeningitis virus (LCMV). This is indicative of both the physiological relevance of the granule pathway for viral defense and the indispensable role of perforin in this process [145]. The importance of perforin has been supported by studies in perforin-deficient mice that are infected with other viruses, including the natural poxvirus pathogen ectromelia [146–148]. Though several studies pointed this out in independent observations, evidence for an indispensable role for granzymes in viral immunity has emerged only in the past few years [147, 148]. The finding that mice lacking both granzymes A and B cluster are as susceptible to ectromelia virus as perforin-deficient animals, underlines the importance of the granzyme system for the cells. It was found that the susceptibility of granzyme-deficient mice to ectromelia virus was not associated with an inability to mount a CTL response to other pathogens [147, 148] or any additional susceptibility to related viruses. This raises an interesting possibility that granzymes might either have additional extracellular functions or have evolved specifically to restrict poxvirus replication. Even serpins that are elaborated by this family of viruses are far less efficient at blocking the granule-mediated, perforin-dependent cell death.

Apart from this, in atherosclerosis, elevated GrB plasma levels are detected in patients with acute coronary syndromes [148]. The plasma levels are significantly higher in patients with unstable carotid plaques compared with those with stable lesions and correspond to an increased incidence of cerebrovascular events [149]. These studies suggest that extracellular GrB activity contributes to ECM degradation and promotes plaque instability and rupture. The causative role for granzymes in atherosclerosis and plaque rupture suggests potential extracellular involvement of this protease in diseases [150]. Studies that use granzyme and/or perforin-KO animal models of atherosclerosis would significantly clarify the intracellular and/or extracellular role of granzymes in promoting atherosclerotic plaque instability. In this regard, Schiller *et al.* [62] examined the role of perforin in atherosclerotic plaque development using the proatherogenic LDL receptor knockout mice model (LDLr-KO). As no difference in plaque size was observed between LDLr-KO mice and perforin double-knockout mice (LDLr/perforin-DKO), it was concluded that atherosclerosis does not involve the granzyme/perforin cytolytic pathway. However, the effect of perforin-independent, extracellular granzyme activity cannot be ruled out in the latter study as the use of perforin-deficient mice alone was insufficient to prove the role of granzymes [151].

Emerging evidence pointed out that granzymes also play a role in controlling inflammation. Granzyme serum levels are elevated in patients with autoimmune diseases and infections, including sepsis thus leading to this proposition. Even though these speculations existed on its role in immune response, it was not until a direct correlation of granzymes with anti-microbial immune response was established [152]. The function of extracellular granzymes in inflammation remains largely unsubstantiated barring few studies which find direct correlation. In one such study it was shown that GrK binds to Gram-negative bacteria and their cell-wall component lipopolysaccharide (LPS) [153]. GrK then synergistically enhances

LPS-induced cytokine release *in vitro* from primary human monocytes and *in vivo* in a mouse model of LPS challenge. Intriguingly, these extracellular effects are independent of GrK catalytic activity. GrK disaggregates LPS from micelles and augments LPS-CD14 complex formation, which might boost monocyte activation by LPS. This unequivocally represents extracellular GrK as an unexpected direct modulator of LPS-TLR4 signaling during the antimicrobial innate immune response [154]. Further comprehensive studies on inflammatory response and granzymes have the potential to open up new avenues in pathophysiological research.

2.5 Substrates and Inhibitors

2.5.1 Substrates and Substrate Specificity

Digestive proteases, such as trypsin, chymotrypsin and elastase, have a very broad range of substrates, and the amino-acid context of the P1 residue of substrates is far less critical than for granzymes. Granzymes have very specific substrate preferences consistent with their role as processing, rather than degradative enzymes. It differs in their primary substrate specificities with four different enzymatic activities detected viz. trypt-ase (cleaving after Arg or Lys), asp-ase (cleaving after Asp), chymase (cleaving after Phe, Trp or Tyr), and met-ase (cleaving after Met, Ile or Leu). Equivalence for these activities can be drawn with the following granzyme genes: tryptase (Grs A and K), asp-ase (Gr B) and met-ase (Gr M). One of the unique features of granzyme specificity is that up to five residues adjacent to the P1 position may influence recognition and cleavage.

The synthetic substrate libraries and substrate phage display together presented the optimal substrate sequence for granzyme B, spanning six subsites. These subsites are P4-P2P with the cleavage site at the Asps-Xaa peptide bond [155]. The Asp substrate specificity of Gr B is highly unusual among proteases with the exception that it is found in caspases, a family of cysteine proteases involved in apoptosis. The definitions of the S1P, S2P, and S3P subsites in granzyme B are consistent with data demonstrating that the enzyme-substrate interactions C-terminal to the scissile bond are catalytically significant and play a role in determining substrate specificity [146].

On comparison with caspases, it has been observed that while granzyme B has deep, tunnel-like features defining S1P, S2P, and S3P subsites, caspases have an extended shallow groove on the surface of the protein representing the S1P subsite [147]. In granzyme B, S1P is a large hydrophobic pocket formed by the disulfide bridge between Cys 58 and Cys 42, the carbon atoms of Lys 40 and Ile 35 side chains. This pocket is large enough to fit a fairly large aromatic tryptophan residue. The other subsite that is represented as S2P has a narrow opening and is in continuation to a network of waters forming the hydrophilic S3P subsite. This narrow corridor, between the main chain of Gly 193, and the side chains of Arg 41 and His 151 is consistent with the specificity for glycine at P2P. The structure of granzyme B has also identified a large hydrophilic S3P subsite which is formed by the main chain atoms of the amino acid pairs, Arg 41 – Gly 43, Val 138 – Gly 142, and Gly 193 – Pro 198. Moreover, this site is covered by the side chains of Tyr 32, Trp 141 and Met 30 and is filled with four well-defined water molecules.

Clear substrate preferences have been identified for granzymes A, D and tryptase-2 which are often defined by their ability to cleave Na-CBZ-L-lysine thiobenzyl ester. Granzyme B, an 'Asp-ase', cleaves at aspartic acid and possibly glutamic acid, whereas granzyme M, a 'Met-ase' cleaves at methionine and finally granzyme H, a chy-mase cleaves after phenyalinine [148, 149, 140]. Gr B is the only mammalian serine protease that prefers acidic side chains [144], a finding of relevance for its role as a pro-apoptotic enzyme as it allows cleavage of Bid and the pro-caspases. Based on these findings, various synthetic compounds including peptide thiobenzyl ester, 7-amino-4-methylcoumarin and paranitroanilide (pNA) derivatives have been tested to determine ideal substrates and cleavage conditions for granzymes. The optimal paranitroanilide substrates for mouse and human granzyme A were identified to be D-Pro-Phe-Arg-pNA and tosyl-Gly-Pro-Arg-pNA respectively [151]. Both these Grs are inhibited by serine protease inhibitors such as di-isopropyl fluorophosphate, phenyl methyl sulfonyl fluoride, bezamidine, aprotinin, leupeptin and soybean trypsin inhibitor. In addition, granzyme A of both species can be blocked by a number of physiological inhibitors, such as 1-protease inhibitor, which produced 85 % inhibition when used at 10 g/ml [152].

With reference to GrM, very few substrates have been reported till date, and the mechanism by which this enzyme recognizes and hydrolyzes its substrates is not well understood. Structural insights into the proteolytic specificity of human GrM (hGrM) were provided by crystal structures of wild-type hGrM, an inactive D86N-GrM mutant with bound peptide substrate, and the complexes with a catalytic product [153]. Structure-based mutagenesis revealed that the N terminus and catalytic triad of hGrM are most essential for proteolytic function. Structural comparisons indicated a large conformational change of the L3 loop upon substrate binding, and suggest that this loop mediates the substrate specificity of hGrM. Based on the complex structure of GrM with its catalytic product, a tetrapeptide chloromethylketone inhibitor was designed and found to specifically block the catalytic activity of hGrM [154]. Co-crystal structure with of the other granzymes with synthetic substrates or inhibitors would assist in providing further insights in the context of specificity.

2.5.2 Granzyme Inhibitors

The regulation of proteolytic enzymes in tissues by endogenous inhibitors is critical to maintaining homeostasis and preventing undesirable damage. Although the trafficking of granzymes within CTLs is designed to minimize leakage of active enzyme out of granules, any stray molecules in the cytoplasm could cause cell death. During granule exocytosis, some granzymes might inadvertently reenter the effector cells. Since CTLs typically kill several targets in succession without harming themselves, the important question is how CTLs protect themselves from their own cytotoxic molecules were studied. One of the most widely known ways is expression of granzyme-specific inhibitors, serpins [146]. Serpins are the largest and most broadly distributed superfamily of protease inhibitors, with more than 1500 family members [146, 147]. Serpins inactivate their targets either by covalently and irreversibly binding to the active site of the enzyme or by forming noncovalent complexes that are tight enough to resist the denaturing conditions of SDS-PAGE [148].

2.5.2.1 Endogenous Inhibitors

The only intracellular inhibitor of human GrB is the nucleocytoplasmic serpin known as proteinase inhibitor-9 (PI-9). PI-9 is expressed by lymphocytes, dendritic, endothelial, mesothelial, and finally mast cells [147, 149, 150] at immune privileged sites such as testis and placenta. This *in vivo* distribution pattern supports the idea that PI-9 protects effector, accessory, and bystander cells from ectopic GrB during an immune response. The expression of PI-9 is induced by inflammatory modulators like lipopolysaccharide, IFN-γ, and IL-1β. Its expression is further enhanced by estrogen and hypoxia because of estrogen responsive elements and hypoxia inducible factor 2 (HIF-2)-binding sites, respectively, in the PI-9 promoter [151, 152].

The mouse counterpart of PI-9 is serine proteinase inhibitor-6 (SPI-6) which is expressed in CTL and NK cells and is upregulated during dendritic cell maturation. Overexpression of SPI-6 in target cells protects them from CTL killing [153]. The persistence of increased numbers of CTLs long after viral clearance in SPI-6 transgenic mice shows that SPI-6 protects CTLs from self-destruction [153]. However contrary to this, GrB-deficient mice do not have increased numbers of CTL after viral infection thus raising questions about the interpretation of the SPI-6 transgenic study [154]. Apart from this, it is conclusively proven that CTLs from mice genetically deficient in SPI-6 have increased cytosolic GrB and reduced viability. One surprising finding in this context is the breakdown of the integrity of cytotoxic granules in SPI-6-deficient CTLs [154], the exact mechanism behind which is unclear.

The overexpression of PI-9 or SPI-6 may be a mechanism by which tumor cells evade the GrB/ perforin pathway as shown in studies with solid tumors and human and mice lymphomas [155]. Though it is difficult to draw a comparison study between serpin expression in tumor cells relative to corresponding normal tissues from these studies. Subset of human lymphoma cell lines expressing PI-9 has been studied recently which further emphasizes its role [87, 88]. In cultured human hepatoma cells, induction of endogenous PI-9 by IFN-γ or estrogen partially blocks CTL and NK-induced apoptosis [156]. Similarly, induction of increasing amounts of endogenous PI-9 by estrogen in a human breast cancer line (MCF-7) progressively increases its resistance to NK-mediated cytolysis [156]. PI-9 expression in pediatric acute lymphoblastic leukemias also correlates with resistance to cytolysis *in vitro*. Most importantly, PI-9 expression is an important determinant of disease-free survival time of melanoma patients following immunotherapy. However, endogenous PI-9 and Bcl-2 expression by some human lymphomas do not confer any resistance to cytolysis by *in vitro*–activated CTLs or NK cells. Many studies carried out the cytotoxicity studies of an *in vitro*–activated cytotoxic lymphocyte considering it to be comparable to the *in vivo* scenario. However, measuring cytotoxicity using highly activated cytotoxic lymphocytes *in vitro* may exaggerate the effectiveness of these cells and thus underestimate the protective capacity of anti-apoptotic molecules. The ability of serpins to make tumors resistant to immune cell destruction most likely depend on the level of expression of the serpin and of other anti-apoptotic molecules, such as Bcl-2 family members and survivin.

2.5.2.2 Viral Granzyme Inhibitors

The first viral inhibitor that was found to inhibit GrB was the pox virus–encoded cytokine response modifier A (CrmA). Both *in vitro* and *in vivo* studies have shown CrmA binding and inhibition of GrB directly. Overexpression of CrmA in target cells inhibits CTL-mediated cell death. However, CrmA also strongly binds and inhibits caspases-1 and −8 and weakly inhibits other caspases like caspase-3; therefore, it is difficult to pinpoint the importance of GrB inhibition in these studies [157]. Parainfluenza virus type 3 specifically inhibits GrB by degrading GrB mRNA in infected T cells [158]. Importantly, GrA transcripts are not affected by this virus and the mechanism of virus-mediated GrB mRNA decay is not known. Human GrB is inhibited by the adenoviral assembly protein (Ad5-100 K) by a mechanism distinct from serpin. In adenovirus-infected cells, Ad5-100 K rapidly complexes with GrB and gets cleaved very slowly at specific sites. GrB that enters the infected target cell upon CTL attack is saturated by the abundant Ad5-100 K protein. Unlike CrmA, which is just an anti-apoptotic factor, Ad5-100 K is also necessary for virus assembly [159], wherein it impedes human GrB but does not inhibit caspases or other apoptotic pathways. Interestingly, the inhibitory activity of Ad5-100 K is specifically directed against human GrB and not its mouse or rat homolog.

Recent studies further show CTLs have become resistant to adenoviruses. In adenovirus-infected cells, the Ad5-100 K-mediated GrB inhibition is released by the action of an orphan granzyme, Gr H by cleaving the former [159]. In general both GrB and GrH target the same adenoviral proteins, DNA-binding protein (DBP) and Ad5-100 K. The direct cleavage of essential viral proteins by granzymes is a novel mechanism by which cytotoxic cells rapidly and directly block viral replication [160]. Moreover, the granzyme-perforin system may also play a significant role in regulating immune cell numbers and function as well as disarming specific intracellular pathogens. The redundancy of granzymes makes sense, given the variety of tasks they need to accomplish. The example of the interplay between GrB, GrH and adenovirus illustrates why multiple granzymes may have evolved to eliminate important pathogens [161]. Additionally, the different specificities of the granzymes allow distinct substrate processing thus leading to synergistic antiviral activity. Viruses have evolved pathways to evade or inhibit granzymes and block apoptosis. This proved to be a unique catalytic specificity of granzymes combined to counter a viral challenge.

2.5.2.3 Synthetic Inhibitors

Design and generation of synthetic inhibitors of granzymes is a powerful tool both for research and potentially for therapeutic applications. The application not only includes the identification of peptide substrate specificity and determination of granzyme function but also immune suppression during autoimmune diseases and organ transplantation. There are several classes of granzyme inhibitors that have been used, including isocoumarin derivatives, peptide chloromethyl ketones, and peptide phosphonates (Table 3.3), but the major limitation has been lack of specificity [162]. Modifications that increase specificity generally diminish efficiency. Thornberry and colleagues [163] reported the identification of a novel

Table 3.3 Granzyme substrates and inhibitors

Granzymes	Endogenous substrates	Endogenous inhibitors	Synthetic inhibitors
Granzyme A	α-Tubulin	K-1-PI	DCI
	PHAP II	Antithrombin III	4-chloro-3-(3 isothioureido-propoxy)isocoumarin
	Nucleolin	Aprotinin	4-chloro-3-ethoxy-7-guanidinoisocoumarin
	Thrombin receptor		Boc-Ala-Ala-Asp-CH2Cl
	pIL-1L		Boc-Ala-Ala-Nle-CH2Cl
	Skeletal muscle proteins (including dystrophin, myosin and nebulin)		Boc-Gly-Leu-Phe-CH2Cl
	Proteoglycans		D-Phe-Pro-Arg-CH2Cl
	Fibronectin		Z-Trp-CH2Cl
	Pro-urokinase Plasminogen activator		Z-LysP(OPh)2
	Myelin basic protein		Z-(4-AmPhGly)P(OPh)2
	Collagen IV		3-(2-furyl)acryloyl-(4-AmPhGly)P(OPh)2
Granzyme B	Pro-caspase 3	α-1-PI	Z-Ala-(4-AmPhGly)P(OPh)2
	Pro-caspase 6	PI 9	Z-Leu-(4-AmPhGly)P(OPh)2
	Pro-caspase 7	SPI 6	Z-Pro-(4-AmPhGly)P(OPh)2
	Pro-caspase 9	rPIT5a	3,3-diphenylpropanoyl-Pro-(4 AmPhGly)P(OPh)2
	Pro-caspase 10	Crm A	PhCH2SO2-Gly-Pro-(4-AmPhGly)P(OPh)2
	PARP	mBM2A	
	DNA-PKcs	Ecotin	
	NuMA		
	Aggrecan		

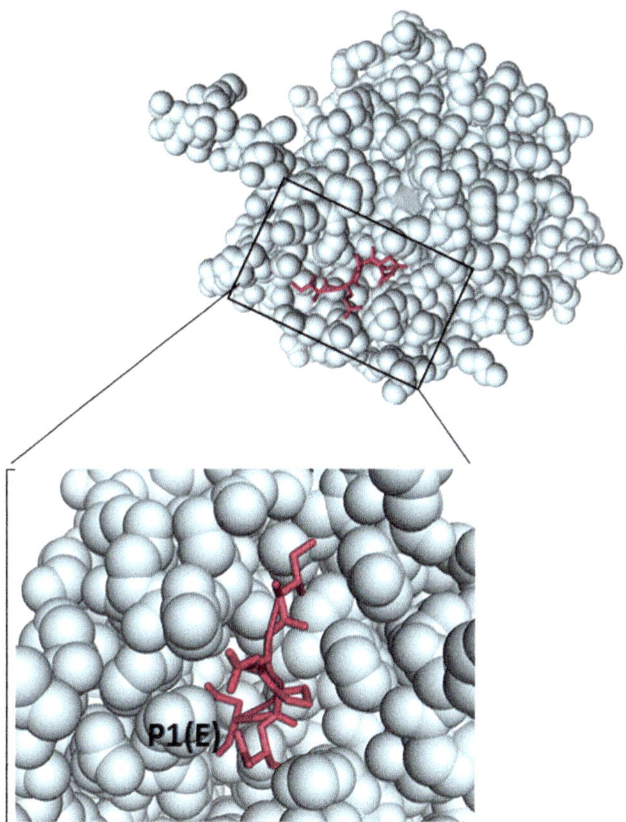

Fig. 3.10 Granzyme inhibition by serpin peptide: Co-crystal structure of granzyme with serpin inhibitor peptide provides details of the interaction at the specificity pocket where the role of Asp at P1 is primary for the interaction as well as inhibition

class of human GrB inhibitors. The key feature of these compounds is a 1, 2, 3-triazole moiety that is crucial for their selectivity and cellular efficacy. Future work with these inhibitors would determine their importance in studying the biology of granzymes.

Recent studies have begun to define multiple cell death pathways activated by individual granzymes and potentially important extracellular roles of these enzymes. The granzymes can trigger at least three distinct pathways, which are just being elucidated with the recent availability of recombinant active forms of many of the granzymes. Co-crystal structures of these Grs with the different inhibitors will thus help in furthering our understanding on the inhibitory mechanism (Fig. 3.10).

2.5.3 Autoinhibition

The phenomenon of auto-inhibition was also identified in Gr C [164]. Although the active-site triad residues adopt canonical conformations, the oxyanion hole is improperly formed, and access to the primary specificity (S1) pocket is blocked through a reversible rearrangement involving Phe-191. The block is specifically due to a shift in the strand preceding the active-site serine that leads Phe-191 filling the S1 pocket. Mutation of a unique Glu–Glu motif at positions 192–193 unlocks the enzyme, which displays chymase activity, and proteomic analysis confirms that activity of the wild-type protease can be released through interactions with an appropriate substrate. The structure of the unlocked enzyme revealed unprecedented flexibility in the 190-strand preceding the active-site serine that results in Phe-191 vacating the S1 pocket [164]. Overall, these observations describe a broadly applicable mechanism of protease regulation that cannot be predicted by template-based modeling or bioinformatic approaches alone.

2.6 Future Perspective

Apart from their active involvement in the induction of cell death and inflammation, granzymes have also been proposed to be responsible for alternative pathogenesis. In this regard, Darrah and Rosen have presented data supporting role of granzymes A, B, and H in autoantigen production, cytokine secretion, and autoimmunity [165]. In addition, they are also involved in cleavage of extracellular proteins, regulation of cell migration, platelet function, and in promoting inflammation [166, 167]. However, as with other putative pathogenic roles for granzymes, animal models showing a direct relationship between granzymes and autoimmunity are required before reaching any definitive conclusions [112].

From a therapeutic point of view, it would be essential to understand how granzymes are induced in immune/non-immune cells during a pathogenic condition. The role of mutations that result in enhanced granzyme production and/or activity can also be investigated. Likewise, it is mandatory that we elucidate how these proteases are regulated, confirm the identity of cells that express granzymes in healthy and diseased tissues. It would also be of great utility to assess how their expression or activity is altered in disease, and determine how perforin and serpins are involved in these physiological and pathological processes. Animal studies in concert with biochemical and clinical studies are necessary to elucidate pathophysiological processes. The study of granzymes H, K, and M with respect to their *in vitro* activity, is still at its infancy and provides great opportunities for future discoveries. Our present understanding of granzyme biology in the context of both cell death and other proteolytic functions requires further investigation to elucidate its specific mechanism.

References

1. Guroff G (1964) A neutral, calcium-activated proteinase from the soluble fraction of rat brain. J Biol Chem 239:149–155
2. Huston RB, Krebs EG (1968) Activation of skeletal muscle phosphorylase kinase by Ca2+ II. Identification of the kinase activating factor as a proteolytic enzyme. Biochemistry 7:2116–2122

3. Dayton WR, Reville WJ, Goll DE, Stromer MH (1976) A calcium(2+) ion-activated protease possibly involved in myofibrillar protein turnover. Partial characterization of the purified enzyme. Biochemistry 15:2159–2167
4. Murachi T, Tanaka K, Hatanaka M, Murakami T (1980) Intracellular Ca2 + −dependent protease (calpain) and its high-molecular-weight endogenousinhibitor (calpastatin). Adv Enzyme Regul 19:407–424
5. Mellgren SI (1980) Alzheimer-type dementia. Possible relation to hypofunction of the cholinergic central nervous system. Tidsskr Nor Laegeforen 100:1355–1356
6. Wheelock MJ (1982) Evidence for two structurally different forms of skeletal muscle Ca2+-activated protease. J Biol Chem 257:12471–12474
7. Saido TC, Sorimachi H, Suzuki K (1994) Calpain: new perspectives in molecular diversity and physiological, pathological involvement. FASEB J 8:814–822
8. Carafoli E, Molinari M (1998) Calpain: a protease in search of a function? Biochem Biophys Res Commun 247:193–203
9. Sorimachi H, Suzuki K (2001) The structure of calpain. J Biochem 129:653–664
10. Huang Y, Wang KKW (2001) The calpain family and human disease. Trend Mol Med 7:355–362
11. Dear TN, Boehm T (1999) Diverse mRNA expression patterns of the mouse calpain genes Capn5, Capn6 and Capn11 during development. Mech Dev 89:201–209
12. Dear TN, Moller A, Boehm T (1999) CAPN11: a calpain with high mRNA levels in testis and located on chromosome 6. Genomics 59:243–247
13. Dear TN, Meier NT, Hunn M, Boehm T (2000) Gene structure, chromosomal localization, and expression pattern of Capn12, a new member of the calpain large subunit gene family. Genomics 68:152–160
14. Blanchard H, Grochulski P, Li Y, Simon J, Arthur C, Davies PL, Elce JS, Cygler M (1997) Structure of a calpain Ca2+-binding domain reveals a novel EF-hand and Ca2+-induced conformational changes. Nat Struct Biol 4:532–538
15. Lin GD, Chattopadhyay D, Maki M, Wang KKW, Carson M, Jin L, Yuen P, Takano E, Hatanaka M, DeLucas LJ, Narayana SVL (1997) Crystal structure of calcium bound domain VI of calpain at 1.9 Å resolution and its role in enzyme assembly, regulation, and inhibitor binding. Nat Struct Biol 4:539–547
16. Hosfield CM, Elce JS, Davies PL, Jia Z (1999) Crystal structure of calpain reveals the structural basis for Ca2+-dependent protease activity and a novel mode of enzyme activation. EMBO J 18:6880–6889
17. Strobl S, Fernandez-Catalan C, Braun M, Huber H, Masumoto H, Nakagawa K, Irie A, Sorimachi H, Bourenkow G, Bartunik H, Suzuki K, Bode W (2000) The crystal structure of calcium-free human m-calpain suggests an electro-static switch mechanism for activation by calcium. Proc Natl Acad Sci U S A 97:588–592
18. Davis TL, Walker JR, Finerty PJ, Mackenzie F, Newman EM, Dhe-Paganon S (2007) The crystal structures of human Calpains 1 and 9 imply diverse mechanisms of action and auto-inhibition. J Mol Biol 366:216–229
19. Moldoveanu T, Gehring K, Green DR (2008) Concerted multi-pronged attack by calpastatin to occlude the catalytic cleft of heterodimeric calpains. Nature 456:404–408
20. Moldoveanu T, Hosfield CM, Lim D, Elce JS, Jia Z, Davies PL (2002) A Ca2+ switch aligns the active site of calpain. Cell 108:649–660
21. Hanna RA, Campbell R, Davies PL (2008) Calcium-bound structure of calpain and its mechanism of inhibition by calpastatin. Nat Struct Biol 456:409–412
22. Moldoveanu T, Gehring K, Green DR (2008) Concerted multi-pronged attack by Calpastatin specifically occludes the catalytic cleft of heterodimeric calpains. Nature 456:404–408
23. Suzuki K, Sorimachi H, Yoshizawa T, Kimbara K, Ishiura S (1995) Calpain: novel family members, activation and physiological function. Biol Chem 376:523–529
24. Arthur JS, Elce JS, Hegadorn C, Williams K, Greer PA (2000) Disruption of the murine calpain small subunit gene, capn4: calpain is essential for embryonic development but not for cell growth and division. Mol Cell Biol 20:4474–4481

25. Shiraha H, Glading A, Chou J, Jia Z, Wells A (2002) Activation of m-calpain(calpain II) by epidermal growth factor is limited by protein kinase A phosphorylation of m-calpain. Mol Cell Biol 22:2716–2727
26. Smith MA, Schnellmann RG (2012) Calpains, mitochondria, and apoptosis. Cardiovasc Res 96:32–37
27. Corbalan-Garcia S, Gomez-Fernandez JC (2010) The C2 domains of classical and novel PKCs as versatile decoders of membrane signals. Biofactors 36:1–7
28. Rizo J, Sudhof TC (1998) C2-domains, structure and function of a universal Ca2+-binding domain. J Biol Chem 273:15879–15882
29. Nakagawa K, Masumoto H, Sorimachi H, Suzuki K (2001) Dissociation of m-calpain subunits after autolysis of the N-terminus of the catalytic subunit, and is not required for activation. J Biochem 130:605–611
30. Sorimachi H, Kinbara K, Kimura S, Takahashi S, Ishiura S, Sasagawa N, Sorimachi N, Shimada H, Tagawa K, Maruyama K, Suzuiki K (1995) Muscle-specific calpain, p94, responsible for limb girdle muscular dystrophy type 2A, associates with connectin through IS2, a p94-specific sequence. J Biol Chem 270:31158–31162
31. Spencer MJ, Guyon JR, Sorimachi H, Potts A, Richard I, Chamberlain J, Dalkilic I, Kunkel LM, Beckmann JS (2002) Stable expression of calpain 3 from a muscle transgenic mice suggests a role for calpain 3 in muscle maturation. Proc Natl Acad Sci U S A 99:8874–8879
32. Koichi Suzuki SH, Kawabata Y, Sorimachi H (2004) Structure, activation, and biology of calpain. DIABETES 53:S12–S18
33. Zimmerman U-JP, Boring L, Park JH, Mukerjee N, Wang KK (2000) The calpain small subunit gene is essential: its inactivation results in embryonic lethality. Life 50:63–68
34. Yoshikawa Y, Mukai H, Hino F, Asada K, Kato I (2000) Isolation of two novel genes, down regulated in gastric cancer. Jpn Can Res 91:459–463
35. Roberts-Lewis JM, Siman R (1993) Spectrin proteolysis in the hippocampus: a biochemical marker for neuronal injury and neuroprotection. Ann N Y Acad Sci 679:78–86
36. Wang KKW, Nath R, Posner A, Rase KD, Burokev-Kilgore M, Hajimoham-madreza I, Robert AW, Marcoux FW, Ye Q, Tankano E, Hatanaka M, Maki M, Caner H, Collins JL, Fergus A, Lee KS, Lunney EA, Hays SJ, Yuen P (1996) An alpha-mercaptoacrylic acid derivative is a selective non-peptidic cell-permeable calpain inhibitor and is neuroprotective. Proc Natl Acad Sci U S A 93:6687–6692
37. Gafni J, Cong X, Chen SF, Gibson BW, Ellerby LM (2009) Calpain-1 cleaves and activates caspase-7. J Biol Chem 284:2544–25449
38. Sorimachi H, Ishiura S, Suzuki K (1997) Structure and physiological function of calpains. Biochem J 328:721–732
39. Richard I, Broux O, Allamand V, Fougerousse F, Chiannilkulchai N, Bourg N, Brenguier L, Devaud C, Pastuard P, Roudaut C, Hillaire D, Passos-Bueno M, Zatz M, Tischfield JA, Fardeau M, Jackson CE, Cohen D, Beckmann JS (1995) Mutations in the proteolytic enzyme calpain 3 cause limb-girdle muscular dystrophy type 2A. Cell 81:27–40
40. Azam M, Andrabi SS, Sahr KE, Kamath L, Kulipoulos A, Chisti AH (2001) Disruption of the mouse u-calpain gene reveals an essential role in platelet function. Mol Cell Biol 21:2213–2220
41. Ono Y, Shimada H, Sorimachi H, Richard I, Saido TC, Beckmann JS, Ishiura S, Suzuki K (1998) Functional defects of a muscle-specific calpain, p94, caused by mutations associated with limb-girdle muscular dystrophy type 2A. J Biol Chem 273:17073–17078
42. Gonen H, Shkedy D, Barnoy S, Kosower NS, Ciechanover A (1997) On the involvement of calpains in the degradation of the tumor suppressor protein p53. FEBS Lett 406:17–22
43. Pariat M et al (1997) Proteolysis by calpains: a possible contribution to degradation of p53. Mol Cell Biol 17:2806–2815
44. Han Y, Weinman S, Boldogh I, Walker RK, Brasier AR (1999) Tumor necrosis factor-α-inducible IκBα proteolysis mediated by cytosolic m-calpain. A mechanism parallel to the ubiquitin-proteasome pathway for nuclear factor-κb activation. J Biol Chem 274:787–794

45. Pianetti S, Arsura M, Romieu-Mourez R, Coffey RJ, Sonenshein GE (2001) Her-2/neu overexpression induces NF-κB via a PI3-kinase/Akt pathway involving calpain-mediated degradation of IκB-α that can be inhibited by the tumor suppressor PTEN. Oncogene 20:1287–1299
46. Lee FY et al (2005) mu-Calpain regulates receptor activator of NF-κB ligand (RANKL)-supported osteoclastogenesis via NF-κB activation in RAW 264.7 cells. J Biol Chem 280:29929–29936
47. Benetti R, Del Sal G, Monte M, Paroni G, Bancolini C, Schneider C (2001) The death substrate Gas2 binds m-calpain and increases susceptibility to p53-dependent apoptosis. EMBO J 20:2702–2714
48. Atencio IA, Ramachandra M, Shabram P, Demers GW (2000) Calpain inhibitor 1 activates p53-dependent apoptosis in tumor cell lines. Cell Growth Differ 11:247–253
49. Bartus RT, Hayward NJ, Elliott PJ, Sawyer SD, Baker KL, Dean RC, Akiyama A, Straub JA, Harbeson SL, Li Z, Powers J (1994) Calpain inhibitor AK295 protects neurons from focal brain ischemia: effects of postocculusion intra-arterial administration. Stroke 25:2265–2270
50. Wang KKW, Yuen PW (1994) Calpain inhibition: an overview of its therapeutic potential. Trend Pharm Sci 15:412–419
51. Patel YM, Lane MD (1999) Role of calpain in adipocyte differentiation. Proc Natl Acad Sci U S A 96:1279–1284
52. Patel YM, Lane M (2000) Mitotic clonal expansion during preadipocyte differentiation: calpain-mediated turnover of p27. J Biol Chem 275:17653–17660
53. Ortho-Malender M, Klannemark M, Svensson MK, Ridderstrale M, Lidgren CM, Groop L (2002) Variants in the calpain-10 gene predispose to insulin resistance and elevated free fatty acid level. Diabetes 51:2658–2664
54. Griffin ME, Marcucci MJ, Cline GW, Bell K, Barucci N, Lee D, Goodyear LJ, Kragen EW, White MF, Shulman GI (1999) Free fatty acid-induced insulin resistance is associated with activation of protein kinase C theta and alterations in the insulin signaling cascade. Diabetes 48:1270–1274
55. Barnoy S, Supino-Rosin L, Kosower N (2000) Regulation of calpain and calpasta-tin in differentiating myoblasts: mRNA levels, protein synthesis and stability. Biochem J 351:413–420
56. Itani S, Zhou Q, Pories WJ, MacDonald KG, Dohm GL (2000) Involvement of protein kinase C in human skeletal muscle resistance and obesity. Diabetes 49:1353–1358
57. Hui-Ju T, Sun G, Weeks DE, Kaushal R, Wolujewicz M, McGarvey ST, Tufa J, Viali S, Deka R (2001) Type 2 diabetes and three calpain-10 gene polymorphisms in Samoans: No evidence of association. AJHG 69:1236–1244
58. Weiwei Gao JL, Hu M, Huang M, Cai S, Zeng Z, Lin B, Cao X, Chen J, Zeng J-Z, Zhou H, Zhang X-K (2013) Regulation of proteolytic cleavage of retinoid X receptor-α by GSK-3β. Carcinogenesis 34:1208–1215
59. Wu HY, Lynch D (2006) Calpain and synaptic function. Mol Neurobiol 33:215–236
60. Wang JC, Zhao Y, Li XD, Zhou NN, Sun H, Sun YY (2012) Proteolysis by endogenous calpain I leads to the activation of calcineurin in human heart. Clin Lab 58:1145–1152
61. Wood DE, Thomas A, Devi LA, Berman Y, Beavis RC, Reed JC, Newcomb EW (1998) Bax cleavage is mediated by calpain during drug-induced apoptosis. Oncogene 17:1069–1078
62. Barnoy S, Glaser T, Kosower NS (1998) The calpain–calpastatin system and protein degradation in fusing myoblasts. Biochim Biophys Acta 1402:52–60
63. Brady CP, Brinkworth RI, Dalton JP, Dowd AJ, Verity CK, Brindley PJ (2000) Molecular modeling and substrate specificity of discrete cruzipain-like and cathepsin L-like cysteine proteinases of the human blood fluke Schistosoma manson. Arch Biochem Biophys 380:46–55
64. Ishiura S, Sugita H, Suzuki K, Imahori K (1979) Studies of a calcium-activated neutral protease from chicken skeletal muscle. II. Substrate specificity. J Biochem 86:579–581
65. Chandra D, Ramana KV, Wang L, Christensen BN, Bhatnagar A, Srivastava SK (2002) Inhibition of fiber cell globulization and hyperglycemia-induced lens opacification by aminopeptidase inhibitor bestatin. Invest Ophthalmol Vis Sci 43:2285–2292

66. Sakai K, Akanuma H, Imahori K, Kawashima S (1987) A unique specificity of a calcium activated neutral protease indicated in histone hydrolysis. J Biochem 101:911–918
67. Stabach PR, Cianci CD, Glantz SB, Zhang Z, Morrow JS (1997) Site-directed mutagenesis of α II spectrin at codon 1175 modulates its μ-calpain susceptibility. Biochemistry (Mosc) 36:57–65
68. Hirao T, Takahashi K (1984) Purification and characterization of a calcium-activated neutral protease from monkey brain and its action on neuropeptides. J Biochem 96:775–784
69. Takahashi K (1990) Calpain substrate specificity. In: Intracellular calcium dependent proteolysis. CRC Press, Boca Raton, pp 571–598
70. duVerle DA, Mamitsuka H (2011) A review of statistical methods for prediction of proteolytic cleavage. Brief Bioinform 13:337–349
71. duVerle DA, Takigawa I, Ono Y, Sorimachi H, Mamitsuka H (2010) CaMPDB: a resource for calpain and modulatory proteolysis. Genome Inf 22:202–213
72. Crooks GE, Hon G, Chandonia JM, Brenner SE (2004) WebLogo: a sequence logo generator. Genome Res 14:1188–1190
73. Nakai K, Kidera A, Kanehisa M (1988) Cluster analysis of amino acid indices for prediction of protein structure and function. Protein Eng 2:93–100
74. Hastie T, Tibshirani R, Friedman J (2009) The elements of statistical learning: data mining, inference, and prediction, 2nd edn. Springer, New York
75. Kiss R, Kovacs D, Tompa A, Perczel A (2008) Local structural preferences of calpastatin, the intrinsically unstructured protein inhibitor of calpain. Biochemistry 47:6936–6945
76. Emori Y, Kawasaki H, Imajoh S, Imahori K, Suzuki K (1987) Endogenous inhibitor for calcium-dependent cysteine protease contains four internal repeats that could be responsible for its multiple reactive sites. Proc Natl Acad Sci U S A 84:3590–3594
77. Maki M, Takano E, Mori H, Kannagi R, Murachi T, Hatanaka M (1987) Repetitive region of calpastatin is a functional unit of the proteinase inhibitor. Biochem Biophys Res Commun 143:300–308
78. Song J, Liu K, Yi J, Zhu D, Liu G, Liu B (2010) Luteolin inhibits lysophosphatidylcholine-induced apoptosis in endothelial cells by a calcium/mitocondrion/caspases dependent pathway. Planta Med 76:433–438
79. Suzuki K (1983) Reaction of calcium-activated neutral protease (CANP) with an epoxysuccinyl derivative (E64c) and iodoacetic acid. J Biochem 93:1305–1312
80. Murray EJ, Grisanti M, Bentley GV, Murray SS (1997) E64d, a membrane-permeable cysteine protease inhibitor, attenuates the effects of parathyroid hormone on osteoblasts in vitro. Metabolism 46:1090–1094
81. Ali MA, Stepanko A, Fan X, Holt A, Schulz R (2012) Calpain inhibitors exhibit matrix metalloproteinase-2 inhibitory activity. Biochem Biophys Res Commun 423:1–5
82. Elce JS (1997) Oxidation inhibits substrate proteolysis by calpain I but not autolysis. J Bio Chem 272:2005–2012
83. Sasaki T, Kishi M, Saito M, Tanaka T, Higuchi N, Kominami E, Katunuma N, Murachi T (1990) Inhibitory effect of di- and tripeptidyl aldehydes on calpains and cathepsins. J Enzyme Inhib 3:195–201
84. Saito Y, Tsubuki S, Ito H, Kawashima S (1990) The structure-function relationship between peptide aldehyde derivatives on initiation of neurite outgrowth in PC12h cells. Neurosci Lett 120:1–4
85. Jensen TJ, Loo M, Pind S, Williams DB, Goldberg AL, Riordan JR (1995) Multiple proteolytic systems, including the proteasome, contribute to CFTR processing. Cell 83:129–135
86. McKernan DP, Guerin MB, O'Brien CJ, Cotter TG (2007) A key role for calpains in retinal ganglion cell death. Invest Ophthalmol Vis Sci 48:5420–5430
87. Baba K, Nishida K (2012) Calpain inhibitor nanocrystals prepared using Nano Spray Dryer B-90. Nanoscale Res Lett 7
88. Momeni HR, Kanje M (2006) Calpain inhibitors delay injury-induced apoptosis in adult mouse spinal cord motor neurons. Neuroreport 17:761–765

89. Polster BM, Aitor Etxebarria GB, Marie Hardwick J, Nicholls DG (2005) Calpain I induces cleavage and release of apoptosis-inducing factor from isolated mitochondria. J Biol Chem 280:6447–6454
90. Tsubuki S, Saito Y, Tomioka M, Ito H, Kawashima S (1996) Differential inhibition of calpain and proteasome activities by peptidyl aldehydes of di-leucine and tri-leucine. J Biochem 119:572–576
91. Saatman KE, Murai H, Bartus RT, Smith DH, Hayward NJ, Perri BR, McIntosh TK (1996) Calpain inhibitor AK295 attenuates motor and cognitive deficits following experimental brain injury in the rat. Proc Natl Acad Sci U S A 93:3428–3433
92. James T, Matzelle D, Bartus R, Hogan EL, Banik NL (1998) New inhibitors of calpain prevent degradation of cytoskeletal and myelin proteins in spinal cord in vitro. J Neurosci Res 2:218–222
93. Wang KK, Nath R, Posner A, Raser KJ, Buroker-Kilgore M, Hajimohammadreza I, Probert WA, Marcoux FW, Ye Q, Takano E, Hatanaka M, Maki M, Caner H, Collins JL, Fergus A, Lee KS, Lunney EA, Hays SJ, Yuen P (1996) An alpha-mercaptoacrylic acid derivative is a selective nonpeptide cell-permeable calpain inhibitor and is neuroprotective. PNAS 93:6687–6692
94. Chatterjee PK, Todorovic Z, Sivarajah A, Mota-Filipe H, Brown PA, Stewart KN, Mazzon E, Cuzzocrea S, Thiemermann C (2005) Inhibitors of calpain activation (PD150606 and E-64) and renal ischemia-reperfusion injury. Biochem Pharmacol 69:1121–1131
95. Chen Q, Paillard M, Gomez L, Ross T, Hu Y, Xu A (2011) Activation of mitochondrial m-calpain increases AIF cleavage in cardiac mitochondria during ischemia – reperfusion. Biochem Biophys Res Commun 415:533–538
96. Ozaki T, Yamashita T, Ishiguro S-i (2008) ERp57-associated mitochondrial m-calpain truncates apoptosis-inducing factor. Biochim Biophys Acta 1783:955–1963
97. Ye H, Cande C, Stephanou NC, Jiang S, Gurbuxani S, Larochette N (2002) DNA binding is required for the apoptogenic action of apoptosis inducing factor. Nat Struct Mol Biol 9:680–684
98. Spencer ML, Theodosiou M, Noonan DJ (2004) NPDC-1, a novel regulator of neuronal proliferation, is degraded by the ubiquitin/proteasome system through a PEST degradation motif. J Biol Chem 279:37069–37078
99. Arrington DD, Van Vleet T, Schnellmann RG (2006) Calpain 10: a mitochondrial calpain and its role in calcium-induced mitochondrial dysfunction. Am J Physiol Cell Physiol 291:C1159–C1171
100. Shumway SD, Maki M, Miyamoto S (1999) The PEST domain of IkBais necessary and sufficient for in vitro degradation bym-calpain. J Biol Chem 274:30874–30881
101. Cregan SP, Dawson VL, Slack RS (2004) Role of AIF in caspase-dependent and caspase-independent cell death. Oncogene 23:2785–2796
102. Jenne DE, Tschopp J (1988) Granzymes, a family of serine protease released from granules of cytolytic T lymphocytes upon T cell receptor stimulation. Immunol Rev 103:53–71
103. Masson DTJ (1987) A family of serine esterases in lytic granules of cytolytic T lymphocytes. Cell 49:679–685
104. Doherty PC (1993) Cell-mediated cytotoxicity. Cell 75:607–612
105. Podack ER, Kupfer A (1991) T-cell effector functions: mechanisms for delivery of cytotoxicity and help. Annu Rev Cell Biol 7:479–504
106. Peters PJ, Geuze HJ, van der Donk HA, Borst J (1990) A new model for lethal hit delivery by cytotoxic T lymphocytes. Immunol Today 11:28–32
107. Tschopp J, Nabholz M (1990) Perforin-mediated target cell lysis by cytolytic T lymphocytes. Annu Rev Immunol 8:279–302
108. Boldin MP, Mett IL, Varfolomeev EE, Chumakov I, Shemer-Avni Y, Camonis JH, Wallach D (1995) Self-association of the "death domains" of the p55 tumor necrosis factor (TNF) receptor and Fas/APO1 prompts signaling for TNF and Fas/APO1 effects. J Biol Chem 270:387–391
109. Nagata S, Golstein P (1995) The Fas death factor. Science 267:1449–1456

110. Murphy ME, Moult J, Bleackley RC, Gershenfeld H, Weissman IL, James MN (1988) Comparative molecular model building of two serine proteinases from cytotoxic T lymphocytes. Proteins 4:190–204
111. Salvesen G, Farley D, Shuman J, Przybyla A, Reilly C, Travis J (1987) Molecular cloning of human cathepsin G: structural similarity to mast cell and cytotoxic T lymphocyte proteinases. Biochemistry 26:2289–2293
112. Granville DJ (2010) Granzymes in disease: bench to bedside. Cell Death Differ 17:565–566
113. Smyth MJ, O'Connor MD, Trapani JA (1996) Granzymes: a variety of serine protease specificities encoded by genetically distinct subfamilies. J Leukoc Biol 5:555–562
114. Smyth MJ, O'Connor M, Trapani JA (1996) Granzymes: a variety of serine protease specificities encoded by genetically distinct subfamilies. J Leukoc Biol 5:555–562
115. Rotonda J, Garcia-Calvo M, Bull HG, Geissler WM, McKeever BM, Willoughby CA, Thornberry NA, Becker JW (2001) The three-dimensional structure of human granzyme B compared to caspase-3, key mediators of cell death with cleavage specificity for aspartic acid in P1. Chem Biol 8:357–368
116. Wu L, Wang L, Hua G, Liu K, Yang X, Zhai Y, Bartlam M, Sun F, Fan Z (2009) Structural basis for proteolytic specificity of the human apoptosis-inducing granzyme M. J Immunol 183:421–429
117. Kaiserman D, Bird CH, Sun J, Matthews A, Ung K, Whisstock JC, Thompson PE, Trapani JA, Bird PI (2006) The major human and mouse granzymes are structurally and functionally divergent. J Cell Biol 175:619–630
118. Harris JL, Peterson EP, Hudig D, Thornberry NA, Craik CS (1998) Definition and redesign of the extended substrate specificity of granzyme B. J Biol Chem 273:27364–27373
119. Caputo A, James MN, Powers JC, Hudig D, Bleackley RC (1994) Conversion of the substrate specificity of mouse proteinase granzyme B. Nat Struct Biol 1:364–367
120. Trapani JA, Smyth MJ (2002) Functional significance of the perforin/granzyme cell death pathway. Nat Rev Immunol 2:735–747
121. Klimpel GR (1996) Immune defenses. In: Baron S (ed) Medical micriobiology, 4th edn. Galveston. ISBN: 0963117211
122. Gogvadze V, Orrenius S, Zhivotovsky B (2006) Multiple pathways of cytochrome c release from mitochondria in apoptosis. Biochim Biophys Acta 1757:639–647
123. Raja SM, Wang B, Dantuluri M, Desai UR, Demeler B, Spiegel K, Metkar SS, Froelich CJ (2002) Cytotoxic cell granule-mediated apoptosis. Characterization of the macromolecular complex of granzyme B with serglycin. J Biol Chem 277:49523–49530
124. Stinchcombe JC, Bossi G, Booth S, Griffiths GM (2001) The immunological synapse of CTL contains a secretory domain and membrane bridges. Immunity 15:751–761
125. Vyas YM, Maniar H, Dupont B (2002) Cutting edge: differential segregation of the SRC homology 2-containing protein tyrosine phosphatase-1 within the early NK cell immune synapse distinguishes noncytolytic from cytolytic interactions. J Immunol 168:3150–3154
126. Hudrisier D, Riond J, Mazarguil H, Gairin JE, Joly E (2001) Cutting edge: CTLs rapidly capture membrane fragments from target cells in a TCR signaling-dependent manner. J Immunol 166:3645–3649
127. Voskoboinik I, Smyth MJ, Trapani JA (2006) Perforin-mediated target-cell death and immune homeostasis. Nat Rev Immunol 6:940–952
128. Metkar SS, Wang B, Aguilar-Santelises M, Raja SM, Uhlin-Hansen L, Podack E, Trapani JA, Froelich CJ (2002) Cytotoxic cell granule-mediated apoptosis: perforin delivers granzyme B-serglycin complexes into target cells without plasma membrane pore formation. Immunity 16:417–428
129. Shi L, Keefe D, Durand E, Feng H, Zhang D, Lieberman J (2005) Granzyme B binds to target cells mostly by charge and must be added at the time as perforin to trigger apoptosis. J Biol Chem 174:5456–5461
130. Thiery J, Keefe D, Saffarian S, Martinvalet D, Walch M, Boucrot E, Kirchhausen T, Lieberman J (2010) Perforin activates clathrin- and dynamin-dependent endocytosis, which is required for plasma membrane repair and delivery of granzyme B for granzyme-mediated apoptosis. Blood 115:1582–1593

131. Calderwood SK, Stevenson MA, Murshid A (2012) Heat shock proteins, autoimmunity, and cancer treatment. Autoimmune Dis 2012:486069
132. Shi L, Mai S, Israels S, Browne K, Trapani JA, Greenberg AH (1997) Granzyme B (GraB) autonomously crosses the cell membrane and perforin initiates apoptosis and GraB nuclear localization. J Exp Med 185:855–866
133. Martinvalet D, Zhu P, Lieberman J (2005) Granzyme A induces caspase-independent mitochondrial damage, a required first step for apoptosis. Immunity 22:355–370
134. ten Cate R, Brinkman DM, van Rossum MA, Lankester AC, Bredius RG, Egeler MR, van Tol MJ, Vossen JM (2002) Macrophage activation syndrome after autologous stem cell transplantation for systemic juvenile idiopathic arthritis. Eur J Pediatr 161:686–686
135. Ravelli A (2002) Macrophage activation syndrome. Curr Opin Rheumatol 14:548–552
136. Ramanan AV, Baildam EM (2002) Macrophage activation syndrome is hemophagocytic lymphohistiocytosis–need for the right terminology. J Rheumatol 29:1105, author reply 1105
137. Adrain C, Murphy BM, Martin SJ (2005) Molecular ordering of the caspase activation cascade initiated by the cytotoxic T lymphocyte/natural killer (CTL/NK) protease granzyme B. J Biol Chem 280:4663–4673
138. Grossman WJ, Revell PA, Lu ZH, Johnson H, Bredemeyer AJ, Ley TJ (2003) The orphan granzymes of humans and mice. Curr Opin Immunol 15:544–552
139. Zhao T, Zhang H, Guo Y, Zhang Q, Hua G, Lu H, Hou Q, Liu H, Fan Z (2007) Granzyme K cleaves the nucleosome assembly protein SET to induce single-stranded DNA nicks of target cells. Cell Death Differ 14:489–499
140. Kelly JM, Waterhouse NJ, Cretney E, Browne KA, Ellis S, Trapani JA, Smyth MJ (2004) Granzyme M mediates a novel form of perforin-dependent cell death. J Biol Chem 279:22236–22242
141. Lu H, Hou Q, Zhao T, Zhang H, Zhang Q, Wu L, Fan Z (2006) Granzyme M directly cleaves inhibitor of caspase-activated DNase (CAD) to unleash CAD leading to DNA fragmentation. J Immunol 177:1171–1178
142. van Domselaar R, Quadir R, van der Made AM, Broekhuizen R, Bovenschen N (2012) All human granzymes target hnRNP K that is essential for tumor cell viability. J Biol Chem 287:22854–22864
143. Kaiserman D, Bird PI (2010) Control of granzymes by serpins. Cell Death Differ 17:586–595
144. Balaji KN, Schaschke N, Machleidt W, Catalfamo M, Henkart PA (2002) Surface cathepsin B protects cytotoxic lymphocytes from self-destruction after degranulation. J Exp Med 196:493–503
145. van Dommelen SL, Sumaria N, Schreiber RD, Scalzo AA, Smyth MJ, Degli-Esposti MA (2006) Perforin and granzymes have distinct roles in defensive immunity and immunopathology. Immunity 25:835–848
146. Wilczynska M, Fa M, Ohlsson PI, Ny T (1995) The inhibition mechanism of serpins. Evidence that the mobile reactive center loop is cleaved in the native protease-inhibitor complex. J Biol Chem 270:29652–29655
147. Law RH, Zhang Q, McGowan S, Buckle AM, Silverman GA, Wong W, Rosado CJ, Langendorf CG, Pike RN, Bird PI, Whisstock JC (2006) An overview of the serpin superfamily. Genome Biol 7:216
148. Sipione S, Simmen KC, Lord SJ, Motyka B, Ewen C, Shostak I, Rayat GR, Dufour JM, Korbutt GS, Rajotte RV, Bleackley RC (2006) Identification of a novel human granzyme B inhibitor secreted by cultured sertoli cells. J Immunol 177:5051–5058
149. Sun J, Bird CH, Sutton V, McDonald L, Coughlin PB, De Jong TA, Trapani JA, Bird PI (1996) A cytosolic granzyme B inhibitor related to the viral apoptotic regulator cytokine response modifier A is present in cytotoxic lymphocytes. J Biol Chem 271:27802–27809
150. Silverman GA, Bird PI, Carrell RW, Church FC, Coughlin PB, Gettins PG, Irving JA, Lomas DA, Luke CJ, Moyer RW, Pemberton PA, Remold-O'Donnell E, Salvesen GS, Travis J, Whisstock JC (2001) The serpins are an expanding superfamily of structurally similar but functionally diverse proteins. Evolution, mechanism of inhibition, novel functions, and a revised nomenclature. J Biol Chem 276:33293–33296

151. Jiang X, Orr BA, Kranz DM, Shapiro DJ (2006) Estrogen induction of the granzyme B inhibitor, proteinase inhibitor 9, protects cells against apoptosis mediated by cytotoxic T lymphocytes and natural killer cells. Endocrinology 147:1419–1426
152. Kanamori H, Krieg S, Mao C, Di Pippo VA, Wang S, Zajchowski DA, Shapiro DJ (2000) Proteinase inhibitor 9, an inhibitor of granzyme B-mediated apoptosis, is a primary estrogen-inducible gene in human liver cells. J Biol Chem 275:5867–5873
153. Phillips T, Opferman JT, Shah R, Liu N, Froelich CJ, Ashton-Rickardt PG (2004) A role for the granzyme B inhibitor serine protease inhibitor 6 in CD8+ memory cell homeostasis. J Immunol 173:3801–3809
154. Medema JP, de Jong J, Peltenburg LT, Verdegaal EM, Gorter A, Bres SA, Franken KL, Hahne M, Albar JP, Melief CJ, Offringa R (2001) Blockade of the granzyme B/perforin pathway through overexpression of the serine protease inhibitor PI-9/SPI-6 constitutes a mechanism for immune escape by tumors. Proc Natl Acad Sci U S A 98:11515–11520
155. Ray M, Hostetter DR, Loeb CR, Simko J, Craik CS (2012) Inhibition of Granzyme B by PI-9 protects prostate cancer cells from apoptosis. Prostate 72:846–855
156. Thomas DC, Xinguo J, David JS (2007) Expression of High Levels of Human Proteinase Inhibitor 9 Blocks Both Perforin/Granzyme and Fas/Fas Ligand-mediated Cytotoxicity. Cell Immunol 245:32–41
157. Cassens U, Lewinski G, Samraj AK, von Bernuth H, Baust H, Khazaie K, Los M (2003) Viral modulation of cell death by inhibition of caspases. Arch Immunol Ther Exp (Warsz) 51:19–27
158. Sieg S, Xia L, Huang Y, Kaplan D (1995) Specific inhibition of granzyme B by parainfluenza virus type 3. J Virol 69:3538–3541
159. Andrade F, Bull HG, Thornberry NA, Ketner GW, Casciola-Rosen LA, Rosen A (2001) Adenovirus L4-100K assembly protein is a granzyme B substrate that potently inhibits granzyme B-mediated cell death. Immunity 14:751–761
160. Andrade F, Casciola-Rosen LA, Rosen A (2003) A novel domain in adenovirus L4-100K is required for stable binding and efficient inhibition of human granzyme B: possible interaction with a species-specific exosite. Mol Cell Biol 23:6315–6326
161. Andrade F, Fellows E, Jenne DE, Rosen A, Young CS (2007) Granzyme H destroys the function of critical adenoviral proteins required for viral DNA replication and granzyme B inhibition. EMBO J 26:2148–2157
162. Kam CM, Hudig D, Powers JC (2000) Granzymes (lymphocyte serine proteases): characterization with natural and synthetic substrates and inhibitors. Biochim Biophys Acta 1477:307–323
163. Thornberry NA, Rano TA, Peterson EP, Rasper DM, Timkey T, Garcia-Calvo M, Houtzager VM, Nordstrom PA, Roy S, Vaillancourt JP, Chapman KT, Nicholson DW (1997) A combinatorial approach defines specificities of members of the caspase family and granzyme B. Functional relationships established for key mediators of apoptosis. J Biol Chem 272:17907–17911
164. Kaiserman D, Buckle AM, Van Damme P, Irving JA, Law RH, Matthews AY, Bashtannyk-Puhalovich T, Langendorf C, Thompson P, Vandekerckhove J, Gevaert K, Whisstock JC, Bird PI (2009) Structure of granzyme C reveals an unusual mechanism of protease autoinhibition. Proc Natl Acad Sci U S A 106:5587–5592
165. Darrah E, Rosen A (2010) Granzyme B cleavage of autoantigens in autoimmunity. Cell Death Differ 17:624–632
166. Hiebert PR, Granville DJ (2012) Granzyme B in injury, inflammation, and repair. Trends Mol Med 18:732–741
167. Anthony DA, Andrews DM, Chow M, Watt SV, House C, Akira S, Bird PI, Trapani JA, Smyth MJ (2010) A role for granzyme M in TLR4-driven inflammation and endotoxicosis. J Immunol 185:1794–1803

Chapter 4
Cathepsins and HtrAs – Multitasking Proteases in Programmed Cell Death

Lalith K. Chaganti, Nitu Singh, and Kakoli Bose

Abstract Apoptosis is a crucial process in embryonic development, adult tissue homeostasis as well as in selective clearance of unwanted or infected cells. It is an energy dependent process generally initiated by cellular damage, stress or several endogenous and extracellular stimuli. During apoptosis, activation of specific proteases results in breakdown of cellular machinery which finally culminates into characteristic morphological and biochemical changes and hence death. For many years, it was believed that caspases were the only enzymes responsible for the proteolytic cascades in apoptosis. However, accumulating evidences indicate that cell death can occur in a programmed fashion in absence of caspase activation. For example, other proteases, such as cathepsins and HtrAs (high temperature requirement protease A), are also involved in the initiation and/or execution of the apoptosis. These proteases are capable of triggering mitochondrial dysfunction with subsequent caspase activation and cellular demise. Cathepsins, a group of proteases enclosed in the lysosomes, have a major role in apoptosis by triggering the death receptor as well as mitochondria-mediated apoptotic pathways. Apart from cathepsins, HtrAs also have a potential role in mediating apoptosis. HtrA family proteins are serine proteases that are involved in important physiological processes, including maintenance of mitochondrial homeostasis, apoptosis and cell signaling. They are involved in the development and progression of several diseases such as cancer, neurodegenerative disorders and arthritis. HtrA proteins are described as potential modulators of programmed cell death and chemotherapy-induced cytotoxicity. This chapter summarizes our current knowledge on the structural and functional aspects of these proteins, with an emphasis on their potential roles in apoptosis.

Funding: A part of work on HtrA family was supported by grants from the Department of Biotechnology (DBT) [BT/PR10552/BRB/10/606/2008] and Department of Science and Technology [SERB/F/108/2012-13], Govt. of India.

L.K. Chaganti • N. Singh • K. Bose (✉)
Integrated Biophysics and Structural Biology Lab, Advanced Centre for Treatment, Research and Education in Cancer (ACTREC), Tata Memorial Centre, Navi Mumbai, Maharashtra 410210, India
e-mail: kbose@actrec.gov.in

K. Bose (ed.), *Proteases in Apoptosis: Pathways, Protocols and Translational Advances*, DOI 10.1007/978-3-319-19497-4_4

Keywords Cathepsins • HtrAs • Apoptosis • Cancer • Neurodegeneration • Inhibitors • Substrates • Catalytic mechanism • Activation • Active-site

1 Cathepsins

Cathepsins belong to a class of globular proteases, and are primarily localized in the lysosomes under normal physiological conditions. In response to stress signals, these proteases are released from the lysosomes into the cytoplasm where they trigger apoptotic cell death either by activation of caspases or inducing release of pro-apoptogenic factors from the mitochondria. The term "cathepsin", which was introduced in 1920, has been derived from the Greek word "*digesting*" as it was originally described as an acidic protease isolated from the stomach mucosa [1]. However, later it was found to comprise three different mechanistic classes of proteases: cysteine, serine, and aspartyl. The cathepsin family mainly embodies cysteine (Cys) proteases which belong to the so-called papain family (C1). In humans 11 of them are reported: cathepsins B, C (J, dipeptidyl peptidase I or DPPI), F, H, K (O2), L, O, S, V or L2, X (P, Y, Z) and W (lymphopain). The other members include cathepsins A and G which are serine proteases and cathepsins D and E which are aspartyl proteases. Depending upon their expression, they are also categorized as ubiquitous cathepsins (cathepsins B, L, H, C, O, X, F) and tissue or cell-specific cathepsins (cathepsins K, V, W, and S) [2].

Cathepsins are kept under tight check to prevent any accidental or uncontrolled proteolytic activity. They are synthesized as inactive pre-proenzymes having a signal peptide and a multi-functional N-terminal region. The processing of the inactive zymogen into a catalytically active form occurs within the lysosome by other active proteases or by auto-proteolysis under specific conditions, such as low pH or presence of glycosaminoglycans [3, 4]. The endogenous regulators of cathepsins are protein inhibitors such as cystatins, stefins [5, 6], thyropins [7] and serpins [8], which bind to the target protease, thereby preventing the substrate hydrolysis. Also, some cathepsins are regulated by the oxidation of Cys residue in the active-site [9].

Although their main function is lysosomal protein recycling, cathepsins do play an essential role in variety of physiological functions such as bone remodeling, wound healing and reproduction [10]. They are also associated with several pathological conditions including cancer, atherosclerosis, rheumatoid arthritis, and bronchial asthma [10, 11]. However, recent studies have discovered a key involvement of cathepsins in apoptosis [12, 13]. So far, a significantly large set of molecules have been identified which are the targets/substrates of these proteases, further inducing cell death in both caspase-dependent and independent manner. The current section will discuss about all the known members of this family; their structure, function, regulation and apoptotic functions.

Classification

The cathepsin protease family consists of at least 15 known members. Based upon the active-site residue, they are sub-divided into three distinct groups such as cysteine, serine and aspartyl proteases.

1.1 Cysteine Cathepsins

The cysteine subfamily of cathepsins with 11 members (B, L, K, S, V, F, H, W, X, C, and O) represents the largest and the best known class of this family. Cysteine cathepsins are majorly papain-like endopeptidases [14], which are located intracellularly in endo-lysosomal vesicles. Cathepsins B and L are expressed constitutively and participate in protein turnover. However, cathepsins S, K, V, F, C, and W are more selectively expressed and exhibit cell-type specific functions. Cathepsins S, F, and V that are involved in antigen processing and presentation, are highly expressed in macrophages, dendritic cells, and/or thymic cortical epithelial cells [15–17]. While cathepsin W is specifically expressed in CD8 and natural killer cells [18], cathepsin O is predominantly expressed in colon cancer cells [19].

Most cysteine cathepsins predominantly exhibit endopeptidase activity, whereas cathepsins C and X function strictly as exopeptidases. Both cathepsins C and H are aminodipeptidases [9], however the latter exhibits endopeptidase activity as well [20]. While cathepsin B play dual role of an endopeptidase as well as a carboxydipeptidase [21], cathepsin X is strictly a carboxymonopeptidase [22].

1.1.1 Structural Assembly

The cysteine cathepsins are synthesized as inactive proenzyme with a signal peptide, propeptide, and a C-terminal catalytic domain (Fig. 4.1). Signal peptide which is of 10–20 amino acids (aa) in length, targets the protease to the endoplasmic reticulum. The middle propeptide, however, is of varying length (36–251aa) which either acts as a scaffold to aid folding of catalytic domain or as a chaperone to direct the proenzyme to the lysosomal compartment. It also acts as a high-affinity reversible inhibitor preventing premature activation of the catalytic domain [23]. The catalytic domains of human cysteine cathepsins are between 214 and 260 aa in length and contain the highly conserved active-site triad comprising a cysteine, histidine, and asparagine residues [24]. Till date, 3D structures of nine cysteine cathepsins are solved with the exception of cathepsins O and W [5, 25]. All except homotetrameric cathepsin C, are relatively small monomers of 20–35 kDa. The papain fold of cysteine cathepsins is shown in Fig. 4.2a using cathepsin L as a typical representative. All the cysteine cathepsins exhibit conserved active-site residues (Cys and His), residues interacting with the main chain of the substrate, (Asn, Gly, and Trp), N-terminus Pro and certain other Cys residues as well. Overall, the structural fold of these cathepsins is highly conserved, which consists of two domains, L and R of similar size. The L-domain contains three α-helices while the R-domain comprises a twisted β-sheet and 2 α-helices. Active-site cysteine is located at the N-terminus of the central helix of the L-domain and histidine is located within the β-barrel residues of R-domain. At the junction between the R and L domains, substrate binding cleft is formed, which contains the protease's

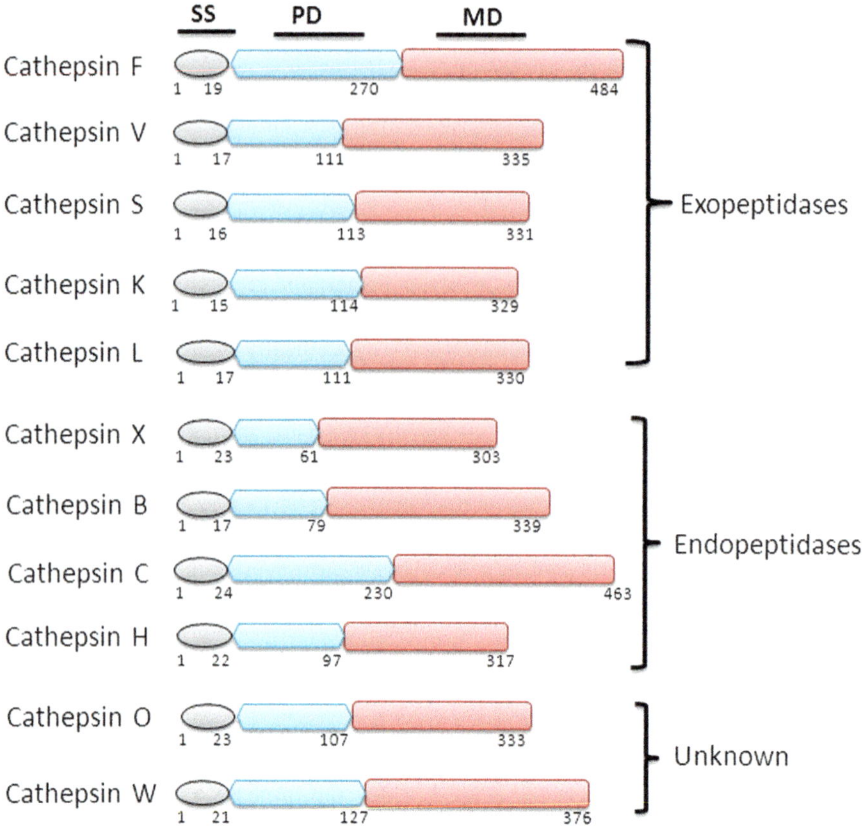

Fig. 4.1 Schematic representation of domain organization of human cysteine cathepsins. SS (signal sequence) is shown as *grey ovals*, PD (prodomain) as *blue hexagons*, MD (mature domain) in *pink rounded rectangles*. Numbers represent length of each domain

catalytic machinery. This cleft is the most cognate part of cathepsins, both in terms of understanding its substrate specificity and in terms of identifying potent and specific inhibitors [26, 27].

Substrate Pocket

S2, S1 and S1′ are the three well defined substrate-binding sites, which involve both the main and side chain contacts between substrate and enzyme residues. P2, P1 and P′ are the corresponding substrate residues, which bind to the enzyme's respective subsites [28]. S2 forms the substrate binding pocket, and determines the substrate specificity of cathepsins whereas S1 and S1′ provide the binding surface. The two loops from the L domain are cross connected with a disulfide bond and their bridged interface provide the surface for side chains of the P3, P1 and P2′ residues of substrate. However, the two R domain loops, form the lid of the β-barrel hydrophobic core and provide the binding surface for the P2 and P1′ residues [29].

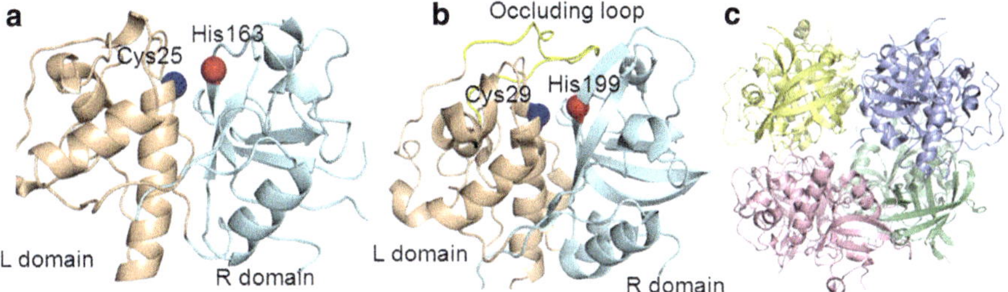

Fig. 4.2 Crystal structure of mature cysteine cathepsins. (**a**) Cartoon representation of native cathepsin L, PDB entry 1ICF. The corresponding domains L and R are shown in *orange* and *cyan* colors, respectively. The side chains of the active-site Cys25 (*blue*) and His163 (*red*) are represented as spheres. (**b**) Cartoon representation of native cathepsin B, PDB entry 1HUC. The corresponding domains L and R are shown in *orange* and *cyan* colors, respectively. The side chains of the active-site Cys29 (*blue*) and His199 (*red*) are represented as *spheres* and occluding loop is highlighted in *yellow*. (**c**) Schematic representation of cathepsin C homotetramer. Each monomer is represented as *yellow*, *blue*, *green* and *pink*, respectively. The figures are generated using PyMOL (DeLano Scientific, USA)

In contrast to the endopeptidases (cathepsins F, L, K, S and V), which possess an extended active-site cleft, the exopeptidases (cathepsins B, C, H and X) have an additional loop, which modifies the active-site cleft and reduces the number of binding sites.

Regulatory Loop

Cathepsin B has an additional regulatory loop (20 residues) known as the 'occluding loop,' which occupies part of the active-site cleft on the primed site (Fig. 4.2b). It essentially blocks the access of substrate to the active-site cleft beyond the S2′ substrate binding site. Three histidine residues of the loop anchor the C-terminal carboxyl group of the substrate. The occluding loop thus plays a decisive role in the exopeptidase activity of cathepsin B [30]. Cathepsin X also has a mini-loop composed of three residues with only one histidine, which is the anchor for the carboxyl group of the C-terminal substrate residue [22]. Cathepsin H, however, has eight mini-chain residues originating from the propeptide of the zymogen, which is attached by a disulphide bond to the catalytic domain of the enzyme and is bound to S2 binding site of the active-site cleft [22]. Thus, the positioning of the mini-chain dictates the aminopeptidase activity of cathepsin H [30].

Unique Assembly

Cathepsin C (dipeptidyl peptidase I) is distinct among the proteases of the papain superfamily due to its oligomeric nature. Indeed, it also has unique structure and mechanism compared to other oligomeric proteolytic complexes such as the proteasome [31], bleomycin hydrolase [32] and tryptase [33]. Unlike other proteases, it exists as a homotetramer with four independent active-sites located on its external surface (Fig. 4.2c). The tetrahedral arrangement of the subunits is the prerequisite for its enzyme activity making it an exceptional oligomeric protease capable of hydrolyzing substrates of different sizes. Each subunit consists of three domains, dual repeats of the papain-like structure containing the catalytic site and a third additional domain. This additional or so called ‘exclusion domain’ has no sequence homology with the rest of the family of papain-like proteases. It contributes essentially to the tetrahedral structure and determines an extension of the active-site cleft. The active-site of cathepsin C is blocked beyond the S2 binding site by the N-terminal residues of the exclusion domain [34].

1.1.2 Activation Mechanism

Most cathepsins are synthesized as inactive zymogens. These pro-cathepsins are activated by proteolytic removal of N-terminal prodomain, which blocks access to the catalytic site. The activation process is triggered by decrease in pH that substantially weakens the interaction between propeptide and catalytic domain [3, 4]. Consequently, the proenzyme adopts a relaxed conformation, in which the propeptide is less tightly bound to the active-site, while the secondary structure remains intact. Finally, the propeptide dissociates from the protease, unfolds, and is proteolytically degraded. The proteolytic removal is brought about by various proteases like pepsin, neutrophil elastase, cathepsins D and other cysteine proteases [29].

1.1.3 Catalytic Mechanism

Cysteine cathepsins contain a catalytic dyad comprising cysteine and histidine. These cysteine cathepsins share a common catalytic mechanism with other endogenous cysteine proteases. The general catalytic mechanism of cysteine proteases is discussed in Chap. 2.

1.1.4 Functions

1.1.4.1 Apoptosis

Cathepsins are well-known for their role in apoptosis and maintenance of tissue homeostasis. Involvement of cathepsins in apoptotic cell death was primarily studied in non-hematopoietic systems. Evidence from these systems provides the basis for understanding the role of cathepsins in both normal and malignant hematopoietic cell death. Lysosome-mediated apoptosis is either death receptor-dependent [35–37] or independent [38]. Apoptotic cell death involving cathepsins requires their release from the lysosomes into the cytosol. This lysosomal membrane permeabilization (LMP) is triggered in response to either extralysosomal or intralysosomal stimuli [38, 39] which include lysosomotropic agents [40, 41], photo damage [42, 43], sphingosine [35], D-galactosamine [44] reactive oxygen species [45]

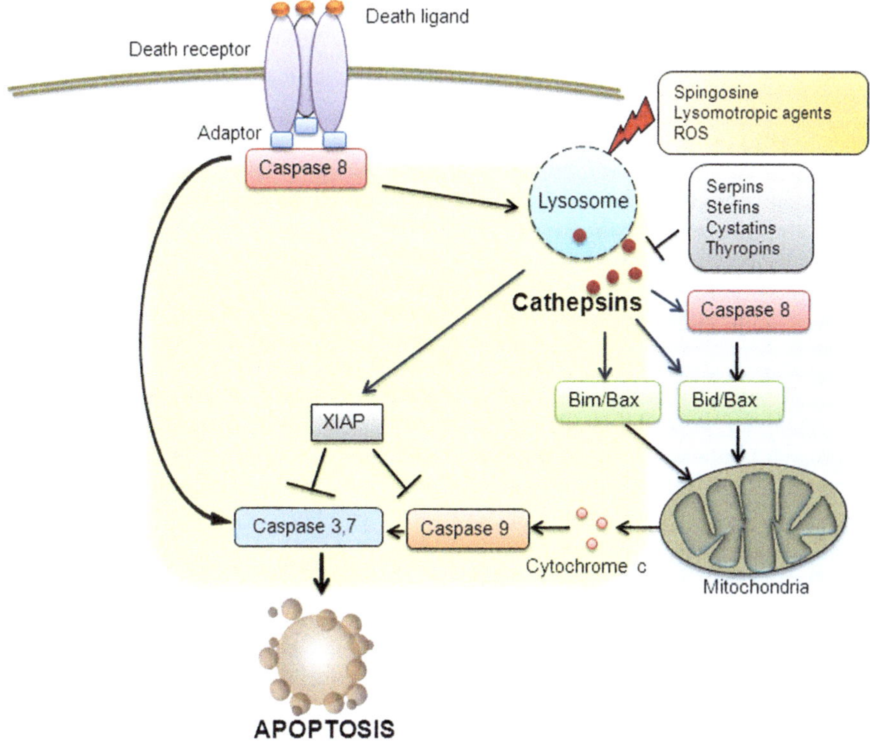

Fig. 4.3 Lysosomal cathepsins in apoptotic signaling cascade. The two major apoptotic pathways, extrinsic and intrinsic, and their major links with the lysosomes, are shown schematically. Upon lysosomal membrane permeabilization (LMP), cathepsins are rapidly released into the cytosol to trigger both the apoptotic pathways. The first pathway includes direct cleavage of Bcl-2 family proteins such as Bid and/or Bak and translocation of these pro-apoptotic proteins to the mitochondrial outer membrane. These processes later induce release of apoptogenic factors like cytochrome c from the mitochondria followed by subsequent activation of downstream caspases and hence apoptosis induction. In addition, anti-apoptotic protein XIAP is also cleaved leading to the activation of executioner caspases. The second pathway involves direct cleavage of caspase-8, followed by cleavage of Bid and/or Bax, translocation of these proteins into the mitochondria and similar downstream events leading to apoptosis. Intracellular cathepsin inhibitors such as serpins, cystatins, thyropins and stefins serve as cellular guardians by inhibiting these cathepsins

and compounds that activate p53-independent apoptosis [45]. Indirect stimulus involves caspase-8, which is activated following binding of a TNFα to its receptor [35–37]. The released cathepsins from lysosomes cleave Bid and the prosurvival Bcl2 homologues, thereby inducing cytochrome c release, formation of a ternary complex with the adaptor protein Apaf-1 and procaspase-9, subsequent activation of caspases and execution of cell death program (Fig. 4.3). On the other side, during TNFα-induced apoptosis, activated caspase-8 induces lysosomal membrane permeabilization (LMP) and promotes release of cathepsins from the lysosomes into the cytosol. Active cathepsins in turn promote the release of cytochrome *c* from mitochondria followed by caspase activation and cell death [35]. In addition, X-chromosome-linked inhibitor of apoptosis (XIAP) is degraded by cysteine cathepsins suggesting that cathepsins can also control apoptosis downstream of mitochondria-mediated apoptotic pathway [36].

Cathepsin C is required for the activation of granzyme serine peptidases that are important in cytotoxic lymphocyte granule-mediated apoptosis. In addition, this cathepsin could be involved in activation of other serine peptidase zymogens such as neutrophil elastase as well [46].

1.1.4.2 Other Functions

Cysteine cathepsins are also responsible for intracellular protein degradation, which is crucial for normal functioning of an organism. Cathepsins B, H and L are expressed ubiquitously in mammalian cells and are considered as 'housekeeping' enzymes essential for normal protein turnover of the cell. Cathepsin C has a role in general protein degradation, and activation of platelet factor XIII. Cathepsins B and L have been implicated in tissue-remodeling events. Cathepsin L is also involved in the degradation of the follicle wall leading to release of mature oocyte [47]. Cathepsin S acts as an intracellular processing enzyme during trafficking and secretion of proteins as well as an extracellular protease during response to damage and tissue remodeling. Another key function of cathepsin S is in antigen processing and presentation, which explains its prominent expression in phagocytic and antigen-presenting cells.

1.1.5 Substrates

1.1.5.1 Synthetic Substrates and Specificity

The specificities of cysteine cathepsins L, V, S, K, F, and B were determined using a tetrapeptide substrate library completely diversified at the P1-P4 positions [48]. The selectivity of the substrate primarily depends on the S2 and S2′ sites. All six human cathepsins generally matched papain in specificity by preferring basic amino acids (R and K) at the P1 position, strictly hydrophobic amino acids at the P2 position, and broader specificities at the P3 and P4 positions. However, these cathepsins demonstrated differences in their chemical characteristics of the favored amino acids at the P2 position and more subtle P3 specificity. At the P2 position, the substrate specificity profile of cathepsin L and V shows a preference for aromatic residues (F, Y, W) over aliphatic amino acids (V, M, L, I), which distinguishes them from cathepsins S and K. Apart from the aliphatic amino acids, cathepsin K has a unique preference for a proline residue in the P2 position. Cathepsin F shows preference for aspartic acid at the P2 and P3 positions, whereas none of the other cathepsins prefer this acidic amino acid residue at either position. Based on these amino acid preferences, a synthetic fluorescently quenched substrate was designed with ortho-aminobenzoic acid (Abz) as fluorophore and N-[[2, 4-dinitrophenyl]ethylenediamine] (EDDnp) as a quencher. The synthetic substrates designed for these cathepsins are Abz-GIVRAK (Dnp)-OH peptide for cathepsin B [49], Abz-HPGGPQ-EDDnp substrate for cathepsin K [50] and Abz-LEQ-EDDnp for cathepsin S as depicted in Table 4.1 [51].

1.1.5.2 Endogenous Substrates

Despite numerous studies highlighting the importance of cathepsins for apoptosis progression in a number of different apoptotic models, only few cathepsin substrates have been identified. Bid is the best characterized cathepsin substrate, initially discovered in a cell free system and later confirmed in a variety of cell models [12]. Moreover, *in vitro* studies demonstrated that a number of cysteine cathepsins, including cathepsins B, L, S and K, cleave Bid, predominantly at Arg65 [41]. In addition to Bid, cathepsins were found to degrade the anti-apoptotic Bcl-2, Bcl-XL, Mcl-1 and XIAP proteins [36]. Thus cathepsins trigger the mitochondrial pathway of apoptosis and activate caspases to execute cell death [43]. In addition, cathepsins are implicated in the regulation of angiogenesis and invasion during cancer progression by degrading E-cadherin, a cell adhesion molecule [52].

1.1.6 Inhibitors

1.1.6.1 Endogenous Inhibitors

Cathepsins have tremendous destructive potential and it is not surprising that a massive lysosomal rupture would cause accidental death. Therefore, one of the important questions is how the activity of cathepsins is regulated and prevented if they leak out of the lysosome. The major regulators of the mature cysteine cathepsins are their endogenous protein inhibitors, cystatin family proteins (stefins, cystatins, and kininogens) [5, 6], thyropins (thyroglobulin type-1 domain inhibitors) [7] and α2-macroglobulin [53]. These are competitive, reversible, tight-binding inhibitors which prevent substrate binding to the active-site. The importance of cystatin inhibitors in preventing cell death under physiological and pathological conditions is demonstrated by cystatin B-deficient mice which display increased apoptosis of cerebellar granule cells [54]. Cystatins inhibit their target enzymes rather unselectively [5], whereas thyropins are much more selective [55], probably reflecting more specialized role *in vivo*. In addition to these inhibitors, some serpins can also inhibit cysteine proteases in cross-class inhibition [8]. The human squamous cell carcinoma antigen-1 (SCCA1) is a potent inhibitor of cathepsins K, L and S [56], whereas hurpin, which is a differentially spliced member of the serpin super family specifically inhibits only cathepsin L [57] . Similarly, serpin endopin 2C selectively inhibits cathepsin L compared to elastase [58]. Thus these inhibitors serve as cellular guardians under normal physiological conditions.

Table 4.1 Substrates, inhibitors and functions of human cathepsins

Protein	Functions	Endogenous substrates in apoptosis	Other endogenous substrates	Synthetic substrates	Endogenous inhibitors	Synthetic inhibitors
Cathepsin B	Apoptosis, Bid activation, protein catabolism, processing of antigens, hormone activation bone turnover	Bid, Bak, Bcl-2, Bcl-xL, Mcl-1, PARP-1, XIAP, E-cadherin	Extracellular matrix	Abz-GIVRAK (Dnp)-OH	Stefin, cystatin, α2-macroglobulin, thyropins	E-64, CA030, CA074
Cathepsin H	Apoptosis, protein turnover, activation of Bid	Bid, Bak, Bcl-2, Bcl-xL, Mcl-1, XIAP, E-cadherin	–	–	Stefin, cystatin, kininogen, α2-macroglobulin, thyropins	E-64
Cathepsin K	Apoptosis, bone remodeling, Bid activation	Bid, Bak, Bcl-2, Bcl-xL, Mcl-1, XIAP,	Type I collagen	Abz-HPGGPQ-EDDnp	Stefin, cystatin, α2-macroglobulin, thyropins, SCCA1	E-64, NSC133452
Cathepsin L	Apoptosis, MHC-II-mediated antigen presentation, protein turnover, Bid activation, tissue remodeling, release of mature oocyte	Bid, Bak, Bcl-2, Bcl-xL, Mcl-1, XIAP, E-cadherin	–	–	Stefin, cystatin, kininogen, α2-macroglobulin, SCCA1, hurpin, serpinendopin 2C	E-64
Cathepsin S	Apoptosis, antigen processing and presentation, Bid activation	Bid, Bak, Bcl-2, Bcl-xL, Mcl-1, XIAP, E-cadherin	–	Abz-LEQ-EDDnp	Stefin, cystatin, α2-macroglobulin, thyropins, SCCA1	–

Cathepsin C	Apoptosis, protein degradation, activation of granzyme B, neutrophil, elastase and factor XIII	–	Dipeptide esters, amides,	–	–	–
Cathepsin F	Apoptosis, MHC-II-mediated antigen presentation	–	–	–	–	–
Cathepsin V	MHC-II-mediated antigen presentation	–	–	–	–	–
Cathepsin X	Phagocytosis, regulation of immune responses	–	–	–	–	–
Cathepsin O	Protein degradation	–	–	–	–	–
Cathepsin W	Cell-mediated cytotoxicity	–	–	–	–	–
Cathepsin D	Apoptosis, antigen presentation, tissue remodeling, protein degradation, activation of Bax, endonuclease 23, procaspase-8 and cysteine cathepsins	Bid, Bax, caspase-8, PARP-1	Cystatin C, kininogen, α_1-antichymotrypsin, lysozyme, antigens, hormones, neuropeptides	MOCAc-Gly-Lys-Pro-Ile-Leu-Phe-Phe-Arg-Leu-Lys(Dnp)-D-Arg-NH_2	–	Pepstatin A, acetyl- and isovaleryl-pepstatin, Diaso-acetyl-norleucine methyl ester

(continued)

Table 4.1 (continued)

Protein	Functions	Endogenous substrates in apoptosis	Other endogenous substrates	Synthetic substrates	Endogenous inhibitors	Synthetic inhibitors
Cathepsin E	Apoptosis, antigen presentation, antitumor activity	TRAIL	–	MOCAc-Gly-Lys-Pro-Ile-Leu-Phe-Phe-Arg-Leu-Lys(Dnp)-D-Arg-NH_2	–	Pepstatin A, acetyl- and isovaleryl-pepstatin,Diaso-acetyl-norleucine methyl ester
Cathepsin G	Modulates coagulation, tissue remodeling induction of platelet activation, chemotaxis of leukocytes and endothelium-dependent vascular relaxation activation of inflammatory cytokine, cardiomyocytes anoikis endothelial cell damage	PARP-1, brm protein	PARs, cytokines and growth receptors	–	–	–
Cathepsin A	Autophagy, elastic fiber formation, associates with β-galactosidase and neuraminidase	–	Endothelin 1 substance P, tachykinins, bradykinin, angiotensin 1, oxytocin, lamp2a	–	–	–

1.1.6.2 Synthetic Inhibitors

E-64 (L-trans-Epoxysuccinyl-leucylamido(4-guanidino)butane), is an active-site directed synthetic inhibitor of cathepsins K, B, H and L [59]. It selectively alkylates the active-site cysteine and remains covalently bound to the enzyme. It also acts as a weak non-selective inhibitor for other members of the cysteine cathepsins. Since it reacts almost exclusively with the reactive site cysteine of papain-like proteases, it has immediately become a widely used indicator of the proteolytic activity of cysteine cathepsins. CA030 (ethyl-ester of epoxysuccinyl-L-Ile-L-Pro-OH), CA074 and their analogs [60, 61] are first specific inhibitors of cathepsin B. These inhibitors bind to the primed side (sites S1′ and S2′) of the active-site cleft in the direction of the substrate. Recently, NSC13345 2-[(2-carbamoylsulphanylacetyl)-amino]benzoic acid, a small-molecule allosteric inhibitor was designed for cathepsin K using high-throughput computational methods [62]. Further studies on identification and design of selective inhibitors for inhibiting catalysis through reversible competition with the substrate, or by covalent modification of catalytic groups is desirable.

1.2 Aspartate Cathepsins

Cathepsins D and E are two major intracellular aspartic proteinases implicated in physiological and pathological degradation of several intra and extracellular proteins. Cathepsin D is ubiquitously expressed in all types of cells, tissues and organs except in lysosome-free erythrocytes [63]. It is synthesized as an inactive pre-procathepsin D in rough endoplasmic reticulum. After removal of signal peptide, the 52 kDa procathepsin D is targeted to intracellular vesicular structures such as lysosomes, endosomes or phagosomes. Under the acidic environment of endosomal or lysosomal compartments, the cleavage of the 44 amino acid N-terminal propeptide results in formation of a 48 kDa single chain intermediate active form (Fig. 4.4). Further proteolytic cleavage finally converts it into the mature active lysosomal protease which is composed of 34 kDa heavy and 14 kDa light chains linked by non-covalent interactions [64–66]. The proteolytically active cathepsin D

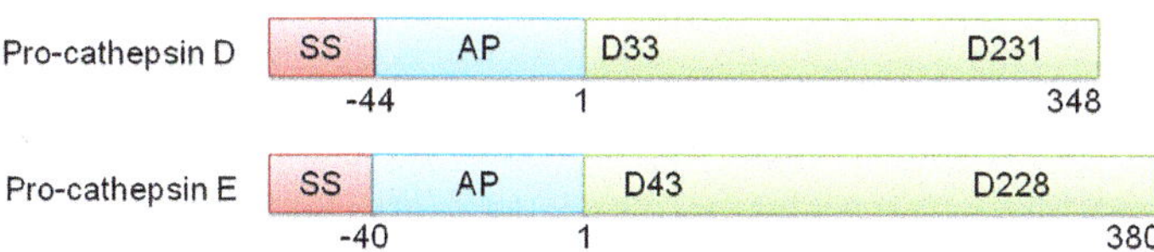

Fig. 4.4 Schematic representation of domain organization of aspartyl cathepsins. SS (signal sequence) is shown in *pink*, AP (activation peptide) in *blue*, mature cathepsin as *green rectangles*. '-44' and '-41' represent deletion of the activation peptide, and '1' corresponds to the first amino acid in mature cathepsins. Catalytic site residues are represented as D (aspartic acid)

plays a crucial role in the control of cell and tissue homeostasis [67–70]. It is also associated with various pathological conditions such as cancer [71], Alzheimer's disease [72] and neuronal ceroid lipofuscinosis [73].

Cathepsin E is however a non-lysosomal cathepsin highly homologous to cathepsin D. It is mainly present in cells of the immune system, including APC (Antigen Presenting Cells) such as lymphocytes, microglia, dendritic cells, Langerhans cells, interdigitating reticulum cells, gastric epithelial cells and osteoclasts [74–76]. Cathepsin E is synthesized as a pre-proenzyme comprising 438 amino acid residues where cleavage of first 18-amino acid propeptide results in the formation of 90 kDa apo-cathepsin E. Under acidic pH, auto-activation of procathepsin E leads to the cleavage of 40-amino acid propeptide and formation of the mature form of cathepsin E (Fig. 4.4). Cathepsin E plays a major role in neurodegeneration associated with brain ischemia and aging [77].

1.2.1 Structural Assembly

The structure of human cathepsins D and E is similar to that of other aspartic proteases (*e.g.* pepsinogen and human immunodeficiency virus protease). Like most aspartic proteases, they are synthesized as inactive zymogens, which auto-activate in an acidic environment. These precursor forms are catalytically inactive towards their natural substrates as the N-terminal pro-sequence folds across the active-site thus preventing accessibility of the substrate to the active-site.

Cathepsin E is the only A1 aspartic protease that exists as a homodimer. The disulfide bridge between two N-terminal Cys34 residues links both the monomers of the molecule. Each monomer has three topologically distinct regions, an N-terminal domain (residues Lys14 to Val157), a C-terminal domain (residues Thr198 to Gln324) and an interdomain comprising six anti-parallel β-sheet that connects the N and C-terminal domains. Active-site residues, Asp43 and Asp228 are contributed by N and C-terminal domains, respectively.

Eder and co-workers determined the crystal structure of the activation intermediate of human cathepsin E at 2.35 Å resolution (Fig. 4.5a). The overall structure follows the general fold of aspartic protease of the A1 family, and the activation intermediate shares features similar to the intermediate-2 of aspartic proteases like pepsin C [78]. In this intermediate, although the pro-sequence is cleaved from the protease it still remains intact after cleavage. The cleaved pro-sequence comprising Gly1 to Arg9 associates with the anti-parallel five β-stranded interdomain, thus forming the outermost sixth strand of this interdomain. Moreover, the N–terminal residues, Lys14 to Glu24, reside in the active-site and occupy the non-primed binding site while the N-terminal extension of the pro-sequence from a neighboring molecule occupies the primed site. Thus, the structure of pro-sequence complex of cathepsin E provides insight into the activation mechanism of aspartic proteases in general.

Crystal structure of native and pepstatin bound mature cathepsin D was solved at 2.5 Å resolution (Fig. 4.5b). Cathepsin D comprises an N-terminal domain

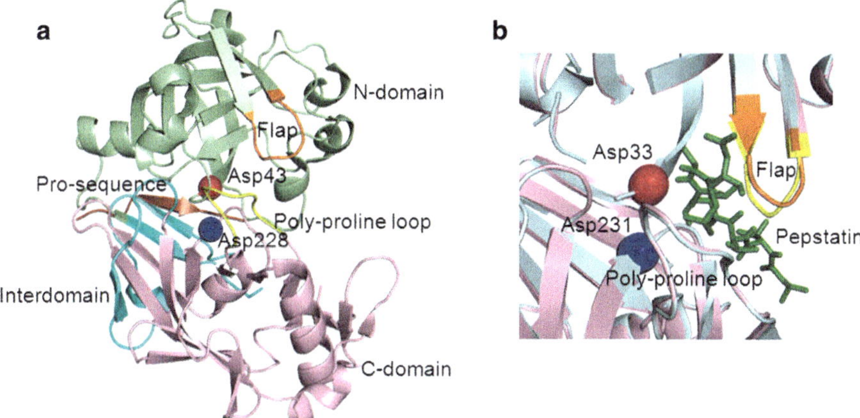

Fig. 4.5 Crystal structure of aspartyl cathepsins. (**a**) Cartoon representation of cathepsin E activation intermediate, PDB entry 1TZS. The N-domain is colored in *green*, C-domain is shown in *pink*, interdomain in *cyan* and pro-sequence in *brown*. Position of active-site aspartic acid residues are represented as *red* and *blue spheres*. Flap (*orange*) and poly-proline (*yellow*) loops are highlighted. (**b**) Close-up view of active-sites for inactive (*cyan*), PDB entry 1LYA and pepstatin bound (*pink*), PDB entry 1LYB. Position of active-site aspartic acid residues are represented as *red* and *blue spheres*. Pepstatin (*green*) is represented as sticks. Flap (*yellow*) is closer to inhibitor in active cathepsin D compared to flap (*orange*) in inactive form. The figures are generated using PyMOL (DeLano Scientific, USA)

(residues 1–188), a C-terminal domain (residues 189–346), and an interdomain that contains anti-parallel β-sheet composed of the N-terminus (residues 1–7), the C-terminus (residues 330–346), and linking residues (160–200). Both N- and C-terminal domains contribute one catalytic aspartic acid residue each (Asp33 and Asp231 respectively) to the active-site. Apart from the active-site residues, these domains have a single carbohydrate group and two disulfide bonds. Two loops, poly-proline loop (loop 8) and flap (loop 2) that are close to the active-site cleft are essential for substrate binding. Pepstatin-bound complex of cathepsin D exhibits structural changes in the flap region where the tip of the flap moves towards the inhibitor, thus promoting activation. However, no sub-domain movements are observed in cathepsin D in complex with inhibitor.

1.2.2 Activation Mechanism

The activation of aspartate cathepsin zymogens is generally initiated by a dramatic conformational rearrangement of the N-terminal pro-sequence. This process is triggered by low pH within the endolysosomal compartments, where several proteolytic processing events generate the mature active cathepsins. In the first step of activation pathway, the pro-sequence gets disordered under acidic pH and the inhibitory salt bridge between the pro-sequence and the active-site is disrupted thus uncovering the active-site. The initial proteolytic event involves removal of pro-sequence to generate single-chain active intermediates. The removal of this pro-sequence is independent of cathepsin D autocatalytic activity and is primarily mediated by cysteine cathepsins other than cathepsins L and B. This single-chain intermediate is further processed into two-chain form, carried out by cysteine cathepsins B or L. The mature form consists of an amino-terminal light and a carboxyl-terminal heavy chain. During this conversion, seven amino acids from interdomain and several more residues from the carboxyl-terminus are removed, thus releasing the active-site for catalysis [79].

1.2.3 Catalytic Mechanism

The catalytic site of aspartyl cathepsins has two aspartic acid residues located in N- and C-terminal domains as shown in Fig. 4.5. In the acidic environment, the carboxyl group of Asp in N-domain undergoes dissociation, whereas that of Asp in C-domain does not. The carboxyl group of the dissociated Asp activates the water molecule and allows proton release from water. On the other hand, the protonated carboxyl group of Asp in C-domain polarizes the carbonyl group of the peptide bond thus facilitating formation of a tetrahedral intermediate and subsequent cleavage of the bond. The reactions are performed by ionizing groups of aspartyl cathepsins and therefore their velocity is pH-dependent. Aspartyl cathepsins show highest activity in an acidic milieu, i.e. between pH 3.5 and 5.5 [79].

1.2.4 Functions

1.2.4.1 Complex Role of Aspartate Cathepsins in Apoptosis

In the classic death paradigm, lysosomes are solely considered to be critical for autophagic or necrotic cell death and the role of lysosomal proteases are limited to non-specific protein degradation. However, in the last decade, it has become evident that the function of lysosome per se in cell death is far more complex. Cathepsin D is known to be involved in both caspase dependent and independent cell death. Studies with lysosomotropic detergents highlight that the key factor in determining the type of cell death is the magnitude of LMP. The release of lysosomal cathepsins directly or indirectly results into mitochondrial dysfunction. Cytoplasmic cathepsin D cleaves Bcl-2 family member, Bid, followed by subsequent release of cytochrome c from mitochondria and activation of caspases-9 and -3. Cathepsin D induces caspase independent cell death in T-cells by triggering the activation of Bax thus leading to selective release of apoptosis inducing factor (AIF) from mitochondria [80]. This mechanism probably involves an excessive calcium influx, and over activation of poly-ADP-ribose polymerase-1 (PARP-1) [81].

Although some studies report that cathepsins can directly induce apoptosis, many others demonstrate that it acts as a mediator of apoptosis induced by several apoptotic agents such as IFN-γ, FAS/APO-1, TNF-α [82–84], oxidative stress [85–87], etoposide [88], adriamycin, 5-flurouracil, cisplatin [88], sphinogosine [89] and staurosporine [90]. Numerous studies have shown that pepstatin A, an aspartate protease inhibitor, could partially delay apoptosis induced by several of the apoptotic agents. Therefore, it seems that cathepsin D plays a key role in apoptosis mediated via its catalytic activity. Moreover, it would be of great interest to design and synthesize some more new specific inhibitors of cathepsin D for research and therapeutic implications. Recently, it has been shown that cathepsin D can also activate procaspase-8, initiating neutrophil apoptosis during the course of inflammation [91]. Interestingly, another study shows the presence of mature cathepsin D in the nucleus during cell death [92], where it has been proposed that nuclear cathepsin D might mediate the proteolytic activation of endonuclease 23 during cryo-necrotic cell death [93]. This raises an interesting possibility of this protease to be involved in more than one apoptotic pathway.

Unlike cathepsin D, the physiological functions of cathepsin E are poorly understood. It was shown that cathepsin E induced apoptosis in human prostate cancer cell lines by proteolysis of tumor necrosis factor related apoptosis-inducing ligand (TRAIL) from the tumor cell surface [94, 95]. This antitumor activity of cathepsin E was also corroborated by *in vivo* studies with mice bearing human and mouse prostrate carcinoma transplants. Administration of nude mice bearing these tumor xenografts with cathepsin E induced tumor growth arrest and apoptosis in tumor cells without any apparent histologic effects on normal tissues and cells [94]. However, the function of cathepsin E in apoptosis needs further in-depth investigation.

Taken together, it appears that aspartate cathepsins trigger apoptosis via multiple pathways in conjunction with key mediators of apoptosis such as cytochrome c, caspases and Bcl-2 family members. The pathways rely either on the catalytic activity of these cathepsins or on their ability to interact with other proteins.

1.2.4.2 Other Functions

One of the major functions of cathepsin D is intracellular catabolism within the lysosomal compartment. Cathepsin D is apparently also involved in the processing of antigens [96], hormones, and neuropeptides. A relatively high concentration of procathepsin D was found in human breast milk suggesting a yet unknown function [97]. Cathepsin D is also reported to be involved in tissue remodeling [98].

Cathepsin E plays an important role in immune function. It is most likely involved in the processing of antigenic peptides during MHC class II-mediated antigen presentation [99]. It is shown to be involved in activation-induced lymphocyte depletion in the thymus, neuronal degeneration and glial cell activation in the brain. Deficiency of cathepsin E in macrophages induces a novel form of lysosomal storage disorder manifesting elevated lysosomal pH with accumulation of major lysosomal membrane glycoproteins such as LAMP-1 and LAMP-2. These cellular alterations are linked to abnormal intracellular trafficking of secretory and cell surface proteins [100].

1.2.5 Substrates

1.2.5.1 Synthetic Substrate and Specificity

Substrate specificity of cathepsins D and E was determined using a 15-residue synthetic peptide library and native protein as substrates [101]. In common with most other aspartic proteinases, cathepsins D and E have an ability to accommodate relatively broad range of residues in the P1 and P1′ positions, but share strong preference for the hydrophobic, β-branched residues. However, it is found that Ile and Val are readily accepted in P1′ but not in the P1 position. Positions P2 and P2′ accept a broad range of amino acids, including charged and polar ones [101]. In synthetic peptide substrates, Glu is the optimum P2′ residue while for cathepsin E, Lys is accepted at P2′. Based on these sequence specificities, a synthetic decapeptide fluorogenic substrate, MOCAc-Gly-Lys-Pro-Ile-Leu-Phe-Phe-Arg-Leu-Lys(Dnp)-D-Arg-NH_2 has been designed for characterizing proteolytic activity of both cathepsins D and E [102].

1.2.5.2 Endogenous Substrates

Till date, only few cytosolic substrates have been identified for aspartate cathepsins. The pro-apoptotic Bcl-2 family member, Bid is the best characterized substrate known so far. Bid was confirmed as a cathepsin substrate in a variety of cell models [41, 103] and it was shown *in vitro* that cathepsin D can cleave Bid into its potent pro-apoptotic t-Bid fragment [41, 84]. It is co-localized with components of the immune system in the endosomes [104] and is capable of generating peptides from ovalbumin [105], lysozyme, sperm whale myoglobin and tetanus toxin [106] that could be presented to T cells by MHC class II molecules [107]. Cathepsin D also specifically hydrolyses papain protease inhibitors, cystatin C, kininogen [108] and α_1-antichymotrypsin [109]. However, there are no reports available on physiological substrates for cathepsin E. It is believed that cathepsin E aids in the antigen presentation by generating peptide from the tetanus toxin [106]. Further investigation is required to identify the potential targets for this class of protease.

1.2.6 Inhibitors

In contrast to other cathepsins (*e.g.* cysteine and serine cathepsins), no endogeneous inhibitors for aspartate cathepsins are known in mammals. Pepstatin is a hexapeptide, competitive and only known natural tight-binding inhibitor of aspartic cathepsins isolated from actinomycites [110]. Pepstatin A and its derivatives, acetyl- and isovaleryl-pepstatin, form multiple non-covalent bonds with the active-site residues of these proteases [110]. Diaso-acetyl-norleucine methyl ester is another synthetic inhibitor of cathepsins D and E, where the reaction is catalyzed with Cu^{2+}

ions [111–113]. These inhibitors provide a certain threshold to prevent inappropriate activity of cathepsins; however, they are less precise than the ones for caspases. This probably explains why cathepsins are more prone to catastrophic events such as necrosis compared to other proteases.

1.3 Serine Protease Cathepsins

Cathepsins A and G, the only identified serine cathepsins have been found to be involved in host immune response and protective functions. Synthesized as zymogens, these cathepsins undergo processing to become mature and active enzymes.

Cathepsin A, also termed as protective protein or carboxypeptidase A, is a ubiquitously expressed lysosomal serine protease with structural homology to yeast (*Saccharomyces cerevisiae*) carboxypeptidase Y. It exhibits carboxypeptidase activity at acidic pH in lysosomes, while acts as a deamidase and an esterase at neutral pH [114, 115]. It functions as a protective protein by associating with β-galactosidase and neuraminidase 1 to form a high molecular weight (700 kDa) multi-enzyme complex and protect them against lysosomal degradation [116]. In addition, cathepsin A hydrolyzes variety of bioactive peptide hormones such as endothelin 1 substance P, tachykinins, bradykinin, angiotensin 1 and oxytocin [117], but its exact physiological function is not fully understood. Cathepsin A is also implicated in autophagy, which occurs due to the digestion of lysosome-associated membrane protein type 2a [118].

Cathepsin G is a neutrophil serine protease that belongs to the S1 class of the serine protease family. It is recognized as degradative enzyme that is capable of killing pathogens and cleaving extracellular matrix components. Cathepsin G plays a potential role in induction of platelet activation [119, 120], induction of chemotaxis of leukocytes [121] and activation of inflammatory cytokine release [122]. It also plays major roles in induction of endothelium-dependent vascular relaxation [123], cardiomyocytes anoikis [124] and endothelial cell damage [125]. Cathepsin G has been shown to affect tumor metastasis by modulating the adherence capacity of tumor cells. It does so by attenuating binding between integrin and extracellular adhesion molecules such as fibronectin [126].

Perspectives

Despite a considerable progress in understanding the role and involvement of cathepsins in apoptosis, we are still at the beginning of harnessing its potential for translational applications. Although at present there are no strategies for therapeutic modulation of cathepsins, its effect on LMP raises the possibility of its application in cell death induction. Therefore, understanding the mechanism of LMP and its consequences is important toward designing novel therapeutic approaches against several life-threatening diseases such as cancer and neurodegenerative disorders.

2 Human HtrA Proteases

HtrA (high-temperature requirement protease A), also known as DegP, was initially identified in *E. coli* as a heat shock-induced envelope-associated serine protease [127]. It generally acts as a molecular chaperone at low temperatures, while, with increase in temperature, it exhibits its proteolytic activity [128]. In humans, HtrAs are involved in numerous cellular processes, ranging from maintenance of mitochondrial homeostasis to cell death in response to stress-inducing agents [129]. In recent years, they have drawn attention as key players in multiple pathways of programmed cell death and are described as potential modulators of chemotherapy-induced cytotoxicity [130, 131]. This multifaceted ability associates them with different pathological conditions such as cancer, neurodegenerative disorders and arthritis including myocardial ischemia/reperfusion injury [129, 132, 133], hence making them therapeutically important.

HtrA proteins belong to the family of S1B class of serine proteases characterized by the presence of a trypsin-like fold domain and have been found to be highly conserved from prokaryotes to humans [131, 134]. Based on their domain architecture that comprise an N-terminal IGFBP- and a Kazal-like module, a trypsin-like fold protease domain, and a C-terminal PDZ domain, four human HtrAs have been identified. They are HtrA1 (L56, PRSS11), HtrA2 (Omi), HtrA3 (PRSP) and HtrA4 [130, 134–138]. Among the HtrA homologs, HtrA1 and HtrA2 are well characterized while very little information on HtrA3 and HtrA4 are currently available.

2.1 HtrA1

HtrA1 was first identified in human fibroblast cells as transformation-sensitive protein whose expression was suppressed in SV-40 transformed fibroblast cell line but not in the normal counterparts [139]. It was initially named as L56, followed by protease serine 11 (PRSS11) and finally as HtrA1 due to its homology to HtrA/Do family of proteases in bacteria [138, 139]. It is ubiquitously expressed in all tissues and organs, with the highest levels in placenta and mature epidermis. Although classified as a secreted protease involved in degradation of extracellular matrix (ECM) proteins, processed forms of HtrA1 are also found in the cytoplasm and in nuclear fractions [140–142]. Several points of evidence confirm that HtrA1 plays a protective role in various malignancies due to its tumor suppressive properties. It has been shown to be down-regulated in malignant melanoma and ovarian cancer. Its over-expression also results in the inhibition of tumor cell growth and proliferation both *in vitro* and *in vivo* [143]. Moreover, transcription of the HtrA1 gene is highly regulated during development as well as in adult tissues, demonstrating that it may not only exert its functions on neoplastic cells but also under normal physiological conditions. HtrA1 has been implicated in pathology of several diseases such as arthritis, age related macular degeneration, Alzheimer's disease and familial ischemic cerebral small vessel disease [142].

2.1.1 Structural Assembly

HtrA1 is a 50 kDa secreted protein composed of a signaling peptide (1–22 aa), an insulin growth factor binding domain or IGFBP (25–111 aa), a Kazal-like protease inhibitor domain (115–135 aa), a conserved serine protease (206–364 aa) domain, and a PDZ domain (382–458 aa) as shown in Fig. 4.6. Daniel and co-workers provided the first structure of N-domain (IGFBP and Kazal) solved at 2.0 Å resolution (Fig. 4.7a). Although this structure preserves the overall IGFBP- and Kazal-like folds, several important structural deviations are also observed [141]. For example, the Kazal-like module differs from the canonical Kazal domains by a longer α-helix and presence of only two rather than three-disulfide bonds. Interestingly, the missing disulfide bond is replaced by H-bonds formed by three arginines (R133, R137 and R141) of Kazal-like module with main-chain carbonyl oxygen of IGFBP, which to an extent preserve the connection between these two modules. These polar interactions efficiently pack the Kazal domain against IGFBP so as to restrict its access to a target protease, thereby categorizing it as a

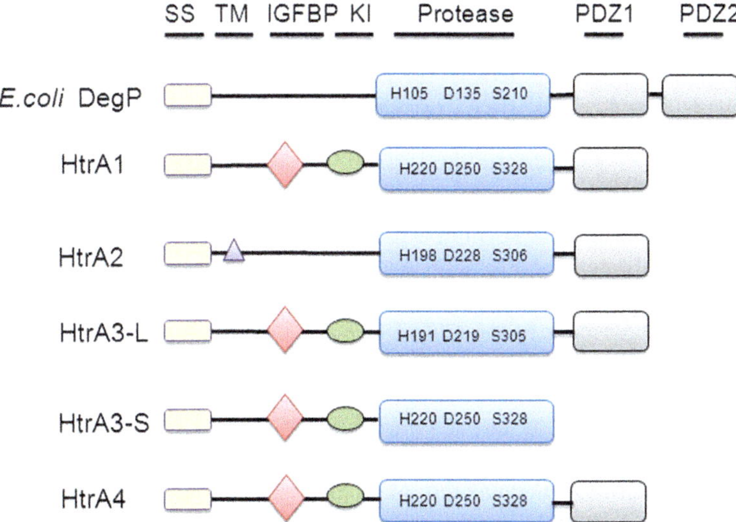

Fig. 4.6 Schematic representation of domain organization of HtrA family of proteases. The protease domain is in *blue rectangles*, PDZ1 and PDZ2 domains in *grey rectangles*, SS (signal sequence) in *yellow rectangles*, TM (transmembrane domain) in *purple triangle*, IGFBP (Insulin-like growth factor binding) in *pink diamond*, and KI (Kazal protease inhibitor domain) in *green oval*. Here, *E. coli* represents *Escherichia coli* and HtrA represents high temperature requirement protease A. Catalytic triad residues are represented as H (histidine), D (aspartate), and S (serine)

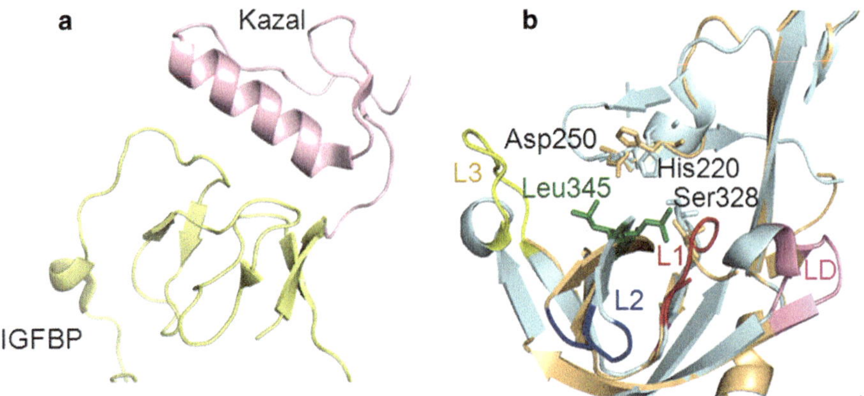

Fig. 4.7 Crystal structure of HtrA1. (**a**) Schematic representation of HtrA1 N-domain. IGFBP and Kazal-like motif are shown in *green* and *pink* colors respectively. (**b**) Close-up view of active-sites for inactive (*orange*), PDB entry 3NUM and substrate bound HtrA1 (*cyan*), PDB entry 3NZ1. The active-site residues are shown in stick model and the functional loops are highlighted. Loops L1, L2, L3, and LD are disordered in incompetent HtrA1 compared to its active form. In inactive protomer, catalytic triad residue His220 is disordered relative to other two active-site residues (Ser328 and Asp250). The S1 specificity pocket in inactive protomer is blocked by Leu345 (*green*) and loop L2. The ribbon diagram has been prepared with PyMOL (DeLano Scientific, USA)

non-inhibitory Kazal domain. Furthermore, the structure-based predictions revealed that both the IGFBP- and the Kazal-like modules neither have any effect on HtrA1 protease activity nor they interact with any of the IGF proteins.

The crystal structure of HtrA1 protease domain is available in different forms, such as unliganded, mutationally inactivated and synthetic inhibitor-bound, thus providing snapshots of its underlying mechanism of activation and regulation. The structure shows that the protease domain is formed by two perpendicular β-barrel lobes (β1–β6 and β7–β12) within which the active-site with the catalytic triad (His220, Asp250 and Ser328) is embedded (Fig. 4.7b). The apo-form of the protease reveals large conformational differences compared to the active form primarily in the arrangement of the activation domain (L1, L2, and LD) and sensor (L3) loops. These subtle structural differences impede the proper functioning of the catalytic triad, oxyanion hole and specificity pocket. Specifically, in an inactive protomer structure, the side chain of the catalytic triad residue His220 is disordered relative to other active-site residues (Ser328 and Asp250) as shown in Fig. 4.7b. The S1 specificity pocket is also completely blocked by the residue L345 as well as loop L2. Moreover, residues 343–348 do not adopt a typical β-strand conformation as observed for analogous residues in active serine proteases. However, in case of competent protomer structure, the catalytic triad residues are positioned such that they support amide bond cleavage. Leu345 and residues 343–345 that are now part of an intact β-strand, define the S1 pocket rather than occluding it. This competent protomer structure is very similar to the HtrA1 protease with a substrate surrogate covalently attached to Ser328 [144].

Low resolution small-angle X-ray scattering or SAXS analysis further provided an overview of the structural organization of HtrA1[141]. Intact HtrA1 represents a saucer like, almost flat trimeric low-resolution molecule with PDZ domains protruding from the protease trimer and the N-domains lying flat against the sides of the protease domains far from the catalytic sites.

2.1.2 Activation Mechanism

Enzymatic activity of HtrA1 is distinct as it is largely independent of ATP or cofactors such as pH, reducing agents and divalent cations. Based upon the inactive and active-state crystal structures, Truebestein and co-workers proposed that HtrA1 protease activity is regulated by substrate-induced remodeling of the active-site. In particular, binding of substrate to the protease domain rearranges the activation domain, and the loop L3 directly interacts with the ligand, leading to rearrangement of the neighboring loops LD and L1/L2 to form fully competent active-site pocket.

Contrary to the induced-fit mechanism, recent structural studies on unliganded HtrA1 display an enzymatically active catalytic site conformation, demonstrating that HtrA1 possibly does not require a substrate for its activation. This model proposes ‘two-state equilibrium’ and a ‘conformational selection’ for activation of HtrA1 [141]. Like other trypsin-like serine proteases, HtrA1 might exist in an equilibrium between active (*E*) and inactive (*E**) states. Substrate (*S*) preferentially binds to the active form to render the formation of enzyme-substrate complex (*E:S*) for catalysis and synthesis of product (*P*) as shown in Eq. (4.1).

$$E^* \leftrightarrow E \overset{+S}{\leftrightarrow} E : S \rightarrow E + P \tag{4.1}$$

In contrast to other *E. coli* homologs (DegS and DegP), the PDZ domain is dispensable for HtrA1 activation. Enzymatic assays with various HtrA1 constructs demonstrated that similar to PDZ, the N-domain is also non-essential for catalytic activity [141]. Therefore, HtrA1 significantly deviates from the bacterial HtrA paradigm where accessory domains regulate the catalytic activity. These observations raise the possibility of existence of an yet unknown allosteric mechanism of HtrA1 activation [141]. Interestingly, in presence of a substrate, HtrA1 protease have been shown to form higher order oligomers (up to 600 kDa) with enhanced processivity [144]. However, further in-depth studies are required to understand the activation and regulatory mechanisms of the protease and its involvement in various physiological processes and diseases.

2.1.3 Catalytic Mechanism

HtrA family proteases contain a catalytic triad comprising serine, histidine and aspartate residues. These HtrAs share a common catalytic mechanism with other endogenous serine proteases. The general catalytic mechanism of serine proteases is discussed in Chap. 3.

2.1.4 Functions

HtrA1 is involved in several important biological and pathological processes such as growth, apoptosis, embryogenesis, invasion, metastasis, and cancer. The following section highlights the reported roles of HtrA1 in physiological functions and diseases.

2.1.4.1 Apoptosis

The exact molecular mechanism by which HtrA1 regulates apoptosis remains largely unknown. Initial studies demonstrated that HtrA1 induced apoptosis in cisplatin treated ovarian cancer cells by decreasing the level of X-linked inhibitor of apoptosis protein (XIAP) and subsequent activation of caspases-3 and/or -7 [130]. Recently it has been observed that HtrA1 over-expression induced apoptosis in esophageal squamous cell carcinoma by blocking the NF-κB (nuclear factor kappa-light chain-enhancer of activated B cells) signaling pathway [145].

HtrA1 is also found to promote anoikis, a programmed cell death that is induced upon cell detachment as a result of inadequate or inappropriate cell-matrix interactions [146]. It does so by attenuating activation of epidermal growth factor receptor EGFR/AKT pathway in a protease-dependent manner. Enhanced expression of HtrA1 in ovarian cancer cells suppressed EGFR/AKT signaling thus leading to increased cell death. However, down regulation of HtrA1 gene attenuated anoikis *in vitro* and also promoted peritoneal dissemination of ovarian cancer cells in mouse model [146].

2.1.4.2 Pathophysiological Functions

Although the normal physiological involvement of HtrA1 is limited, its role in pathobiology of different diseases is better studied.

(a) **Cancer**

Extensive functional studies have been carried out on pathobiology of HtrA1 and its role in cancer. It has been found to be absent or significantly reduced in various cancers including gastric [131], breast [143], ovarian [131, 143, 147], endometrial [148] and hepatocellular carcinomas [149], indicating that its down regulation is associated with cancer progression. HtrA1 is also reported to be absent in melanomas [150] and mesothelioma [151] and over-expression of it in a mouse model inhibited cell proliferation. HtrA1 is further associated with the occurrence and development of esophageal cancer, where it participates in invasion and metastasis [131, 143, 146, 150]. Moreover, over-expression of HtrA1 in ovarian cancer cell line (OV202) has been shown to promote cell sensitivity to cisplatin-induced apoptosis which could be reversed by increased expression of XIAP or X-linked inhibitor of apoptosis protein [130].

(b) **Osteoarthritis**

HtrA1 is involved in degradation of extracellular matrix (ECM) proteins, such as decorin [152, 153], biglycan [21], fibronectin [152, 154] that are essential in matrix remodeling, cartilage catabolism, and arthritis. Proteolysis of fibronectin by HtrA1 releases peptide factors that stimulate MMP (matrix metallo-proteins) production, which culminates in further proteolysis and turnover of the extracellular matrix. It thus leads to progressive degeneration of ECM that protects the joints from osteoarthritis. These results are further appreciated by the studies which correlate the role of HtrA1 in regulating bone homeostasis through BMP (bone morphogenic protein) and TGF-β (transforming growth factor- β) signaling [153].

(c) **Alzheimer's Disease**

Apart from its involvement in apoptosis, HtrA1 also has a protective role against the development of Alzheimer's disease (AD), a neurodegenerative disorder characterized by progressive memory loss and cognitive decline [155]. HtrA1 also degrades various fragments of amyloid precursor protein, including β-40 and β-42. Inhibition of HtrA1 proteolytic activity results in accumulation of β-amyloid in the supernatant of human astrocytoma cell line [155].

2.1.5 Substrates

2.1.5.1 Synthetic Substrate and Substrate Specificity

Mixture-based peptide library screens determined the specificity and selectivity of HtrA1 in both the primed and unprimed cleavage sites [141, 144]. The primary sequence specificity of the unprimed cleavage site indicates that the protease has strongest selectivity for non-polar aliphatic amino acids (V, L, A, I, and M) at the P1 position. These screenings led to the design of a fluorescently quenched peptide substrate Mca (methoxycoumarin)–IRRVSYSF-Dnp (dinitrophenyl) KK [156]. Mechanistically, the P1 valine fits in the S1 specificity pocket, which is formed by the side chains of Lys346 and of Ile323 while P2 and P4 residues interact with the L3 loop of HtrA1 [144]. For the prime sites, HtrA1 shows preference for polar residues (R, D) in P1′ and P3′ positions and non-polar residues (P, F) in P2′ and P4′. Along with the peptide based reporter substrates, fluorescently-labeled full length substrates have also been used to measure protease activity of HtrA1 which include resorufin or FITC- (fluoroisothiocynate) labeled casein. Substrate specificity for HtrA1 remains unaltered either in presence or on deletion of C-terminal PDZ domain, suggesting that this accessory domain does not influence substrate interactions.

2.1.5.2 Endogenous Substrates

Although evidences show involvement of HtrA1 in apoptosis, very little is known about its endogenous substrates. The substrates identified for HtrA1 majorly belong to the class of extracellular proteins. These include components of matrix, such as type II collagen and fibronectin, or components of cartilage, such as aggrecan, decorin, fibromodulin [153, 154], clusterin, ADAM9, vitronectin, and α2-macroglobulin (Table 4.2). Some cell surface proteins like talin-1, fascin and chloride intracellular channel protein 1 are also found to be its substrates [157]. In addition, *in vitro* studies have shown that HtrA1 degrades various amyloid precursor protein fragments, including amyloid-β and its precursor C99 peptide, both involved in maintenance of metabolic balance and prevention of plaque formation in the brain of Alzheimer's disease patients [155]. A recent report suggests that it cleaves TGF-β receptors such as TβRI and TβRII, resulting in downregulation of TGF-β signaling [158]. Furthermore, XIAP is also identified as a potential substrate of HtrA1 [130].

2.1.6 Inhibitors

Till date, no endogenous inhibitors are found for the human HtrA homologs. A specific synthetic inhibitor, 1-(3-cyclohexyl-2-propionyl)-pyrrolidine-2-carboxylic acid 5-(3-cyclohexyl-ureido)-1-dihydroxyboranyl-pentyl)-amide from Novartis has been reported which is depicted in Table 4.2 [154]. In addition, di-isopropyl-fluorophosphate (DFP) is an irreversible synthetic active-site inhibitor that forms a covalent bond with the catalytic serine and occupies the S1 specificity pocket. It acts as an excellent probe to investigate conformational states of active-sites due to its specificity to enzymatically competent active-sites, as shown for *E. coli* DegS [159]. Interestingly, alpha-1-anti-trypsin has also been reported to be associated with HtrA1, suggesting that it could be considered as a potential inhibitor of the protease.

2.2 HtrA2

HtrA2/Omi is a mitochondrial pro-apoptotic serine protease with multitasking ability. It was first identified as an Inhibitor of Apoptosis (IAP) binding protein due to its IAP recognizing Reaper-like motif (AVPS) similar to human Smac/DIABLO [160] and Drosophila death proteins Reaper, Grim, Hid and Sickle [161–163]. It is predominantly localized in the mitochondria; the proform, possessing the transmembrane region (TM) is anchored in the inner membrane while the proteolytic and PDZ domains are exposed to the intermembrane space (IMS). Mature form of HtrA2 (lacking the TM region) largely resides in the IMS as a soluble protein [164]. The protein has also been detected in the endoplasmic reticulum and nucleus in relatively smaller amounts [134, 136, 140]. Nuclear DNA damage,

Table 4.2 Substrates, inhibitors and functions of human HtrA family proteases

Protein	Functions	Endogenous substrates in apoptosis	Other endogenous substrates	Synthetic substrates	Endogenous inhibitors	Synthetic inhibitors
HtrA1	Apoptosis, caspase and EGFR/AKT activation, blocking the NF-κB, anoikis, matrix remodeling, cartilage catabolism, bone homeostasis, Alzheimer's disease, TGF-β receptor signaling	XIAP	TGFβI, TGFβII, Amyloid β, C99,type II collagen, aggrecan, fibronectin, decorin, fibromodulin, clusterin, ADAM9, vitronectin, α2-macroglobulin, talin-1, fascin	Mca-IRRVSYSF (Dnp) KK, FITC-β-casein	–	DFP, 1-(3-cyclohexyl-2-propionyl)-pyrrolidine-2-carboxylic acid 5-(3-cyclohexyl-ureido)-1-dihydroxyboranyl-pentyl)-amid, alpha-1-anti-trypsin
HtrA2	Apoptosis, caspase and Fas activation, mitochondrial homeostasis, Alzheimer's disease, chaperone function	XIAP, cIAP1, cIAP2, PEA15, HAX1, FLIP, Fas, actin, α-, β-tubulin, vimentin	Mulan, WARTS kinase, eIF-4G1, EF-1R, HADH2, ERAB, KIAA1967 and KIAA0251, PDHB, IDH3A, HSPA8, Amyloid β, p73	Mca-IRRVSYSF (Dnp) KK, FITC-β-casein	–	UCF-101
HtrA3	Apoptosis, embryo implantation, formation of placenta, trophoblast invasion, TGF-β receptor signaling	–	TGFβ	–	–	–
HtrA4	Apoptosis, TGF-β receptor signaling, suppresses cell-cell fusion	–	TGFβ, syncytin-1	–	–	–

death-receptor activation and numerous other apoptotic stresses trigger translocation of the mature protease into the cytosol, where it contributes to apoptosis through both caspase-dependent and independent mechanisms [165]. In addition, HtrA2/Omi might function as a protein quality control factor in the mitochondrial intermembrane compartment similar to prokaryotic HtrAs in the periplasmic space [132, 138].

2.2.1 Structural Assembly

HtrA2/Omi is expressed as a 49 kDa proenzyme that is targeted primarily to the IMS [166], where it is attached through its N-terminal transmembrane anchor [166]. During maturation, the first 133 amino acids from the N-terminus gets cleaved and upon apoptotic stimulation, it is released from IMS into the cytosol as a 36 kDa mature protease [161–163, 166]. This cleavage exposes an internal tetrapeptide motif (AVPS) that binds to IAPs. Mature HtrA2 has a short N-terminal region and well defined serine protease and PDZ domains (Fig. 4.6). Crystal structure of mature form of inactive HtrA2 was solved at 2.1 Å, which provides a broad overview of the global structural organization of the inactive protease [167]. It has pyramidal trimeric (110 kDa) architecture with the short N-terminal region at the top and PDZ domains residing at the base of the pyramid (Fig. 4.8a). Each monomer comprises 7 α-helices and 19 β-strands which fold into a compact globular structure. The active-site pocket is surrounded by several activation and regulatory loops to accommodate

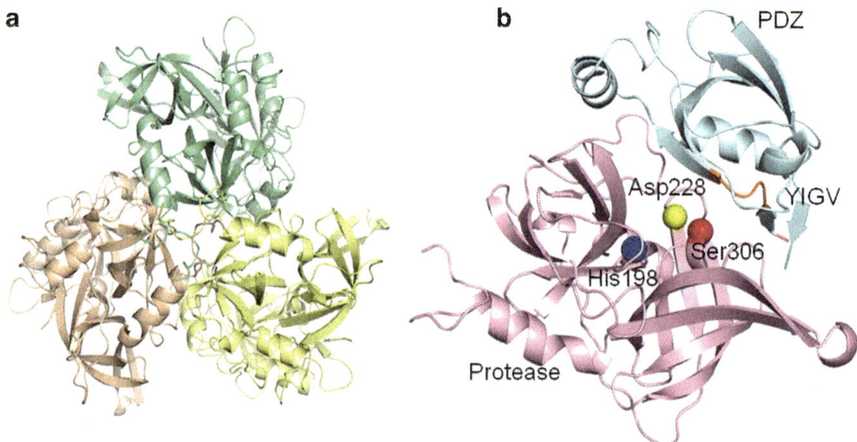

Fig. 4.8 Crystal structure of HtrA2. (**a**) Schematic representation of HtrA2 trimer (PDB entry 1LCY). Each monomeric subunit is represented in *green*, *orange* and *yellow*. (**b**) Cartoon representation of HtrA2 monomer. Serine protease and PDZ domains are colored in *pink* and *cyan* respectively. The position of catalytic triad residues: His198 (*blue*), Asp228 (*yellow*) and Ser306 (*red*) are shown as *spheres*. The position of canonical peptide binding groove 'YIGV' is represented in *orange*. The figures are generated using PyMOL (DeLano Scientific, USA)

the catalytic triad residues (Ser306, His198 and Asp228) in the hydrophobic core of the serine protease domain (Fig. 4.8b). Trimerization is mediated through extensive intermolecular hydrophobic and Van der Waals interactions involving aromatic residues (Tyr147, Phe149 and Phe256) primarily from the N-terminal region. The core serine protease domains that are arranged 25 Å above the base of the pyramid are gated by C-terminal PDZ domains. This, along with its trimeric structural arrangement, restricts HtrA2's accessibility to substrate molecules thus leading to its low basal activity. PDZ is attached covalently to SPD through a flexible linker sequence which regulates HtrA2 activity though subtle conformational changes. Canonical PDZ peptide-binding pocket with consensus 'GLGF-motif' is substituted by 'YIGV' in HtrA2 which is highly buried in the intimate interface between the PDZ and the protease domains. This peptide binding groove is also loosely occupied by two hydrophobic residues, Pro225 and Val226 that are located between strands 5 and 6 of the protease domain. Therefore, the binding groove is unavailable for interaction with other proteins in this 'closed' conformation.

2.2.2 Activation Mechanism

From the structural insights, Li and co-workers proposed a working model for HtrA2 activation [167], where the relative intra-molecular PDZ-protease movement was considered the primary regulatory factor. According to this model, substrate binding at 'YIGV' groove induces a huge conformational change at the PDZ-protease interface which removes the inhibitory effect of PDZ from the active-site. This structural rearrangement leads to significant increase in activity thus emphasizing intramolecular PDZ-protease crosstalk to be pivotal in HtrA2 activation. The model also hypothesized that in the monomeric HtrA2 variant, complete collapse of PDZ on protease leads to its inactivity, which can be subsequently rescued through removal of the PDZ domain. Since the YIGV groove is deeply embedded within the hydrophobic core where the residues are intertwined with each other through several intramolecular interactions, accessibility of ligands to this site was limited. Recently, Bose and co-workers revisited the model for HtrA2 activation and identified a complex allosteric mechanism involving a series of conformational changes leading to ligand binding and subsequent substrate cleavage [168, 169]. This revised model proposes the importance of N-terminal region, oligomerization, and intricate intermolecular PDZ-protease interaction in proper active-site formation, enzyme-substrate complex stabilization, and hence HtrA2 functions (Fig. 4.9). The model also highlights the involvement of N-terminal-mediated allostery in transforming HtrA2 to a proteolytically competent state in a PDZ-independent yet synergistic manner [168, 170]. Overall, the mechanism of HtrA2 activation seems to be distinct from the bacterial counterparts as well as its homolog HtrA1.

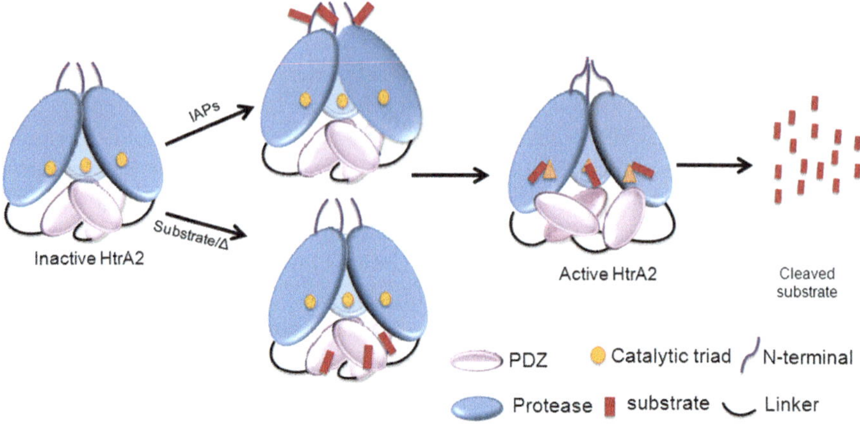

Fig. 4.9 Model for HtrA2 activation adapted from Chaganti *et al.*, 2013 (*168*). Binding of N-/C-terminal substrates or increase in temperature leads to intricate intermolecular PDZ-protease interactions. These intermolecular interactions lead to subtle rearrangements of the regulatory loops to form the active enzyme

2.2.3 Functions

2.2.3.1 Apoptosis

Apoptotic insults such as nuclear DNA damage and death receptor activation trigger the translocation of mature HtrA2 into the cytosol, where it contributes to apoptosis through both caspase-dependent and independent pathways (Fig. 4.10). The following section highlights both these signaling mechanisms in detail.

(a) ***Caspase-dependent mechanism***

In response to various apoptotic stimuli, mature HtrA2 is released into the cytosol, where it interacts with XIAP through the AVPS motif and relieves caspase-9 inhibition [163]. The homotrimeric HtrA2 may also interact via its PDZ domain with a trimeric assembly of TNFR1 or Fas [167, 171]. The Fas-ligand-induced trimerization activates the 'death domain' present in the cytoplasmic region of each Fas monomer thereby initiating caspase-8-dependent apoptotic pathway [172]. This hypothesis is further supported by observation of increased FasL expression after myocardial ischemia/reperfusion [171].

(b) ***Caspase-independent mechanism***

HtrA2 via its serine protease activity induces cell death, independent of molecules such as caspase-9 and Apaf-1 [162]. It does so by degrading several cellular proteins such as IAPs through its serine protease activity [173, 174]. HtrA2 promotes cell death specifically by binding to and cleavage of the death effector domains (DED) of FLIP [175] and anti-apoptotic PED/PEA-15 [176]. Furthermore,

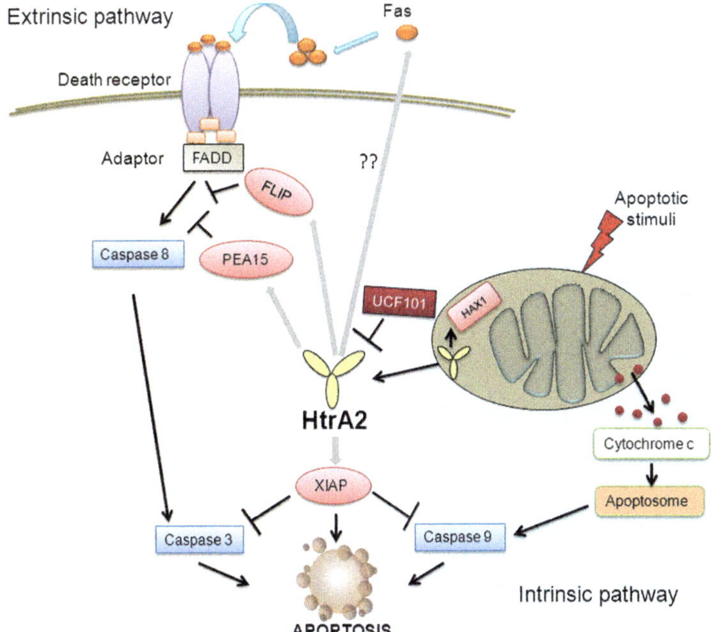

Fig. 4.10 HtrA2 mediated apoptotic pathway. Upon apoptotic stimuli, mature HtrA2 is released from the mitochondrial intermembrane space into the cytosol. HtrA2 interacts with anti-apoptotic protein, XIAP thus relieving its inhibitory action on caspases, thereby facilitating caspase-dependent apoptosis. Homotrimeric HtrA2 via its PDZ domain interacts with a trimeric assembly of Fas, thus initiating death-receptor pathway of apoptosis. HtrA2 through its serine protease activity also induces caspase-independent cell death by binding and cleaving anti-apoptotic proteins FLIP, PEA15 and HAX1

caspase-independent pro-apoptotic property is manifested by its ability to cleave important cytoskeletal proteins such as actin, α-, β-tubulin and vimentin [177]. Moreover, upon induction of apoptosis, HtrA2 translocates to the nucleus where it cleaves p73. Proteolytically modified p73 thereby stimulates transcription of the BAX gene, whose protein product exhibits pro-apoptotic function [140].

2.2.3.2 Other Functions

HtrA2 acts as a regulator of mitochondrial homeostasis under physiological conditions facilitating cell survival rather than cell death. Alteration in the HtrA2 proteolytic activity leads to accumulation of unfolded proteins in mitochondria, dysfunction of mitochondrial respiration, and also loss of mitochondrial competence [161, 178, 179]. The mnd2 (motor neuron degeneration 2) mice carrying a missense mutation Ser276Cys, as well as the knockout mice carrying a homologous deletion of the HtrA2 gene exhibit phenotypes with features typical for the Parkinsonian syndrome [178, 180]. Several points of evidences show that it can also function as a chaperone protein. It prevents aggregation of amyloid β42, a major element of neurotoxic deposits in brains of Alzheimer's disease patients by keeping the peptide in monomeric state [181]. Recently, it was shown that HtrA2 is involved in maintaining the levels of Mulan (Mitochondrial ubiquitin ligase activator of NF-kB) protein under normal conditions as well as during oxidative stress [182]. Upregulation of Mulan, in the absence of HtrA2, leads to the degradation and removal of Mfn2 (mitofusin 2) protein leading to mitochondrial dysfunction and mitophagy in muscle cells. This mechanism of maintaining the levels of Mulan defines a new function of HtrA2 in mitochondrial homeostasis [182].

2.2.4 Substrates

2.2.4.1 Synthetic Substrate and Substrate Specificity

Substrate specificity of HtrA2 was determined using degenerate peptide libraries to identify the optimal substrate sequence for cleavage by HtrA2 [156]. This is defined by the residue positions at carboxyl-terminal (P1′, P2′, P3′ and P4′) and amino-terminal (P1, P2 and P3) to the HtrA2/Omi cleavage site. The strongest selectivity at P1 position is for non-polar aliphatic amino acids (V, I, and M). At the P2 and P3 positions, R is selected most strongly with a secondary selection for other polar residues. The major selectivity at P1′ is for A and S and at P2′ for Y and F [156]. Based on this amino acid specificity, a fluorescently quenched peptide substrate, H2-Opt (Mca-IRRVSYSF (Dnp) KK) was designed [156] which accommodates optimal HtrA2/Omi amino acid preferences at each position as discussed above.

Ligand specificity at PDZ domain was determined to characterize its binding properties using the peptide libraries fused to the C- or N- terminus of a phage coat protein [183]. This specificity is defined by the residue positions from the C- to the N-terminus (0, −1, −2, and −3). From C-terminal libraries, the strong selectivity at position 0 was found to be for aliphatic hydrophobic residues (V, L, I and A) that are typical of most PDZ domain ligands [184]. Positions −1, −2, and −3 were preferred by the large hydrophobic residues (W and F). At position −1, W was dominant while F was preferred at position −3. From N-terminal libraries, G or D residue was preferred at position 0 while hydrophobic residues predominated at −1, −2, and −3 positions as observed in case of C-terminal peptide library as well. Thus, PDZ of HtrA2 binds to both C-terminal and internal peptide sequences that contain extended hydrophobic residues [183].

2.2.4.2 Endogenous Substrates

Despite substantial evidence of involvement of HtrA2/Omi and its proteolytic activity in apoptosis, little is known about its cytosolic targets. The most studied substrates of HtrA2 are human IAP family of proteins, including XIAP, cellular

IAP (cIAP1) and cIAP2 (Table 4.2). HtrA2 mediates apoptosis by binding and degrading IAPs, causing the release and subsequent activation of caspases [163]. Besides the IAP proteins, HtrA2 interacts with DED (death effector domain) of cytosolic PED/PEA-15 [176] and FLIP [171, 175], promote their degradation and activate apoptosis by Fas-mediated pathway. WARTS kinase, a serine–threonine kinase that plays important roles in mitotic and post-mitotic cell cycle regulation [185], binds directly to the PDZ domain of HtrA2 through its C-terminal region and is proteolyzed by activated HtrA2. Thus, this protein is not only a regulator but also a downstream target of HtrA2. HS-1 associated protein X-1 (HAX-1), a mitochondrial anti-apoptotic protein is also reported as a substrate of HtrA2 [186]. HtrA2 induced cell death in etoposide [187], cisplatin [186] and H_2O_2 [188], treated mouse embryonic fibroblasts (MEF) cells by degrading HAX-1. Therefore cleavage of HAX-1 by HtrA2 might be an early event, and defines a potential new pro-apoptotic pathway initiated in the mitochondria [186].

In another report, proteome-wide study using mass spectrometry identified 10 potential targets of HtrA2. It includes cytoskeletal proteins such as actin, tubulin-α, $-\beta$ and vimentin, eukaryotic translation initiation factor-4 gamma 1 (eIF-4γ1) and elongation factor-1 alpha (EF-1α). HADH2 (3-hydroxyacyl-coenzyme A dehydrogenase type 2) or ERAB (ER-associated amyloid-β binding protein), KIAA1967 and KIAA0251 were also among the potential targets. These identified and validated substrates highlight the involvement of HtrA2 at various levels of apoptotic process involving cytoskeleton, translation initiation complex, and organelle destruction [177]. Mitochondrial proteomic analysis from 293 cells identified three potential substrates of HtrA2 that impact metabolism and ATP production. Out of this, two substrates are the key proteins of Kreb's cycle, PDHB (Pyruvate dehydrogenase E1 component beta subunit) and IDH3A (Isocitrate dehydrogenase [NAD] subunit alpha) and the third identified protein was a chaperone heat shock 70 kDa protein 8, HSPA8 [189]. Moreover, amyloid precursor protein (APP) is directly and efficiently cleaved by HtrA2 serine protease *in vitro* and *in vivo*. Recently, Mulan was identified a potential substrate of HtrA2 in mitochondrial intermembrane space [182].

2.2.5 Inhibitors

Till date, no specific endogenous inhibitors for HtrA2 are known in mammals. UCF-101 (5-[5-(2-nitrophenyl) furfuryliodine] 1, 3-diphenyl-2-thiobarbituric acid), a highly selective competitive, reversible synthetic inhibitor with an IC_{50} (i.e. half maximal inhibitory concentration value) of 9.5 μM has been synthesized [190]. This inhibitor has a natural fluorescence and can be used to monitor its ability to enter mammalian cells thus making it suitable for *in vivo* experiments. UCF-101 shows a profound effect on the activity of HtrA2 and could substantially inhibit its ability to induce caspase-independent apoptosis in caspase-9 (-/-) null fibroblasts [190, 191]. UCF-101 can therefore be used as a tool to dissect the two different functions (caspase-dependent *versus* caspase-independent) of HtrA2 and their respective contribution to apoptosis in various biological systems.

2.3 *HtrA3*

HtrA3 was initially identified as a pregnancy-related serine protease [192] and is found to share a high degree of similarity in sequence and domain organizations with HtrA1 as shown in Fig. 4.6 [139, 193]. It is a nuclear-encoded mitochondrial protease whose localization depends on the presence of the Mac25 domain [133, 194]. In mammalian cells, it is present in two isoforms: the long 49 kDa (HtrA3-L) and the short PDZ-lacking 39 kDa, (HtrA-S) variants resulting from alternative splicing [195]. However, there are no reports characterizing their respective biological functions in the cell. Processed forms of HtrA3 (i.e. removal of N-terminal Mac25 domain) was also found in the cytoplasm [195]. The highest expression of HtrA3 is observed in heart tissues and reproductive organs such as ovary, uterus and placenta [193]. HtrA3 mRNA levels are down regulated in several cancers including human endometrial, ovarian and lung cancers [195, 196].

2.3.1 Structural Assembly

The open reading frame (ORF) of HtrA3-L (long form) mRNA encodes a polypeptide of 453 aa with mass of approximately 49 kDa. It contains a signal peptide (1–17aa), IGFBP (29–94 aa), a Kazal-type inhibitor domain (89–126 aa), a serine protease domain (176–341 aa) with the catalytic triad His191-Asp219-Ser305 and one PDZ domain (384–440 aa) as illustrated in Fig. 4.6. Domain with homology to the IGFBP and a Kazal-type inhibitor motif shares identity with Mac25 protein. The ORF of HtrA3-S (short form) mRNA encodes a polypeptide of 357 amino acids with mass of about 38 kDa. Unlike HtrA3-L, HtrA3-S lacks the PDZ domain and the last seven C-terminal residues of HtrA3-S (APSLAVH) are completely different from the corresponding (DWKKRFI) residues in HtrA3-L [193].

High precision PDZ domain structure of HtrA3 in complex with a peptide ligand ($FGRWV_{COOH}$) was obtained by X-ray crystallography [197]. PDZ fold consists of a five-stranded β-sandwich (β1–β5) capped by two α-helices (α1, α3) [198]. In addition, there are short β-strands at the N- and C-termini. Similar to PDZ domains of HtrA2/Omi [183] and bacterial DegP as well as DegS [199, 200], β1–β2 loops of the PDZ domain form a well-defined α-helix, but the orientation of the helix relative to the rest of the domain varies.

2.3.2 Functions

2.3.2.1 Apoptosis

As HtrA1 and 3 share a high degree of domain homology, they may also share functional similarities. It is believed that HtrA3 promotes activation of the intrinsic mitochondria-mediated apoptotic pathway, that is dependent on its release into the cytoplasm [195]. In support of this, down regulation of HtrA3 expression attenuates cisplatin and etoposide induced cytotoxicity in lung cancer cell lines while re-expression of proteolytic active HtrA3 promotes etoposide and cisplatin-induced cytotoxicity [195]. HtrA3 mRNA expression is also down-regulated in ovarian cancer and granulosa cell tumor cell lines, suggesting that HtrA3 may act as a tumor suppressor [201]. However, further experimental evidences are required to understand the exact role of HtrA3 in apoptosis.

2.3.2.2 Other Functions

HtrA3 binds to several members of the TGF-β protein family, including BMP4 (bone morphogenetic protein 4), TGF-β1, TGF- β2, GDF5 (Growth differentiation factor 5), and suppresses signal transduction mediated by these extracellular cytokines [202]. It plays a significant role during embryo implantation and formation of placenta in mammals during the early stage of pregnancy [203–205]. HtrA3, due to its proteolytic activity, negatively regulates trophoblast invasion during placental development [206].

2.3.3 Substrates and Inhibitors

Till date, no endogenous substrates and inhibitors are identified for HtrA3. It is believed that as HtrA1 and 3 share a high degree of domain homology, they may share similar substrates in the cell as well [139, 202].

2.4 HtrA4

HtrA4 is the least characterized member of the HtrA protease family. It is a nuclear-encoded secreted protein almost restricted to the placenta, although a small amount of expression is also observed in thyroid and uterus [207]. It contains a signal peptide (1–31 aa), a IGFBP domain (36–99 aa), a Kazal-type inhibitor motif (8–154), a serine protease domain (202–362 aa) with the catalytic triad His218-Asp248-Ser326 and one PDZ domain (384–474 aa) as shown in Fig. 4.6 [138]. There are no structures available for any of the domains of HtrA4 so far.

Similar to other human HtrA family proteins, HtrA4 regulates TGF-β receptor signaling pathway [207]. HtrA4 binds to the surface-associated (SU) subunit of syncytin-1 through its PDZ domain. This decreases the surface level of syncytin-1 thereby suppressing syncytin-1-mediated cell-cell fusion [208]. Further investigations are required to understand the structural assembly and role of HtrA4 in apoptosis and associated pathophysiological conditions.

2.5 *Perspectives*

Although a clear picture of structural and functional relationship of human HtrA1 and 2 has emerged in the recent years, a lot more needs to be done especially on HtrA3 and 4. It would be intriguing to know whether the mechanism of action and mode of regulation of these non-caspase proteases are unique or conserved across the family. Moreover, identification of new binding partners or substrates will address key questions about the biological roles of these proteases as well as their contribution to various pathological conditions. Given their involvement in prevention of neurodegenerative diseases and tumorigenesis, as well as role in protein quality control, HtrAs have tremendous potential as targets for therapeutic intervention. Therefore, determination of structures of substrate-bound HtrAs and understanding of their intricate regulatory mechanisms will be a step forward toward clinical applications.

References

1. Willstaetter R, Bamann E (1929) Ueber die Proteasen der Magenschleimhaut. Hoppe Seylers Z Physiol Chem 180:127–143
2. Turk B, Turk D, Turk V (2000) Lysosomal cysteine proteases: more than scavengers. Biochim Biophys Acta Protein Struct Mol Enzymol 1477:98–111
3. Rozman J, Stojan J, Kuhelj R, Turk V, Turk B (1999) Autocatalytic processing of recombinant human procathepsin B is a bimolecular process. FEBS Lett 459:358–362
4. Jerala R, Zerovnik E, Kidric J, Turk V (1998) pH-induced conformational transitions of the propeptide of human cathepsin L: a role for a molten globule state in zymogen activation. J Biol Chem 273:11498–11504
5. Turk B, Turk V, Turk D (1997) Structural and functional aspects of papain-like cysteine proteinases and their protein inhibitors. Biochemistry 3–4:141–150
6. Turk V, Bode W (1991) The cystatins: protein inhibitors of cysteine proteinases. FEBS Lett 285:213–219
7. Lenarcic B, Bevec T (1998) Thyropins–new structurally related proteinase inhibitors. Biol Chem 2:105–111
8. Turk B, Turk D, Salvesen GS (2002) Regulating cysteine protease activity: essential role of protease inhibitors as guardians and regulators. Curr Pharm Des 18:1623–1637
9. Turk D, Janjic V, Stern I, Podobnik M, Lamba D, Dahl SW, Lauritzen C, Pedersen J, Turk V, Turk B (2001) Structure of human dipeptidyl peptidase I (cathepsin C): exclusion domain added to an endopeptidase framework creates the machine for activation of granular serine proteases. EMBO J 23:6570–6582
10. Gelb BD, Shi GP, Chapman HA, Desnick RJ (1996) Pycnodysostosis, a lysosomal disease caused by cathepsin K deficiency. Science 5279:1236–1238
11. Vasiljeva O, Reinheckel T, Peters C, Turk D, Turk V, Turk B (2007) Emerging roles of cysteine cathepsins in disease and their potential as drug targets. Curr Pharm Des 4:387–403
12. Stoka V, Turk B, Schendel SL, Kim T-H, Cirman T, Snipas SJ, Ellerby LM, Bredesen D, Freeze H, Abrahamson M, Brömme D, Krajewski S, Reed JC, Yin X-M, Turk V, Salvesen GS (2001) Lysosomal protease pathways to apoptosis: clevage of bid, not pro-caspase, is the most likely route. J Biol Chem 276:3149–3157

13. Boya P, Gonzalez-Polo R-A, Poncet D, Andreau K, Vieira HLA, Roumier T, Perfettini J-L, Kroemer G (2003) Mitochondrial membrane permeabilization is a critical step of lysosome-initiated apoptosis induced by hydroxychloroquine. Oncogene 22:3927–3936
14. Barrett AJ, Rawlings ND, Jr JFW (eds) (1998) Handbook of proteolytic enzymes. Academic, London
15. Riese RJ, Chapman HA (2000) Cathepsins and compartmentalization in antigen presentation. Curr Opin Immunol 12:107–113
16. Tolosa E, Li W, Yasuda Y, Wienhold W, Denzin LK, Lautwein A, Driessen C, Schnorrer P, Weber E, Stevanovic S, Kurek R, Melms A, Bromme D (2003) Cathepsin V is involved in the degradation of invariant chain in human thymus and is overexpressed in myasthenia gravis. J Clin Invest 112:517–526
17. Shi G-P, Bryant RAR, Riese R, Verhelst S, Driessen C, Li Z, Bromme D, Ploegh HL, Chapman HA (2000) Role for cathepsin F in invariant chain processing and major histocompatibility complex class II peptide loading by macrophages. J Exp Med 191:1177–1186
18. Linnevers C, Smeekens SP, Brömme D (1997) Human cathepsin W, a putative cysteine protease predominantly expressed in CD8+ T-lymphocytes. FEBS Lett 405:253–259
19. Velasco G, Ferrando AA, Puente XS, Sánchez LM, López-Otín C (1994) Human cathepsin O. Molecular cloning from a breast carcinoma, production of the active enzyme in Escherichia coli, and expression analysis in human tissues. J Biol Chem 269:27136–27142
20. Guncar G, Podobnik M, Pungercar J, Strukelj B, Turk V, Turk D (1998) Crystal structure of porcine cathepsin H determined at 2.1 A resolution: location of the mini-chain C-terminal carboxyl group defines cathepsin H aminopeptidase function. Structure 1:51–61
21. Musil D, Zucic D, Turk D, Engh RA, Mayr I, Huber R, Popovic T, Turk V, Towatari T, Katunuma N (1991) The refined 2.15 A X-ray crystal structure of human liver cathepsin B: the structural basis for its specificity. EMBO J 9:2321–2330
22. Gunčar G, Klemenčič I, Turk B, Turk V, Karaoglanovic-Carmona A, Juliano L, Turk D (2000) Crystal structure of cathepsin X: a flip–flop of the ring of His23 allows carboxy-monopeptidase and carboxy-dipeptidase activity of the protease. Structure 8:305–313
23. Wiederanders B (2003) Structure-function relationships in class CA1 cysteine peptidase propeptides. Acta Biochim Pol 3:691–713
24. Polgar L, Halasz P (1982) Current problems in mechanistic studies of serine and cysteine proteinases. Biochem J 1:1–10
25. McGrath ME (1999) The lysosomal cysteine proteases. Annu Rev Biophys Biomol Struct 28:181–204
26. Turk D, Guncar G, Podobnik M, Turk B (1998) Revised definition of substrate binding sites of papain-like cysteine proteases. Biol Chem 2:137–147
27. Turk D, Guncar G (2003) Lysosomal cysteine proteases (cathepsins): promising drug targets. Acta Crystallogr Sect D 59:203–213
28. Barrett AJ (1994) [1] Classification of peptidases. In: Alan JB (ed) Methods in enzymology. Academic Press, New York, pp 1–15
29. Turk V, Stoka V, Vasiljeva O, Renko M, Sun T, Turk B, Turk D (2012) Cysteine cathepsins: from structure, function and regulation to new frontiers. Biochim Biophys Acta (BBA) – Proteins Proteomics 1824:68–88
30. Jenko S, Dolenc I, Gunčar G, Doberšek A, Podobnik M, Turk D (2003) Crystal structure of stefin A in complex with cathepsin H: N-terminal residues of inhibitors can adapt to the active sites of endo- and exopeptidases. J Mol Biol 326:875–885
31. Lowe J, Stock D, Jap B, Zwickl P, Baumeister W, Huber R (1995) Crystal structure of the 20S proteasome from the archaeon T. acidophilum at 3.4 A resolution. Science 268:533–539
32. Joshua-Tor L, Xu H, Johnston S, Rees D (1995) Crystal structure of a conserved protease that binds DNA: the bleomycin hydrolase, Gal6. Science 269:945–950
33. Pereira PJB, Bergner A, Macedo-Ribeiro S, Huber R, Matschiner G, Fritz H, Sommerhoff CP, Bode W (1998) Human [beta]-tryptase is a ring-like tetramer with active sites facing a central pore. Nature 392:306–311

34. Mølgaard A, Arnau J, Lauritzen C, Larsen S, Petersen G, Pedersen J (2007) The crystal structure of human dipeptidyl peptidase I (cathepsin C) in complex with the inhibitor Gly-Phe-CHN2. Biochem J 401:645–650
35. Guicciardi ME, Deussing J, Miyoshi H, Bronk SF, Svingen PA, Peters C, Kaufmann SH, Gores GJ (2000) Cathepsin B contributes to TNF-α–mediated hepatocyte apoptosis by promoting mitochondrial release of cytochrome c. J Clin Invest 106:1127–1137
36. Foghsgaard L, Wissing D, Mauch D, Lademann U, Bastholm L, Boes M, Elling F, Leist M, Jäättelä M (2001) Cathepsin B acts as a dominant execution protease in tumor cell apoptosis induced by tumor necrosis factor. J Cell Biol 153:999–1010
37. Liu N, Raja SM, Zazzeroni F, Metkar SS, Shah R, Zhang M, Wang Y, Brömme D, Russin WA, Lee JC, Peter ME, Froelich CJ, Franzoso G, Ashton-Rickardt PG (2003) NF-κB protects from the lysosomal pathway of cell death. EMBO J 22:5313–5322
38. Turk B, Stoka V, Rozman-Pungercar J, Cirman T, Droga-Mazovec G, Oreic K, Turk V (2002) Apoptotic pathways: involvement of lysosomal proteases. Biol Chem 383:1035–44
39. Guicciardi ME, Leist M, Gores GJ (2004) Lysosomes in cell death. Oncogene 23:2881–2890
40. Li W, Yuan X, Nordgren G, Dalen H, Dubowchik GM, Firestone RA, Brunk UT (2000) Induction of cell death by the lysosomotropic detergent MSDH. FEBS Lett 470:35–39
41. Cirman T, Orešić K, Mazovec GD, Turk V, Reed JC, Myers RM, Salvesen GS, Turk B (2004) Selective disruption of lysosomes in HeLa cells triggers apoptosis mediated by cleavage of bid by multiple papain-like lysosomal cathepsins. J Biol Chem 279(5):3578–3587
42. Reiners JJ Jr, Caruso JA, Mathieu P, Chelladurai B, Yin XM, Kessel D (2002) Release of cytochrome c and activation of pro-caspase-9 following lysosomal photodamage involves Bid cleavage. Cell Death Differ 9:934–944
43. Boya P, Andreau K, Poncet D, Zamzami N, Perfettini J-L, Metivier D, Ojcius DM, Jäättelä M, Kroemer G (2003) Lysosomal membrane permeabilization induces cell death in a mitochondrion-dependent fashion. J Exp Med 197:1323–1334
44. Katunuma N, Murata E, Le QT, Hayashi Y, Ohashi A (2004) New apoptosis cascade mediated by lysosomal enzyme and its protection by epigallo-catechin gallate. Adv Enzym Regul 44:1–10
45. Erdal H, Berndtsson M, Castro J, Brunk U, Shoshan MC, Linder S (2005) Induction of lysosomal membrane permeabilization by compounds that activate p53-independent apoptosis. Proc Natl Acad Sci U S A 102:192–197
46. Pham CTN, Ley TJ (1999) Dipeptidyl peptidase I is required for the processing and activation of granzymes A and B in vivo. Proc Natl Acad Sci 96:8627–8632
47. Robker RL, Russell DL, Espey LL, Lydon JP, O'Malley BW, Richards JS (2000) Progesterone-regulated genes in the ovulation process: ADAMTS-1 and cathepsin L proteases. Proc Natl Acad Sci 97:4689–4694
48. Choe Y, Leonetti F, Greenbaum DC, Lecaille F, Bogyo M, Brömme D, Ellman JA, Craik CS (2006) Substrate profiling of cysteine proteases using a combinatorial peptide library identifies functionally unique specificities. J Biol Chem 281:12824–12832
49. Cotrin SS, Puzer L, de Souza Judice WA, Juliano L, Carmona AK, Juliano MA (2004) Positional-scanning combinatorial libraries of fluorescence resonance energy transfer peptides to define substrate specificity of carboxydipeptidases: assays with human cathepsin B. Anal Biochem 335:244–252
50. Alves MFM, Puzer L, Cotrin SS, Juliano MA, Juliano L, Brömme D, Carmona AK (2003) S3 to S3′ subsite specificity of recombinant human cathepsin K and development of selective internally quenched fluorescent substrates. Biochem J 373:981–986
51. Oliveira M, Torquato RJS, Alves MFM, Juliano MA, Brömme D, Barros NMT, Carmona AK (2010) Improvement of cathepsin S detection using a designed FRET peptide based on putative natural substrates. Peptides 31:562–567
52. Gocheva V, Zeng W, Ke D, Klimstra D, Reinheckel T, Peters C, Hanahan D, Joyce JA (2006) Distinct roles for cysteine cathepsin genes in multistage tumorigenesis. Genes Dev 20:543–556

53. Mason RW (1989) Interaction of lysosomal cysteine proteinases with alpha 2-macroglobulin: conclusive evidence for the endopeptidase activities of cathepsins B and H. Arch Biochem Biophys 2:367–374
54. Pennacchio LA, Bouley DM, Higgins KM, Scott MP, Noebels JL, Myers RM (1998) Progressive ataxia, myoclonic epilepsy and cerebellar apoptosis in cystatin B-deficient mice. Nat Genet 20:251–258
55. Bevec T, Stoka V, Pungercic G, Dolenc I, Turk V (1996) Major histocompatibility complex class II-associated p41 invariant chain fragment is a strong inhibitor of lysosomal cathepsin L. J Exp Med 183:1331–1338
56. Schick C, Pemberton PA, Shi G-P, Kamachi Y, Çataltepe S, Bartuski AJ, Gornstein ER, Brömme D, Chapman HA, Silverman GA (1998) Cross-class inhibition of the cysteine proteinases cathepsins K, L, and S by the serpin squamous cell carcinoma antigen 1: a kinetic analysis†. Biochemistry 37:5258–5266
57. Welss T, Sun J, Irving JA, Blum R, Smith AI, Whisstock JC, Pike RN, von Mikecz A, Ruzicka T, Bird PI, Abts HF (2003) Hurpin is a selective inhibitor of lysosomal cathepsin L and protects keratinocytes from ultraviolet-induced apoptosis†. Biochemistry 42:7381–7389
58. Hwang S-R, Stoka V, Turk V, Hook V (2006) Resistance of cathepsin L compared to elastase to proteolysis when complexed with the serpin endopin 2C, and recovery of cathepsin L activity. Biochem Biophys Res Commun 340:1238–1243
59. Hanada K, Tamai M, Yamagishi M, Ohmura S, Sawada J, Tanaka I (1978) Isolation and characterization of E-64, a new thiol protease inhibitor. Agric Biol Chem 42:523–528
60. Towatari T, Nikawa T, Murata M, Yokoo C, Tamai M, Hanada K, Katunuma N (1991) Novel epoxysuccinyl peptides A selective inhibitor of cathepsin B, in vivo. FEBS Lett 280:311–315
61. Murata M, Miyashita S, Yokoo C, Tamai M, Hanada K, Hatayama K, Towatari T, Nikawa T, Katunuma N (1991) Novel epoxysuccinyl peptides selective inhibitors of cathepsin B, in vitro. FEBS Lett 280:307–310
62. Novinec M, Korenč M, Caflisch A, Ranganathan R, Lenarčič B, Baici A (2014) A novel allosteric mechanism in the cysteine peptidase cathepsin K discovered by computational methods. Nat Commun 5
63. Yamamoto K (1999) Cathepsin E and cathepsin D. In: Turk V (ed) Proteases new perspectives. Birkhäuser, Basel, pp 59–71
64. Erickson AH, Conner GE, Blobel G (1981) Biosynthesis of a lysosomal enzyme. Partial structure of two transient and functionally distinct NH2-terminal sequences in cathepsin D. J Biol Chem 21:11224–11231
65. Conner GE, Richo G (1992) Isolation and characterization of a stable activation intermediate of the lysosomal aspartyl protease cathepsin D. Biochemistry 4:1142–1147
66. Gieselmann V, Hasilik A, von Figura K (1985) Processing of human cathepsin D in lysosomes in vitro. J Biol Chem 5:3215–3220
67. Koblinski JE, Dosescu J, Sameni M, Moin K, Clark K, Sloane BF (2002) Interaction of human breast fibroblasts with collagen I increases secretion of procathepsin B. J Biol Chem 277:32220–32227
68. Laurent-Matha V, Maruani-Herrmann S, Prebois C, Beaujouin M, Glondu M, Noel A, Alvarez-Gonzalez ML, Blacher S, Coopman P, Baghdiguian S, Gilles C, Loncarek J, Freiss G, Vignon F, Liaudet-Coopman E (2005) Catalytically inactive human cathepsin D triggers fibroblast invasive growth. J Cell Biol 168:489–499
69. Heylen N, Vincent LM, Devos V, Dubois V, Remacle C, Trouet A (2002) Fibroblasts capture cathepsin D secreted by breast cancer cells: possible role in the regulation of the invasive process. Int J Oncol 20:761–767
70. Laurent-Matha V, Farnoud MR, Lucas A, Rougeot C, Garcia M, Rochefort H (1998) Endocytosis of pro-cathepsin D into breast cancer cells is mostly independent of mannose-6-phosphate receptors. J Cell Sci 111(Pt 17):2539–2549
71. Vetvicka V, Vektvickova J, Fusek M (1994) Effect of human procathepsin D on proliferation of human cell lines. Cancer Lett 2:131–135

72. Zhou W, Scott SA, Shelton SB, Crutcher KA (2006) Cathepsin D-mediated proteolysis of apolipoprotein E: possible role in Alzheimer's disease. Neuroscience 143:689–701
73. Siintola E, Partanen S, Strömme P, Haapanen A, Haltia M, Maehlen J, Lehesjoki A-E, Tyynelä J (2006) Cathepsin D deficiency underlies congenital human neuronal ceroid-lipofuscinosis. Brain 129:1438–1445
74. Sakai H, Saku T, Kato Y, Yamamoto K (1989) Quantitation and immunohistochemical localization of cathepsins E and D in rat tissues and blood cells. Biochim Biophys Acta Gen Subj 991:367–375
75. Kageyama T, Moriyama A, Kato T, Sano M, Yonezawa S (1996) Determination of cathepsins D and E in various tissues and cells of rat, monkey, and man by the assay with β-endorphin and substance P as substrates. Zool Sci 13:693–698
76. Finzi G, Cornaggia M, Capella C, Fiocca R, Bosi F, Solcia E, Samloff IM (1993) Cathepsin E in follicle associated epithelium of intestine and tonsils: localization to M cells and possible role in antigen processing. Histochemistry 3:201–211
77. Nakanishi H, Tominaga K, Amano T, Hirotsu I, Inoue T, Yamamoto K (1994) Age-related changes in activities and localizations of cathepsins D, E, B, and L in the rat brain tissues. Exp Neurol 126:119–128
78. Khan AR, Cherney MM, Tarasova NI, James MNG (1997) Structural characterization of activation/'intermediate 2/' on the pathway to human gastricsin. Nat Struct Mol Biol 4:1010–1015
79. Alina M, Marek G, Alicja K, Łukasz M (2008) Human cathepsin D. Folia Histochem Cytobiol 46:23–38
80. Roberts L, Adjei P, Gores G (1999) Cathepsins as effector proteases in hepatocyte apoptosis. Cell Biochem Biophys 30:71–88
81. Yu S-W, Wang H, Poitras MF, Coombs C, Bowers WJ, Federoff HJ, Poirier GG, Dawson TM, Dawson VL (2002) Mediation of poly(ADP-ribose) polymerase-1-dependent cell death by apoptosis-inducing factor. Science 297:259–263
82. Deiss LP, Galinka H, Berissi H, Cohen O, Kimchi A (1996) Cathepsin D protease mediates programmed cell death induced by interferon-gamma, Fas/APO-1 and TNF-alpha. EMBO J 15:3861–3870
83. Demoz M, Castino R, Cesaro P, Baccino FM, Bonelli G, Isidoro C (2002) Endosomal-lysosomal proteolysis mediates death signalling by TNFalpha, not by etoposide, in L929 fibrosarcoma cells: evidence for an active role of cathepsin D. Biol Chem 383:1237–1248
84. Heinrich M, Neumeyer J, Jakob M, Hallas C, Tchikov V, Winoto-Morbach S, Wickel M, Schneider-Brachert W, Trauzold A, Hethke A, Schutze S (2004) Cathepsin D links TNF-induced acid sphingomyelinase to Bid-mediated caspase-9 and -3 activation. Cell Death Differ 11:550–563
85. Ollinger K (2000) Inhibition of cathepsin D prevents free-radical-induced apoptosis in rat cardiomyocytes. Arch Biochem Biophys 373:346–351
86. Kagedal K, Johansson U, Ollinger K (2001) The lysosomal protease cathepsin D mediates apoptosis induced by oxidative stress. FASEB J 15:1592–1594
87. Takuma K, Kiriu M, Mori K, Lee E, Enomoto R, Baba A, Matsuda T (2003) Roles of cathepsins in reperfusion-induced apoptosis in cultured astrocytes. Neurochem Int 42:153–159
88. Emert-Sedlak L, Shangary S, Rabinovitz A, Miranda MB, Delach SM, Johnson DE (2005) Involvement of cathepsin D in chemotherapy-induced cytochrome c release, caspase activation, and cell death. Mol Cancer Ther 4:733–742
89. Kagedal K, Zhao M, Svensson I, Brunk UT (2001) Sphingosine-induced apoptosis is dependent on lysosomal proteases. Biochem J 2:335–343
90. Johansson AC, Steen H, Ollinger K, Roberg K (2002) Cathepsin D mediates cytochrome c release and caspase activation in human fibroblast apoptosis induced by staurosporine. Cell Death Differ 10:1253–1259
91. Conus S, Perozzo R, Reinheckel T, Peters C, Scapozza L, Yousefi S, Simon HU (2008) Caspase-8 is activated by cathepsin D initiating neutrophil apoptosis during the resolution of inflammation. J Exp Med 205:685–698

92. Zhao S, Aviles ER Jr, Fujikawa DG (2010) Nuclear translocation of mitochondrial cytochrome c, lysosomal cathepsins B and D, and three other death-promoting proteins within the first 60 minutes of generalized seizures. J Neurosci Res 88:1727–1737
93. Grdovic N, Vidakovic M, Mihailovic M, Dinic S, Uskokovic A, Arambasic J, Poznanovic G (2010) Proteolytic events in cryonecrotic cell death: proteolytic activation of endonuclease P23. Cryobiology 60:271–280
94. Kawakubo T, Okamoto K, Iwata J-I, Shin M, Okamoto Y, Yasukochi A, Nakayama KI, Kadowaki T, Tsukuba T, Yamamoto K (2007) Cathepsin E prevents tumor growth and metastasis by catalyzing the proteolytic release of soluble TRAIL from tumor cell surface. Cancer Research 67:10869–10878
95. Yasukochi A, Kawakubo T, Nakamura S, Yamamoto K (2010) Cathepsin E enhances anticancer activity of doxorubicin on human prostate cancer cells showing resistance to TRAIL-mediated apoptosis. Biol Chem 8:947–958
96. Mohamadzadeh M, Mohamadzadeh H, Brammer M, Sestak K, Luftig RB (2004) Identification of proteases employed by dendritic cells in the processing of protein purified derivative (PPD). J Immune Based Ther Vaccines 1:8
97. Vetvicka V, Vagner J, Baudys M, Tang J, Foundling SI, Fusek M (1993) Human breast milk contains procathepsin D–detection by specific antibodies. Biochem Mol Biol Int 5:921–928
98. Saftig P, Hetman M, Schmahl W, Weber K, Heine L, Mossmann H, Koster A, Hess B, Evers M, von Figura K et al (1995) Mice deficient for the lysosomal proteinase cathepsin D exhibit progressive atrophy of the intestinal mucosa and profound destruction of lymphoid cells. EMBO J 15:3599–3608
99. Sealy L, Mota F, Rayment N, Tatnell P, Kay J, Chain B (1996) Regulation of cathepsin E expression during human B cell differentiation in vitro. Eur J Immunol 26:1838–1843
100. Tsukuba T, Okamoto K, Yamamoto K (2012) Cathepsin E is critical for proper trafficking of cell surface proteins. J Oral Biosci 54:48–53
101. Arnold D, Keilholz W, Schild H, Dumrese T, Stevanović S, Rammensee H-G (1997) Substrate specificity of cathepsins D and E determined by N-terminal and C-terminal sequencing of peptide pools. Eur J Biochem 249:171–179
102. Yasuda Y, Kageyama T, Akamine A, Shibata M, Kominami E, Uchiyama Y, Yamamoto K (1999) Characterization of new fluorogenic substrates for the rapid and sensitive assay of cathepsin E and cathepsin D. J Biochem 125:1137–1143
103. Blomgran R, Zheng L, Stendahl O (2007) Cathepsin-cleaved Bid promotes apoptosis in human neutrophils via oxidative stress-induced lysosomal membrane permeabilization. J Leukoc Biol 81:1213–1223
104. Guagliardi LE, Koppelman B, Blum JS, Marks MS, Cresswell P, Brodsky FM (1990) Co-localization of molecules involved in antigen processing and presentation in an early endocytic compartment. Nature 343:133–139
105. Diment S (1990) Different roles for thiol and aspartyl proteases in antigen presentation of ovalbumin. J Immunol 145:417–422
106. Hewitt EW, Treumann A, Morrice N, Tatnell PJ, Kay J, Watts C (1997) Natural processing sites for human cathepsin E and cathepsin D in tetanus toxin: implications for T cell epitope generation. J Immunol 159:4693–4699
107. Van Noort JM, Jacobs MJM (1994) Cathepsin D, but not cathepsin B, releases T cell stimulatory fragments from lysozyme that are functional in the context of multiple murine class II MHC molecules. Eur J Immunol 24:2175–2180
108. Lenarčič B, Krašovec M, Ritonja A, Olafsson I, Turk V (1991) Inactivation of human cystatin C and kininogen by human cathepsin D. FEBS Lett 280:211–215
109. Pimenta D, Chen V, Chao J, Juliano M, Juliano L (2000) α1-Antichymotrypsin and Kallistatin hydrolysis by human cathepsin D. J Protein Chem 19:411–418
110. Ueno E, Sakai H, Kato Y, Yamamoto K (1989) Activation mechanism of erythrocyte cathepsin E. Evidence for the occurrence of the membrane-associated active enzyme. J Biochem 105:878–882

111. Tarasova NI, Szecsi PB, Foltmann B (1986) An aspartic proteinase from human erythrocytes is immunochemically indistinguishable from a non-pepsin, electrophoretically slow moving proteinase from gastric mucosa. Biochim Biophys Acta 1:96–100
112. Hill J, Montgomery DS, Kay J (1993) Human cathepsin E produced in E. coli. FEBS Lett 326:101–104
113. Keilová H, Tomášek V (1972) Effect of pepsin inhibitor from Ascaris lumbricoides on cathepsin D and E. Biochim Biophys Acta (BBA) Enzymology 284:461–464
114. Galjart NJ, Morreau H, Willemsen R, Gillemans N, Bonten EJ, d'Azzo A (1991) Human lysosomal protective protein has cathepsin A-like activity distinct from its protective function. J Biol Chem 266:14754–14762
115. Itoh K, Takiyama N, Kase R, Kondoh K, Sano A, Oshima A, Sakuraba H, Suzuki Y (1993) Purification and characterization of human lysosomal protective protein expressed in stably transformed Chinese hamster ovary cells. J Biol Chem 268:1180–1186
116. D'Azzo A, Hoogeveen A, Reuser AJ, Robinson D, Galjaard H (1982) Molecular defect in combined beta-galactosidase and neuraminidase deficiency in man. Proc Natl Acad Sci U S A 79:4535–4539
117. Hiraiwa M (1999) Cathepsin A/protective protein: an unusual lysosomal multifunctional protein. Cell Mol Life Sci CMLS 56:894–907
118. Cuervo AM, Mann L, Bonten EJ, d'Azzo A, Dice JF (2003) Cathepsin A regulates chaperone-mediated autophagy through cleavage of the lysosomal receptor. EMBO J 1:47–59
119. Sambrano GR, Huang W, Faruqi T, Mahrus S, Craik C, Coughlin SR (2000) Cathepsin G activates protease-activated receptor-4 in human platelets. J Biol Chem 275:6819–6823
120. Selak MA, Smith JB (1990) Cathepsin G binding to human platelets. Evidence for a specific receptor. Biochem J 1:55–62
121. Sun R, Iribarren P, Zhang N, Zhou Y, Gong W, Cho EH, Lockett S, Chertov O, Bednar F, Rogers TJ, Oppenheim JJ, Wang JM (2004) Identification of neutrophil granule protein cathepsin G as a novel chemotactic agonist for the G protein-coupled formyl peptide receptor. J Immunol 173:428–436
122. Uehara A, Muramoto K, Takada H, Sugawara S (2003) Neutrophil serine proteinases activate human nonepithelial cells to produce inflammatory cytokines through protease-activated receptor 2. J Immunol 170:5690–5696
123. Glusa E, Adam C (2001) Endothelium-dependent relaxation induced by cathepsin G in porcine pulmonary arteries. Br J Pharmacol 133:422–428
124. Sabri A, Alcott SG, Elouardighi H, Pak E, Derian C, Andrade-Gordon P, Kinnally K, Steinberg SF (2003) Neutrophil cathepsin G promotes detachment-induced cardiomyocyte apoptosis via a protease-activated receptor-independent mechanism. J Biol Chem 278:23944–23954
125. Iacoviello L, Kolpakov V, Salvatore L, Amore C, Pintucci G, de Gaetano G, Donati MB (1995) Human endothelial cell damage by neutrophil-derived cathepsin G: role of cytoskeleton rearrangement and matrix-bound plasminogen activator inhibitor-1. Arterioscler Thromb Vasc Biol 15:2037–2046
126. Yui S, Tomita K, Kudo T, Ando S, Yamazaki M (2005) Induction of multicellular 3-D spheroids of MCF-7 breast carcinoma cells by neutrophil-derived cathepsin G and elastase. Cancer Sci 96:560–570
127. Lipinska B, Fayet O, Baird L, Georgopoulos C (1989) Identification, characterization, and mapping of the Escherichia coli htrA gene, whose product is essential for bacterial growth only at elevated temperatures. J Bacteriol 171:1574–1584
128. Spiess C, Beil A, Ehrmann M (1999) A temperature-dependent switch from chaperone to protease in a widely conserved heat shock protein. Cell 97:339–347
129. Zurawa-Janicka D, Skorko-Glonek J, Lipinska B (2010) HtrA proteins as targets in therapy of cancer and other diseases. Expert Opin Ther Targets 14:665–679
130. He X, Khurana A, Maguire JL, Chien J, Shridhar V (2012) HtrA1 sensitizes ovarian cancer cells to cisplatin-induced cytotoxicity by targeting XIAP for degradation. Int J Cancer 5:10295–10235

131. Chien J, Aletti G, Baldi A, Catalano V, Muretto P, Keeney GL, Kalli KR, Staub J, Ehrmann M, Cliby WA, Lee YK, Bible KC, Hartmann LC, Kaufmann SH, Shridhar V (2006) Serine protease HtrA1 modulates chemotherapy-induced cytotoxicity. J Clin Invest 7:1994–2004
132. Clausen T, Kaiser M, Huber R, Ehrmann M (2011) HTRA proteases: regulated proteolysis in protein quality control. Nat Rev Mol Cell Biol 12:152–162
133. Chien J, Campioni M, Shridhar V, Baldi A (2009) HtrA serine proteases as potential therapeutic targets in cancer. Curr Cancer Drug Targets 9:451–468
134. Faccio L, Fusco C, Chen A, Martinotti S, Bonventre J, Zervos A (2000) Characterization of a novel human serine protease that has extensive homology to bacterial heat shock endoprotease HtrA and is regulated by kidney Ischemia. J Biol Chem 275:2581–2588
135. Akhurst RJ, Derynck R (2001) TGF-beta signaling in cancer–a double-edged sword. Trends Cell Biol 11:s44–s51
136. Gray C, Ward R, Karran E, Turconi S, Rowles A, Viglienghi D, Southan C, Barton A, Fantom K, West A, Savopoulos J, Hassan N, Clinkenbeard H, Hanning C, Amegadzie B, Davis J, Dingwall C, Livi G, Creasy C (2000) Characterization of human HtrA2, a novel serine protease involved in the mammalian cellular stress response. Eur J Biochem 267:5699–5710
137. Oka C, Tsujimoto R, Kajikawa M, Koshiba-Takeuchi K, Ina J, Yano M, Tsuchiya A, Ueta Y, Soma A, Kanda H, Matsumoto M, Kawaichi M (2004) HtrA1 serine protease inhibits signaling mediated by Tgfbeta family proteins. Development 5:1041–1053
138. Clausen T, Southan C, Ehrmann M (2002) The HtrA family of proteases: implications for protein composition and cell fate. Mol Cell 10:443–455
139. Zumbrunn J, Trueb B (1996) Primary structure of a putative serine protease specific for IGF-binding proteins. FEBS Lett 398:187–192
140. Marabese M, Mazzoletti M, Vikhanskaya F, Broggini M (2008) HtrA2 enhances the apoptotic functions of p73 on bax. Cell Death Differ 5:849–858
141. Eigenbrot C, Ultsch M, Lipari MT, Moran P, Lin SJ, Ganesan R, Quan C, Tom J, Sandoval W, van Lookeren Campagne M, Kirchhofer D (2012) Structural and functional analysis of HtrA1 and its subdomains. Structure 6:1040–1050
142. Hara K, Shiga A, Fukutake T, Nozaki H, Miyashita A, Yokoseki A, Kawata H, Koyama A, Arima K, Takahashi T, Ikeda M, Shiota H, Tamura M, Shimoe Y, Hirayama M, Arisato T, Yanagawa S, Tanaka A, Nakano I, Ikeda S-I, Yoshida Y, Yamamoto T, Ikeuchi T, Kuwano R, Nishizawa M, Tsuji S, Onodera O (2009) Association of HTRA1 mutations and familial ischemic cerebral small-vessel disease. N Engl J Med 17:1729–1739
143. Chien J, Staub J, Hu S, Erickson-Johnson M, Couch F, Smith D, Crowl R, Kaufmann S, Shridhar V (2004) A candidate tumor suppressor HtrA1 is downregulated in ovarian cancer. Oncogene 23:1636–1644
144. Truebestein L, Tennstaedt A, Mönig T, Krojer T, Canellas F, Kaiser M, Clausen T, Ehrmann M (2011) Substrate-induced remodeling of the active site regulates human HTRA1 activity. Nat Struct Mol Biol 18:386–388
145. Xia J, Wang F, Wang L, Fan Q (2013) Elevated serine protease HtrA1 inhibits cell proliferation, reduces invasion, and induces apoptosis in esophageal squamous cell carcinoma by blocking the nuclear factor-κB signaling pathway. Tumor Biol 34:317–328
146. He X, Ota T, Liu P, Su C, Chien J, Shridhar V (2010) Downregulation of HtrA1 promotes resistance to anoikis and peritoneal dissemination of ovarian cancer cells. Cancer Res 8:3109–3018
147. Narkiewicz J, Klasa-Mazurkiewicz D, Zurawa-Janicka D, Skorko-Glonek J, Emerich J, Lipinska B (2008) Changes in mRNA and protein levels of human HtrA1, HtrA2 and HtrA3 in ovarian cancer. Clin Biochem 41:561–569
148. Bowden MA, Di Nezza-Cossens LA, Jobling T, Salamonsen LA, Nie G (2006) Serine proteases HTRA1 and HTRA3 are down-regulated with increasing grades of human endometrial cancer. Gynecol Oncol 103:253–260
149. Zhu F, Jin L, Luo TP, Luo GH, Tan Y, Qin XH (2010) Serine protease HtrA1 expression in human hepatocellular carcinoma. Hepatobiliary Pancreat Dis Int 9:508–512

150. Baldi A, De Luca A, Morini M, Battista T, Felsani A, Baldi F, Catricalà C, Amantea A, Noonan D, Albini A, Natali P, Lombardi D, Paggi M (2002) The HtrA1 serine protease is down-regulated during human melanoma progression and represses growth of metastatic melanoma cells. Oncogene 21:6684–6688
151. Baldi A, Mottolese M, Vincenzi B, Campioni M, Mellone P, Di Marino M, di Crescenzo V, Visca P, Menegozzo S, Spugnini E, Citro G, Ceribelli A, Mirri A, Chien J, Shridhar V, Ehrmann M, Santini M, Facciolo F (2008) The serine protease HtrA1 is a novel prognostic factor for human mesothelioma. Pharmacogenomics 9:1069–1077
152. Hadfield KD, Rock CF, Inkson CA, Dallas SL, Sudre L, Wallis GA, Boot-Handford RP, Canfield AE (2008) HtrA1 inhibits mineral deposition by osteoblasts: requirement for the protease and PDZ domains. J Biol Chem 283:5928–5938
153. Tsuchiya A, Yano M, Tocharus J, Kojima H, Fukumoto M, Kawaichi M, Oka C (2005) Expression of mouse HtrA1 serine protease in normal bone and cartilage and its upregulation in joint cartilage damaged by experimental arthritis. Bone 37:323–336
154. Grau S, Richards P, Kerr B, Hughes C, Caterson B, Williams A, Junker U, Jones S, Clausen T, Ehrmann M (2006) The role of human HtrA1 in arthritic disease. J Biol Chem 281:6124–6129
155. Grau S, Baldi A, Bussani R, Tian X, Stefanescu R, Przybylski M, Richards P, Jones SA, Shridhar V, Clausen T, Ehrmann M (2005) Implications of the serine protease HtrA1 in amyloid precursor protein processing. Proc Natl Acad Sci U S A 17:6021–6026
156. Martins L, Turk B, Cowling V, Borg A, Jarrell E, Cantley L, Downward J (2003) Binding specificity and regulation of the serine protease and PDZ domains of HtrA2/Omi. J Biol Chem 278:49417–49427
157. An E, Sen S, Park SK, Gordish-Dressman H, Hathout Y (2010) Identification of novel substrates for the serine protease HTRA1 in the human RPE secretome. Invest Ophthalmol Vis Sci 7:3379–3386
158. Graham JR, Chamberland A, Lin Q, Li XJ, Dai D, Zeng W, Ryan MS, Rivera-Bermúdez MA, Flannery CR, Yang Z (2013) Serine protease HTRA1 antagonizes transforming growth factor-β signaling by cleaving its receptors and loss of HTRA1 <italic> in vivo </italic> enhances bone formation. PLoS ONE 8, e74094
159. Sohn J, Grant R, Sauer R (2007) Allosteric activation of DegS, a stress sensor PDZ protease. Cell 131:572–583
160. Wu G, Chai J, Suber TL, Wu J-W, Du C, Wang X, Shi Y (2000) Structural basis of IAP recognition by Smac/DIABLO. Nature 408:1008–1012
161. Martins L, Iaccarino I, Tenev T, Gschmeissner S, Totty N, Lemoine N, Savopoulos J, Gray C, Creasy C, Dingwall C, Downward J (2002) The serine protease Omi/HtrA2 regulates apoptosis by binding XIAP through a reaper-like motif. J Biol Chem 277:439–444
162. Hegde R, Srinivasula S, Zhang Z, Wassell R, Mukattash R, Cilenti L, DuBois G, Lazebnik Y, Zervos A, Fernandes-Alnemri T, Alnemri E (2002) Identification of Omi/HtrA2 as a mitochondrial apoptotic serine protease that disrupts inhibitor of apoptosis protein-caspase interaction. J Biol Chem 277:432–438
163. Verhagen A, Silke J, Ekert P, Pakusch M, Kaufmann H, Connolly L, Day C, Tikoo A, Burke R, Wrobel C, Moritz R, Simpson R, Vaux D (2002) HtrA2 promotes cell death through its serine protease activity and its ability to antagonize inhibitor of apoptosis proteins. J Biol Chem 277:445–454
164. Kadomatsu T, Mori M, Terada K (2007) Mitochondrial import of Omi: the definitive role of the putative transmembrane region and multiple processing sites in the amino-terminal segment. Biochem Biophys Res Commun 2:516–521
165. Vande Walle L, Lamkanfi M, Vandenabeele P (2008) The mitochondrial serine protease HtrA2/Omi: an overview. VandeWalle 2008(5):453–460
166. Suzuki Y, Imai Y, Nakayama H, Takahashi K, Takio K, Takahashi R (2001) A serine protease, HtrA2, is released from the mitochondria and interacts with XIAP, inducing cell death. Mol Cell 8:613–621

167. Li W, Srinivasula S, Chai J, Li P, Wu J, Zhang Z, Alnemri E, Shi Y (2002) Structural insights into the pro-apoptotic function of mitochondrial serine protease HtrA2/Omi. Nat Struct Biol 9:436–441
168. Chaganti LK, Kuppili RR, Bose K (2013) Intricate structural coordination and domain plasticity regulate activity of serine protease HtrA2. FASEB J 8:3054–3066
169. Bejugam PR, Kuppili RR, Singh N, Gadewal N, Chaganti LK, Sastry GM, Bose K (2012) Allosteric regulation of serine protease HtrA2 through novel non-canonical substrate binding pocket. PLoS ONE 8, e55416
170. Singh N, D'Souza A, Cholleti A, Sastry GM, Bose K (2014) Dual regulatory switch confers tighter control on HtrA2 proteolytic activity. FEBS J 281:2456–2470
171. Bhuiyan MS, Fukunaga K (2008) Activation of HtrA2, a mitochondrial serine protease mediates apoptosis: current knowledge on HtrA2 mediated myocardial ischemia/reperfusion injury. Cardiovasc Ther 26:224–232
172. Boldin MP, Mett IL, Varfolomeev EE, Chumakov I, Shemer-Avni Y, Camonis JH, Wallach D (1995) Self-association of the "death domains" of the p55 tumor necrosis factor (TNF) receptor and Fas/APO1 prompts signaling for TNF and Fas/APO1 effects. J Biol Chem 1:387–391
173. Yang Q, Church-Hajduk R, Ren J, Newton M, Du C (2003) Omi/HtrA2 catalytic cleavage of inhibitor of apoptosis (IAP) irreversibly inactivates IAPs and facilitates caspase activity in apoptosis. Genes Dev 17:1487–1496
174. Srinivasula SM, Gupta S, Datta P, Zhang Z, Hegde R, Cheong N, Fernandes-Alnemri T, Alnemri ES (2003) Inhibitor of apoptosis proteins are substrates for the mitochondrial serine protease Omi/HtrA2. J Biol Chem 34:31469–31472
175. Bhuiyan MS, Fukunaga K (2007) Inhibition of HtrA2/Omi ameliorates heart dysfunction following ischemia/reperfusion injury in rat heart in vivo. Eur J Pharmacol 557:168–177
176. Trencia A, Fiory F, Maitan M, Vito P, Barbagallo A, Perfetti A, Miele C, Ungaro P, Oriente F, Cilenti L, Zervos A, Formisano P, Beguinot F (2004) Omi/HtrA2 promotes cell death by binding and degrading the anti-apoptotic protein ped/pea-15. J Biol Chem 279:46566–46572
177. Vande Walle L, Van Damme P, Lamkanfi M, Saelens X, Vandekerckhove J, Gevaert K, Vandenabeele P (2007) Proteome-wide identification of HtrA2/Omi substrates. J Proteome Res 6:1006–1015
178. Jones JM, Datta P, Srinivasula SM, Ji W, Gupta S, Zhang Z, Davies E, Hajnoczky G, Saunders TL, Van Keuren ML, Fernandes-Alnemri T, Meisler MH, Alnemri ES (2003) Loss of Omi mitochondrial protease activity causes the neuromuscular disorder of mnd2 mutant mice. Nature 6959:721–727
179. Krick S, Shi S, Ju W, Faul C, Tsai SY, Mundel P, Bottinger EP (2008) Mpv17l protects against mitochondrial oxidative stress and apoptosis by activation of Omi/HtrA2 protease. Proc Natl Acad Sci U S A 37:14106–14111
180. Martins LM, Morrison A, Klupsch K, Fedele V, Moisoi N, Teismann P, Abuin A, Grau E, Geppert M, Livi GP, Creasy CL, Martin A, Hargreaves I, Heales SJ, Okada H, Brandner S, Schulz JB, Mak T, Downward J (2004) Neuroprotective role of the Reaper-related serine protease HtrA2/Omi revealed by targeted deletion in mice. Mol Cell Biol 22:9848–9862
181. Kooistra J, Milojevic J, Melacini G, Ortega J (2008) A new function of human HtrA2 as an amyloid-beta oligomerization inhibitor. J Alzheimers Dis 2:281–294
182. Cilenti L, Ambivero CT, Ward N, Alnemri ES, Germain D, Zervos AS (2014) Inactivation of Omi/HtrA2 protease leads to the deregulation of mitochondrial Mulan E3 ubiquitin ligase and increased mitophagy. Mol Cell Res 1843:1295–1307
183. Zhang Y, Appleton B, Wu P, Wiesmann C, Sidhu S (2007) Structural and functional analysis of the ligand specificity of the HtrA2/Omi PDZ domain. Protein Sci 16:2454–2471
184. Appleton BA, Zhang Y, Wu P, Yin JP, Hunziker W, Skelton NJ, Sidhu SS, Wiesmann C (2006) Comparative structural analysis of the Erbin PDZ domain and the first PDZ domain of ZO-1. Insights into determinants of PDZ domain specificity. J Biol Chem 31:22312–22320

185. Kuninaka S, Nomura M, Hirota T, Iida S-I, Hara T, Honda S, Kunitoku N, Sasayama T, Arima Y, Marumoto T, Koja K, Yonehara S, Saya H (2005) The tumor suppressor WARTS activates the Omi / HtrA2-dependent pathway of cell death. Oncogene 34:5287–5298
186. Cilenti L, Soundarapandian M, Kyriazis G, Stratico V, Singh S, Gupta S, Bonventre J, Alnemri E, Zervos AS (2004) Regulation of HAX-1 anti-apoptotic protein by Omi/HtrA2 protease during cell death. J Biol Chem 279:50295–50301
187. Jin S, Kalkum M, Overholtzer M, Stoffel A, Chait BT, Levine AJ (2003) CIAP1 and the serine protease HTRA2 are involved in a novel p53-dependent apoptosis pathway in mammals. Genes Dev 17:359–367
188. Ding X, Patel M, Shen D, Herzlich AA, Cao X, Villasmil R, Klupsch K, Tuo J, Downward J, Chan C-C (2009) Enhanced HtrA2/Omi expression in oxidative injury to retinal pigment epithelial cells and murine models of neurodegeneration. Invest Ophthalmol Vis Sci 50:4957–4966
189. Johnson F, Kaplitt M (2009) Novel mitochondrial substrates of omi indicate a new regulatory role in neurodegenerative disorders. PLoS ONE 4, e7100
190. Cilenti L, Lee Y, Hess S, Srinivasula S, Park KM, Junqueira D, Davis H, Bonventre JV, Alnemri ES, Zervos AS (2003) Characterization of a novel and specific inhibitor for the pro-apoptotic protease Omi/HtrA2. J Biol Chem 13:11489–11494
191. Bhuiyan MS, Fukunaga K (2009) Mitochondrial serine protease HtrA2/Omi as a potential therapeutic target. Curr Drug Targets 10:372–383
192. Nie GY, Li Y, Minoura H, Batten L, Ooi GT, Findlay JK, Salamonsen LA (2003) A novel serine protease of the mammalian HtrA family is up-regulated in mouse uterus coinciding with placentation. Mol Hum Reprod 5:279–290
193. Nie GY, Hampton A, Li Y, Findlay JK, Salamonsen LA (2003) Identification and cloning of two isoforms of human high-temperature requirement factor A3 (HtrA3), characterization of its genomic structure and comparison of its tissue distribution with HtrA1 and HtrA2. Biochem J 1:39–48
194. Beleford D, Liu Z, Rattan R, Quagliuolo L, Boccellino M, Baldi A, Maguire J, Staub J, Molina J, Shridhar V (2009) Methylation induced gene silencing of HtrA3 in smoking-related lung cancer. Clin Cancer Res 2:389–409
195. Beleford D, Liu Z, Rattan R, Quagliuolo L, Boccellino M, Baldi A, Maguire J, Staub J, Molina J, Shridhar V (2010) Methylation induced gene silencing of HtrA3 in smoking-related lung cancer. Clin Cancer Res 2:398–409
196. Narkiewicz J, Klasa-Mazurkiewicz D, Zurawa-Janicka D, Skorko-Glonek J, Emerich J, Lipinska B (2008) Changes in mRNA and protein levels of human HtrA1, HtrA2 and HtrA3 in ovarian cancer. Clin Biochem 7–9:561–569
197. Runyon ST, Zhang Y, Appleton BA, Sazinsky SL, Wu P, Pan B, Wiesmann C, Skelton NJ, Sidhu SS (2007) Structural and functional analysis of the PDZ domains of human HtrA1 and HtrA3. Protein Sci 11:2454–2471
198. Sheng M, Sala C (2001) PDZ domains and the organization of supramolecular complexes. Annu Rev Neurosci 24:1–29
199. Krojer T, Garrido-Franco M, Huber R, Ehrmann M, Clausen T (2002) Crystal structure of DegP (HtrA) reveals a new protease-chaperone machine. Nature 416:455–459
200. Wilken C, Kitzing K, Kurzbauer R, Ehrmann M, Clausen T (2004) Crystal structure of the DegS stress sensor: how a PDZ domain recognizes misfolded protein and activates a protease. Cell 117:483–494
201. Bowden MA, Drummond AE, Fuller PJ, Salamonsen LA, Findlay JK, Nie G (2010) High-temperature requirement factor A3 (Htra3): a novel serine protease and its potential role in ovarian function and ovarian cancers. Mol Cell Endocrinol 327:13–18
202. Tocharus J, Tsuchiya A, Kajikawa M, Ueta Y, Oka C, Kawaichi M (2004) Developmentally regulated expression of mouse HtrA3 and its role as an inhibitor of TGF-beta signaling. Dev Growth Differ 3:257–274
203. Nie G, Findlay JK, Salamonsen LA (2005) Identification of novel endometrial targets for contraception. Contraception 4:272–281

204. Nie G, Li Y, Hale K, Okada H, Manuelpillai U, Wallace EM, Salamonsen LA (2006) Serine peptidase HTRA3 is closely associated with human placental development and is elevated in pregnancy serum. Biol Reprod 2:366–374
205. Bowden MA, Li Y, Liu Y-X, Findlay JK, Salamonsen LA, Nie G (2008) HTRA3 expression in non-pregnant rhesus monkey ovary and endometrium, and at the maternal-fetal interface during early pregnancy. Reprod Biol Endocrinol 6:22
206. Singh H, Makino SI, Endo Y, Nie G (2011) Inhibition of HTRA3 stimulates trophoblast invasion during human placental development. Hum Reprod 4:748–757
207. Inagaki A, Nishizawa H, Ota S, Suzuki M, Inuzuka H, Miyamura H, Sekiya T, Kurahashi H, Udagawa Y (2012) Upregulation of HtrA4 in the placentas of patients with severe pre-eclampsia. Placenta 11:919–926
208. Wang LJ, Cheong ML, Lee Y-S, Lee M-T, Chen H (2012) High-temperature requirement protein A4 (HtrA4) suppresses the fusogenic activity of syncytin-1 and promotes trophoblast invasion. Mol Cell Biol 18:3707–3717

Chapter 5
Proteases in Apoptosis: Protocols and Methods

Saujanya Acharya, Raja Reddy Kuppili, Lalith K. Chaganti, and Kakoli Bose

Abstract Proteases in apoptosis have evolved as major drug targets in the past few decades. Development in this direction has been brought about by efficient design and refinement of the various platforms of protease assays. These can be broadly categorized into general assays, that characterize kinetics and biochemistry of apoptotic proteases, and the more specific assays devoted to discern proteases involved in apoptosis. Together, these two approaches comprise a flawless two-pronged approach to understand the role of proteases in apoptosis and their therapeutic applications. This chapter lays down a comprehensive account of different experimental procedures spanning the use of *in vitro* purified proteases to those that monitor enzyme activity and its apoptotic effect in fixed or live cells. In this regard, fluorescence based platforms are the workhorse of fast, accurate, easy-to-use and high throughput screening amenable procedures. Therefore, they form the majority of techniques, among others, covered in this chapter. Apart from the popular methods currently in use, this chapter also provides a bird's eye view of the future of the protease assays with special mention of protease activatable prodrugs and protease engineering.

Keywords Apoptosis • Proteases • Assays

1 Introduction

Proteases are a class of enzymes that catalyze the cleavage of their substrates through the formation of hydrogen, electrostatic and hydrophobic bonds [1, 2]. They form an integral part of any biological system performing a range of diverse functions such as cell proliferation, differentiation, immune response, neuronal development, angiogenesis, coagulation and apoptosis [3]. They have also been implicated in the process of disease development; a few of them

S. Acharya • R.R. Kuppili • L.K. Chaganti • K. Bose (✉)
Integrated Biophysics and Structural Biology Lab, Advanced Centre for Treatment, Research and Education in Cancer (ACTREC), Tata Memorial Centre, Navi Mumbai, Maharashtra 410210, India
e-mail: kbose@actrec.gov.in

K. Bose (ed.), *Proteases in Apoptosis: Pathways, Protocols and Translational Advances*, DOI 10.1007/978-3-319-19497-4_5

include cardiovascular diseases, infectious diseases and cancer [4]. Apoptosis, a biologically important phenomenon that maintains cellular homeostasis, is deregulated in many pathological conditions such as cancer, neurodegenerative and autoimmune diseases. Several critical proapoptotic proteases such as caspases, calpains and cathepsins are the key mediators of this biological process [5–11].

Apoptosis is distinctively characterized by a series of morphological changes that include cell shrinkage, chromatin condensation, nuclear disintegration and finally the formation of membrane bound apoptotic bodies [12, 13]. DNA cleavage has been proved to be a hallmark of apoptosis through early studies, and researchers have pointed to the role of endonucleases in the event [14]. However, establishing a central role for nuclear cleavage in initiation and promotion of apoptosis has been slightly more difficult owing to the fact that characteristic apoptotic changes other than nucleosomal degradation have been observed in cytoplasts lacking nuclei [15]. Since then the focus has shifted away from endonucleases as major players of apoptosis. At the same time, proteases have steadily gained importance through a multitude of evidence that comes in the form of biochemical and physiological studies. A number of experiments describe the proteolytic cleavage of cellular proteins such as lamins, poly ADP-ribose polymerase (PARP) and histones in the early stage of apoptosis [16–20]. Several other reports demonstrate the use of protease inhibitors to prevent apoptosis [21, 22] such as some viral proteins that have been used to target pro-apoptotic proteases [23, 24]. In addition, gene knockout experiments have also brought to the forefront the importance of this class of enzymes in regulating apoptosis [25]. The wealth of information gained from such studies has highlighted proteases as invaluable therapeutic targets, and research in this direction over the past few years has triggered the search for small molecule inhibitors or mimetics.

Till date, more than a dozen protease inhibitor drugs have been cleared for clinical use which target angiotensin converting enzyme (ACE), HIV protease and the proteasome complex [4]. Captopril, an ACE inhibitor has been used to treat hypertension while HIV and HCV inhibitors have successfully been used against AIDS [26–29]. Other key examples include Argatroban, the thrombin inhibitor for treatment of coagulation disorders and DPP4 (Sitagliptin) against diabetes [30–33].

Proteases with established roles in the apoptotic pathway are also routinely exploited to develop drugs against a wide spectrum of chronic diseases. For example, a recent study suggests that inhibition of cathepsin B by proteinase inhibitor E-64 induces oxidative stress followed by mitochondria mediated apoptosis in filarial parasites leading to their death [34]. This suggests that filarial cathepsin B is a potential chemotherapeutic target for lymphatic filariasis. Bortezomib (VELCADE) a proteasome inhibitor, causes G_2–M cell cycle arrest followed by apoptosis and is currently approved for treatment of multiple myeloma and mantle cell non-hodgkin's lymphoma (NHL) [35–37]. Many such examples highlight the involvement of apoptosis related proteases in diverse disease states and their evolving status as major drug targets.

With over 500 proteases having been identified so far, the task of targeting them becomes so challenging that it is necessary to understand the biology of these proteases; primarily how they bind and process their substrates. Moreover, new proteases are regularly screened for industrial use based on the mechanism of hydrolysis. This has led to the improvement of traditional assays and a spurt in the development of new assay formats. In this chapter, we will provide a comprehensive account of different assay formats for investigating general protease activity as well as identifying putative proteases involved in apoptosis. These include cell based studies, conventional biophysical assays as well as the more recently developed and extremely sensitive high throughput fluorescent assays.

2 General Protease Assays

The basic properties of a protease are investigated by mainly studying its enzyme activity and response to inhibitors. These studies offer us a wealth of information required to classify and define a protease. Some of these parameters include mechanism of catalysis, specificity for a particular site, change in activity owing to varying assay conditions and the extent of exo/endopeptidase activity [38]. Keeping this in mind, it becomes imperative to understand the biochemistry and the kinetics of enzyme action. In this context, inhibition studies provide an insight into how the protease interacts with its substrates, whether its action can be regulated by the inhibitor and if so how they can be used to target the enzyme. Besides, they also serve as an effective means of classifying a protease, since proteases with similar catalytic mechanism can be targeted by a common group of inhibitors [38]. A new protease thus can be easily grouped into any of the known seven classes of proteases based on the enzyme inhibition assay.

2.1 Use of Enzyme Inhibitors to Identify Proteases

The use of exogenous small molecule protease inhibitors is aimed at studying the reduction in catalytic activity of its target protease. The inhibitor brings about this change by binding to the regulatory unit or by modifying the active site amino acid(s) [39]. The inhibition assay explores the kinetic behavior of the protease with and without the inhibitor. Traditionally, *in vitro* experiments utilize purified proteases or extracts from individual organisms, tissues and cells so as to deduce the kinetic properties and constants explained in Sect. 2.2. Since these kinetic concepts are integral toward understanding enzyme action and its inhibition, they have been elaborated in the subsequent sections.

2.2 Understanding Enzyme Kinetics

A simple enzyme (E) catalyzed reaction in which a substrate (S) is transformed into a product (P) through an intermediate enzyme – substrate complex (ES), proceeds as follows [40]:

$$E + S \overset{k_1}{\leftrightarrow} ES \xrightarrow{k_2 \text{ or } k_{cat}} E + P \tag{5.1}$$

Where, k_1 is the rate constant of the forward reaction. k_2 or k_{cat} is the rate constant of the forward reaction yielding P and releasing E for the next round of catalysis. To study enzyme kinetics, it is imperative to measure the initial rate (v_0) which is defined as the time (t) dependent formation of the product.

$$v_o = \left(\frac{\mathrm{d}\,[P]}{\mathrm{d}t}\right)_0 = k_2\,[\mathrm{ES}] \tag{5.2}$$

This equation contains a measurable variable v_0, a kinetic parameter k_2 or k_{cat}, and another unknown variable [ES]. Assuming steady state kinetics, the mathematical description that can be used to depict the kinetic behavior of enzymes in terms of substrate concentration is the Michaelis-Menten equation as shown in Eq. (5.3) (Fig. 5.1a).

$$v_0 = v_{max}\,[\mathrm{S}] \,/\, (K_M + [\mathrm{S}]) \tag{5.3}$$

Where, v_{max} is the maximal reaction rate, [S] is the substrate concentration, K_M is the Michaelis-Menten constant i.e. equal to the substrate concentration at which

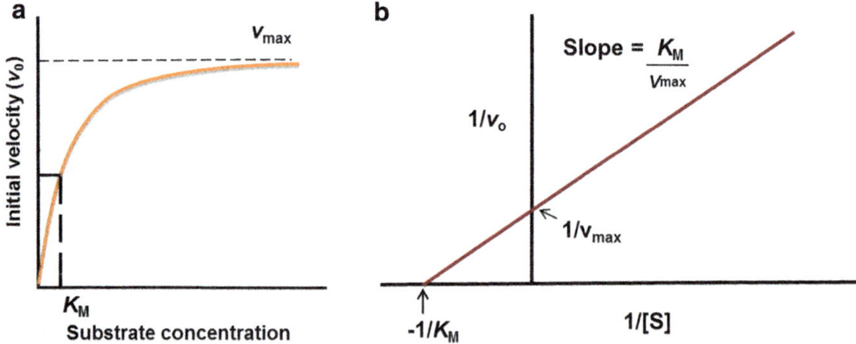

Fig. 5.1 (**a**) The Michaelis-Menten plot: A plot of the reaction velocity (v_0) as a function of the substrate concentration [S] for an enzyme that obeys Michaelis-Menten kinetics shows that the maximal velocity (v_{max}) is approached asymptotically. The Michaelis-Menten constant (K_M) is the substrate concentration yielding a velocity of $v_{max}/2$. (**b**) The Lineweaver-Burk plot: A plot of $1/v_0$ versus 1/[S] gives a straight line from which $1/v_{max}$ and K_M/v_{max} can be calculated from the y-intercept and the slope respectively

the reaction rate is half its maximum value. K_M indicates the binding strength of an enzyme to its substrate with a lower value corresponding to better efficiency. k_{cat}/K_M gives a measure of substrate specificity and is defined as the catalytic efficiency [40, 41].

2.2.1 Determining the Different Kinetic Parameters

In practice, the plot of v_0 versus [S] is not very useful in determining the value of v_{max} because finding v_{max} at very high substrate concentrations is often difficult. This can be circumvented by any of the three common linearization methods to obtain estimates for K_M and v_{max} such as Lineweaver – Burke, Eadie – Hofstee and Hanes plots. The Lineweaver Burk method employs the double-reciprocal plot of $1/v_0$ versus 1/[S] to give a straight line. In contrast to nonlinear plots, changes in enzyme kinetics, for example, due to the action of an inhibitor, are readily apparent in linear plots. The Michaelis-Menten equation can now be rewritten as a **Lineweaver – Burk** as shown below in Eq. (5.4) [40–42].

$$\frac{1}{v_0} = \frac{K_M}{v_{max}[S]} + \frac{1}{v_{max}} \tag{5.4}$$

As shown in Fig. 5.1b, both K_M and v_{max} can be obtained from the slope and intercepts of the straight line. Although useful and widely employed in enzyme kinetics and inhibition studies, the Lineweaver – Burk plot has the disadvantage of often being less accurate.

2.2.2 Quantitating Enzyme Activity

Most of the kinetic experiments designed use spectrophotometry to calculate kinetic parameters from absorbance. The initial rate of reaction (v_0) corresponds to the slope calculated from the linear portion of the curve. This can be expressed as a change in absorbance per unit time for product (colour) formation and is defined as ΔA_λ/min as shown in Eq. (5.5) [43].

$$v_0\left(\text{min}^{-1}\right) = \frac{\text{final absorbance} - \text{initial absorbance}}{\text{final time} - \text{initial time}} \tag{5.5}$$

However, it is more useful to express the rate in terms of the actual amount of product formed per unit time. To begin with, the absorbance value can be converted to concentration using the **Beer – Lambert law** (Eq. 5.6).

$$c = \frac{A}{\varepsilon.l} \tag{5.6}$$

Where, c (μM/min) is the concentration of absorbing material, A is the absorbance measured at suitable wavelength, l is the path length, and ε is the extinction

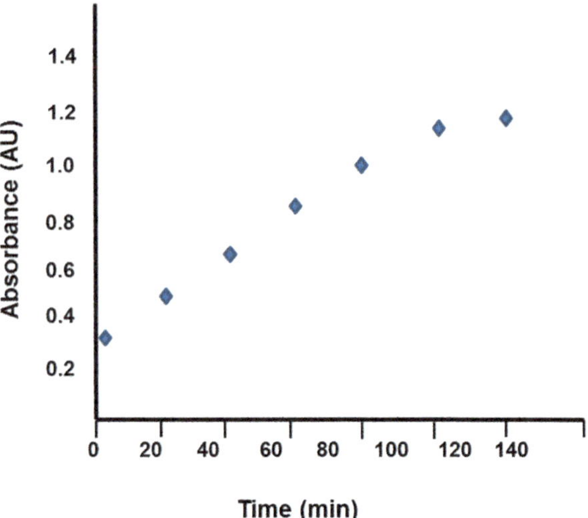

Fig. 5.2 The plot of absorbance versus time: The plot uses the time course absorbance measurement (abs/min) for each substrate concentration to determine the initial velocity

coefficient. Concentration (c) is multiplied with the volume of reaction in order to convert this concentration to the total amount of product formed [43]. The result is finally in terms of nmol/min or μmol/min as seen in Eq. (5.7).

$$\mathrm{c}\left(\frac{\mathrm{nM}}{\mathrm{min}}\right) \times \text{total volume of reaction (ml)}$$
$$= \text{total amount of coloured product formed (nmol/ min)} \qquad (5.7)$$

For each substrate concentration, the plot of absorbance versus time is drawn to determine the initial velocity (Fig. 5.2). Subsequently, a simple hyperbolic plot of v_0 (in nmol/min) as a function of [S] is drawn. A double-reciprocal graph of $1/v_0$ vs. 1/[S] (Lineweaver–Burk or any other method described earlier) is then used for a linear plot. The K_M and v_{max} values for the reaction are determined as described previously.

2.2.3 Determining Enzyme Inhibition Constants

The inhibitor constant or K_I values are determined by performing a series of experiments with varying amounts of inhibitor. For each concentration of inhibitor [I] used, a K_M denoted by $K_M{}^{+I}$ is derived which yields a line with a slope of K_M/K_I and a *y*-intercept of K_M (Fig. 5.3a) when plotted. The type of inhibition can be easily determined by comparing the K_M and v_{max} of a catalytic reaction in presence and absence of the inhibitor. The K_I values for competitive, noncompetitive and uncompetitive inhibition can be determined by using Eqs. (5.8), (5.9) and (5.10) respectively [43]. The three types of inhibition have been described briefly in Sect. 2.3.

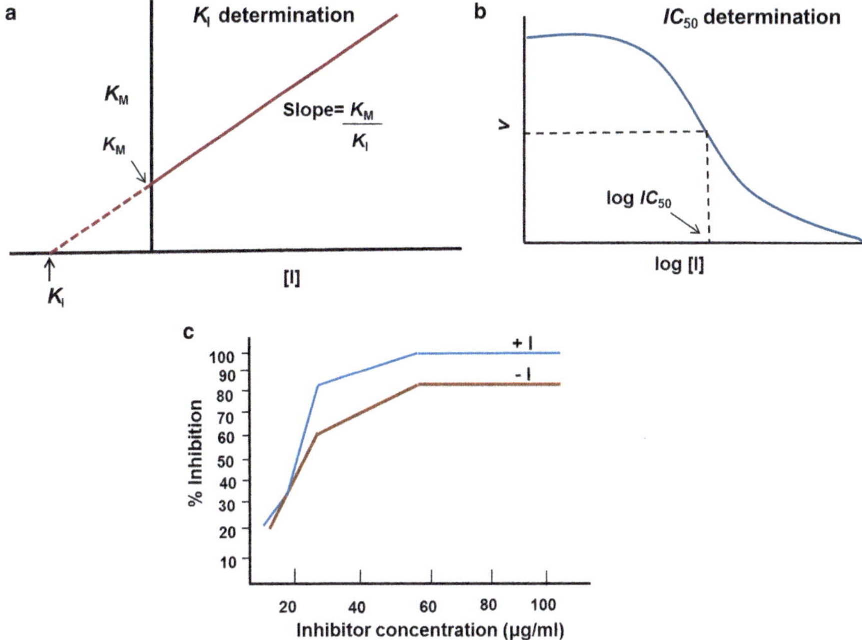

Fig. 5.3 Determining enzyme inhibition constants: (**a**) The inhibitor constant or K_I is calculated from the slope (K_M/K_I) of a plot of K_M versus the inhibitor concentration. (**b**) The concentration response curve obtained from the plot of rate of reaction against inhibitor concentration (log [I]) allows the determination inhibitor IC_{50} for an enzyme inhibitor. (**c**) The efficiency of an inhibitor is graphically represented by showing the percentage inhibition as a function of inhibitor concentration

$$K_M{}^{+I} = K_M{}^{-I}\left(1 + \frac{I}{K_I}\right) \tag{5.8}$$

$$v_{max}{}^{+I} = v_{max}{}^{-I} / \left(1 + \frac{I}{K_I}\right) \tag{5.9}$$

$$K_{M+1} = K_M / \left(1 + \frac{I}{K_I}\right) \tag{5.10}$$

The IC_{50} value i.e. the concentration of inhibitor required to decrease the rate of reaction by half is determined through a series of experiments performed with saturating substrate and varying inhibitor concentrations. The amount of inhibitor (log [I]) is plotted against the rate of the reaction (v) obtained from each individual experiment to generate a sigmoidal curve. The point of inflection of this curve corresponds to the logarithm of the inhibitor concentration i.e. IC_{50} (Fig. 5.3b), and the efficiency is obtained by calculating the percentage inhibition using Eq. (5.11). Percentage inhibition versus substrate concentration can then be represented as a graph as shown in Fig. 5.3c [44].

$$\% \ inhibition = \frac{rate \ without \ inhibitor - rate \ with \ inhibitor}{rate \ without \ inhibitor} \times 100 \quad (5.11)$$

An additional measure of the robustness of the enzyme inhibition assay is given by the Z′ factor which should ideally be above 0.7.

$$Z' = 1 - \frac{3 \times (SD_+ + SD_-)}{|\mu_+ - \mu_-|} \quad (5.12)$$

where, μ_+ and μ_- represent the mean of the positive and negative control signals, while SD_+ and SD_- are standard deviations of the mean values for the positive and negative controls.

2.3 Enzyme Inhibition Assay

Background

Activity assays require measuring the kinetic parameters of uninhibited reactions as described in the previous section [45]. Moreover, the evaluation of enzyme activity forms an inseparable part of an inhibition assay. Generally, an experimental set up that studies the effect of an inhibitor on the catalytic activity of a protease is termed as an enzyme inhibition assay [38, 46, 47]. Therefore, the guidelines for this experimental set up will also be covered in this section.

An inhibitor, in principle, is any low molecular weight chemical compound that decreases the rate of a catalytic reaction when added in presence of a competing substrate. Inhibitors may act in a reversible or an irreversible manner [39]. While irreversible inhibition results in permanently disabling the enzyme, reversible inhibition results in a temporary reduction in enzyme activity.

There are three basic mechanisms of reversible inhibition – competitive, noncompetitive, and uncompetitive [39, 48]. These inhibitors usually bind to the catalytic site, regulatory site and to the enzyme substrate complex respectively. A competitive inhibitor binds at the catalytic site thus decreasing the ability of the enzyme to bind with its substrate thereby raising the apparent K_M and K_I value with no change in the v_{max}. A noncompetitive inhibitor on the other hand, binds to a site apart from the active site and acts by reducing the turnover rate (k_{cat}) of the reaction. It thus lowers the apparent v_{max} value, without any effect on the K_M and K_I values. Lastly, an uncompetitive inhibitor with no structural similarity to the substrate binds to the enzyme-substrate complex instead of the free enzyme (Fig. 5.4). This interaction brings about structural distortion of the active and allosteric sites of the enzyme-substrate complex which prevents catalysis.

Before setting up an inhibition assay, preparatory assays have to be carried out to predetermine the optimal temperature, pH, buffer component concentrations and other experimental conditions at which the protease is active. For an enzymatic

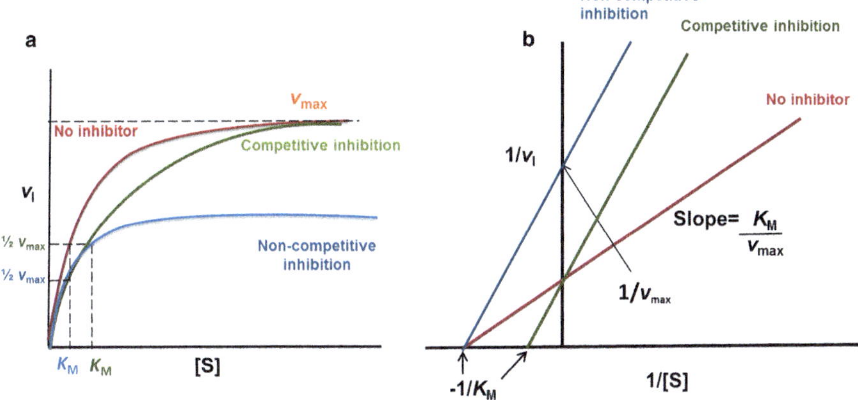

Fig. 5.4 Three types of enzyme inhibition: The two figures represent the kinetic parameters of an enzymatic reaction with and without inhibitors in the form of a Michaelis-Menten and a Lineweaver-Burk plot respectively. The type of inhibition can be predicted by the change in K_M or v_{max} values as is evident on the x and y axes respectively. While, a competitive inhibitor increases the K_M, a noncompetitive inhibitor decreases the v_{max} value keeping K_M unchanged

assay involving competitive inhibitors, it is essential to run the reaction under initial velocity conditions with substrate concentrations at or below the K_M value [38, 49]. The IC_{50} and/or K_I values are determined based on assay design. This section describes the development and validation of assays for proteases with competitive or reversible inhibitors.

General Methodology

The flowchart for the experiment is as shown in Fig. 5.5.

Materials

1. Enzyme: Protease being studied.
2. Inhibitor: Dissolved in a suitable solvent.
3. Substrate: The substrate chosen for an experiment is usually a dye conjugated chemical compound which on being processed by the protease releases the dye that absorbs at a particular wavelength. For example azocasein is casein conjugated to a dye, which on hydrolysis releases a yellow compound with an absorbance at 440 nm. This spectrophotometric data of its absorbance is used to calculate the enzyme activity.
4. Other reagents:

   (i) Assay buffer (with pH appropriately maintained), pH upto 5.0 is suitable for azocasein.
   (ii) Trichloroacetic acid (TCA): stopping reagent.

5. Microfuge tubes, tips.

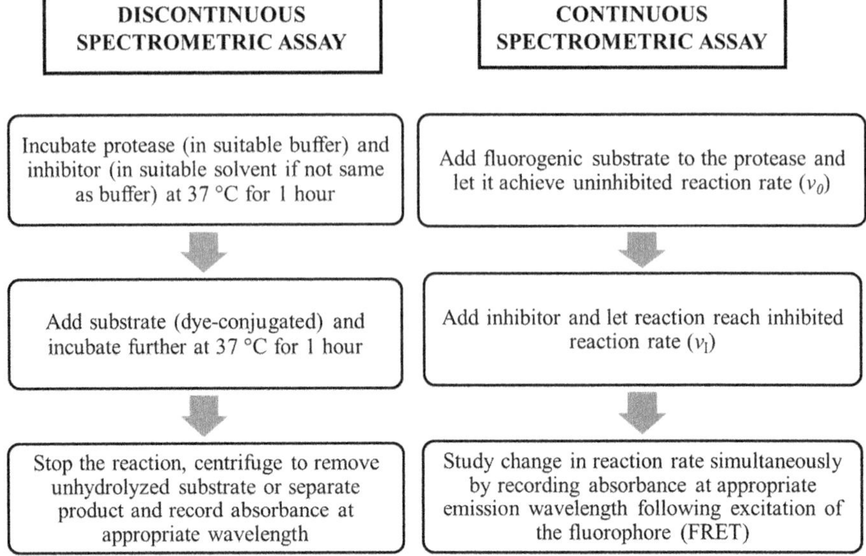

Fig. 5.5 General methodology for discontinuous and continuous spectrophotometric assay

Instrumentation

1. Centrifuge.
2. Spectrophotometer (discontinuous assays) or plate reader (for continuous assays).

Method

1. Description of the reaction components:

   Protease control: To ensure that the protease extract does not absorb at the same wavelength as the released dye (product of hydrolysis).

   Inhibitor solvent control (if the solvent is not as same as the assay buffer): This ensures that the solvent has no effect on the enzyme activity. If the inhibitor is dissolved in the same buffer as the reaction, the control for the inhibitor vehicle can be eliminated.

   Blank: This contains none of the other components except the substrate in suitable buffer. It is used to normalize absorbance data from the test reactions to calculate enzyme activity.

2. Perform all the reactions simultaneously in the order mentioned. The total reaction volume in each tube is 1.5 ml. It is preferable to run reactions in triplicates.

Components	Inhibitory activity	Protease activity	Protease control	Inhibitor solvent control	Blank
Buffer	470 μl	480 μl	480 μl	470 μl	1000 μl
Protease	20 μl	20 μl	20 μl	20 μl	–
Inhibitor	10 μl	–	–	–	–
Inhibitor solvent	–	–	–	10 μl	–
Incubated at appropriate temperature for a suitable time period (e.g. 37 °C for 1 h)					
Reaction stopping agent (TCA)	–	–	0.5 ml	–	–
Substrate	0.5 ml	0.5 ml	0.5 ml	0.5 ml	0.5 ml
	Incubate				
Reaction stopping agent	0.5 ml	0.5 ml	–	0.5 ml	–
Centrifuge at 6500 g, 5 min					
Record absorbance					

The set up described above is typical for a discontinuous spectrometric assay.

Data Analysis

Calculations for v_0, v_{max}, K_M, k_{cat}, IC_{50}, percentage inhibition and other kinetic parameters are described in Sect. 2.2.

Precautions

1. Time and temperature need to be standardized for each protease as mentioned before.
2. For initial velocity calculations, keep experimental conditions constant. Signal strength should clearly be able to indicate at least 10 % product formation or 10 % product loss. Background correction should be done by measuring signal strength at zero time point.
3. For determining K_M and v_{max} values, use different substrate concentrations to generate a saturation curve. Use initial velocity conditions.
4. For measuring IC_{50} and percentage of inhibition, use substrate concentration below or equal to the K_M in a competitive inhibition. Maintain steady state conditions. Keep enzyme and substrate conditions constant and use well-spaced inhibitor concentrations to obtain sufficient data points (half below and half above the IC_{50} value) to generate accurate models for data fitting. Since, IC_{50} value is greatly affected by experimental conditions this should be kept in mind while designing the assay. The minimum percentage of inhibition should be over 50 % for reporting it as cutoff IC_{50}.
5. Determining suitable storage conditions of the enzyme to ensure maximum activity and stability.
6. Perform preliminary optimization assays with varying amounts of the reagents such as salts, EDTA (Ethylenediaminetetraacetic acid), reducing agents, detergents and BSA (bovine serum albumin) to set up a robust assay.
7. Buffer source needs to be carefully chosen after trying out different conditions.

Protease Characterization Using Enzyme Activity and Inhibition: An Example For characterization of one such novel protease, four fractions from Langostilla *(Pleuroncodes planipes)* extracts were probed for protease activity [38]. This was done using a discontinuous assay with azocasein as the substrate. To assess the class of this novel protease, general inhibitors pre-defined for each broad group of proteases were used for the inhibition studies. Of all the inhibitors used, treatment with PMSF (Phenylmethylsulfonyl fluoride) showed 32 % inhibition indicating the presence of serine proteases in the extract. Trypsin, a known serine protease used as a control for comparative study showed 96 % inhibition. Other fractions showed inhibition with EDTA, EGTA and 1, 10 phenanthroline which clearly indicated the presence of metalloproteases in those fractions of Langostilla extract [38].

2.4 Fluorescent Methods to Detect Enzyme Activity

A normal proteolytic reaction involves the cleavage of the scissile amide bond of the substrate. Amino acids on the N and carboxy terminal of the scissile bond are called the P and P' sites respectively, as shown in Fig. 5.6 [40, 50]. The most widely used principles to monitor an enzyme-catalyzed reaction are UV-visible spectrophotometry and fluorimetry [49]. In the current scenario, fluorescence based

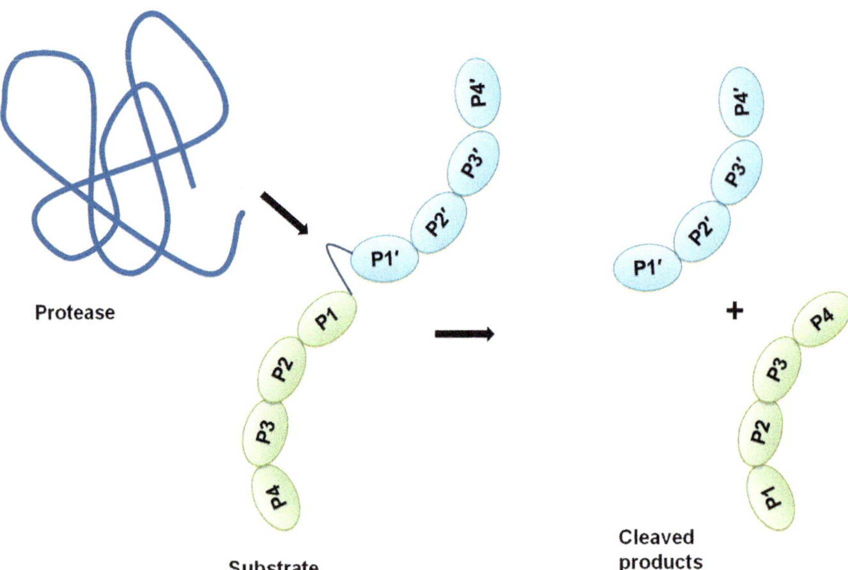

Fig. 5.6 A schematic representation of a typical proteolytic cleavage: Amino acids on the N and carboxy terminal of the scissile bond are called the P and P′ sites respectively

approaches are widely used in assays for proteases because of their sensitivity and suitability for high throughput screening [51, 52]. Fluorescence is the property by which chromophores can absorb light at a certain wavelength and emit at a longer one and is utilized in biophysical techniques such as FRET (Förster resonance energy transfer) and FP (Fluorescence polarization) [53, 54]. The assays discussed under this section generally focus on the interaction of a protease with its target substrate and measures change in enzymatic activity using fluorescent probes.

These assays are designed to quantitate enzymatic activity in cell extracts or purified proteins [55]. They employ substrates or a reaction product that are highly fluorescent. The fluorescence can then be studied using a fluorescence microscope, fluorimeter, a 96-well plate reader, a flow cytometer or UV illuminator. The fluorescence spectra of substrate and product ideally should not overlap. However, a separation step is included in case it happens. Fortunately, many substrates have low intrinsic fluorescence or are metabolized to products that have longer- excitation or emission wavelength [53, 54].

A wide variety of fluorescent moieties can be utilized, such as coumarin derivatives (hydroxyl or amino substituted): 7-amino-4-methylcoumarin (AMC), 7-amino-4-(trifluoromethyl) coumarin (AFC), 3-cyano-7-hydroxycoumarin, 7-hydroxy-4-trifluoromethylcoumarin and 6, 8-difluoro-7-hydroxy-4-methylcoumarin. These are highly soluble and give brilliant blue fluorescence [56, 57]. Aromatic amines like AMC are fully deprotonated at physiological pH which reduces variability in the fluorescence spectra due to pH-dependent protonation/deprotonation. Other highly sensitive fluorescent substrates are those derived from fluoresceins, rhodamines and resorufins [58–60]. Longer wavelength fluorophores are preferred because they reduce background noise and autofluorescence to facilitate detection.

A highly efficient detection system for discontinuous assays can also be developed by derivatizing reaction products with suitable reagents followed by a separation step in order to generate a product-specific fluorescent signal. For example, fluorescamine or *o*-phthaldialdehyde reacts with primary amines in the presence of 2-mercaptoethanol to form fluorescent products [61]. Thus, it can detect the rate of a peptidase reaction by measuring the increase in the concentration of free amines in solution. Enzyme activity can also be studied utilizing substrates that yield insoluble chromogenic products at the site of reaction. Halogenated indolyl derivatives are the most popular of these substrates [62]. Initially colourless or mildly fluorescent blue in the reaction mixture, these compounds are soon oxidized by nitro blue tetrazolium (NBT) or potassium ferricyanide to an indigo precipitate.

There are a few basic approaches to design fluorescent based assays. These include the presence of a single fluorophore (using fluorescently quenched peptide or substrate) or a pair of fluorophores (FRET) in the experimental setup and can be grouped as follows:

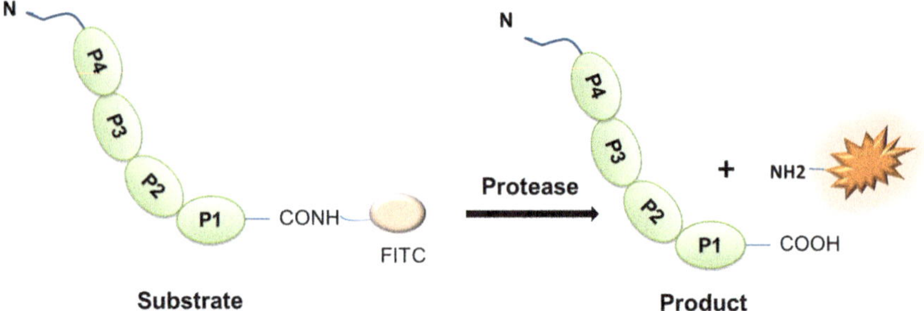

Fig. 5.7 A representative protease assay using fluorescently quenched peptide or substrate: The quenched dye conjugated to the substrate is released on cleavage by the protease

2.4.1 Protease Assays Using Fluorescently Quenched Peptide or Substrate

In this protease assay, change in fluorescent intensity of a chemically quenched dye is monitored on being released by substrate cleavage. The substrates used in this format are prepared by covalently conjugating a fluorescent dye. As a result, their absorption maxima are usually shortened significantly. Protease activity cleaves the substrate to release the free dye which restores its fluorescence [51, 52, 63]. The increase in fluorogenic signal is directly proportional to enzyme activity and can be quantitated (Fig. 5.7). Fluoresceinated casein or FITC (fluorescein isothiocyanate) β-casein is one of the most prominent dye conjugate used to detect and quantitate enzyme activity based on this principle [64]. Hydrolysis relieves this quenching conjugate, yielding intense green fluorescent peptides measured at λex/λem (excitation/emission) = 485/530 nm. Such an approach allows for fast and accurate detection of signal. However, the major disadvantage of this approach is that it does not work for proteases that do not allow room for modifications in substrates on the P1′ site, for example carboxypeptidases [65]. The basic protocol for setting up an *in vitro* enzymatic assay using FITC-casein as a fluorogenic substrate has been summarized as follows [66, 67]:

Materials

1. Fluorogenic substrate: FITC-conjugated casein.
2. Assay buffer: Buffer and its working concentration needs to be optimized for the protease being investigated (for example, 50–100 mM Tris buffer at pH 7.4 for trypsin/chymotrypsin).
3. Milli-Q-purified water, PBS buffer, FITC standard (25 μM).
4. Protease enzyme: concentration as required.
5. Microwell plates, tips.

Instrumentation

1. Fluorometer/Fluorescent microplate reader (excitation and emission wavelengths will depend on the property of the fluorogenic substrate).
2. Orbital shaker.
3. Centrifuge.

Method

Set up appropriate positive and negative controls for the experiment.

1. Dilute FITC standard in assay buffer to obtain a series of different concentrations for example 0, 0.05, 0.1, 0.15, 0.2, 0.25 nmol in each well. Make up the volume to 100 μl with assay buffer. This will be used for the preparation of FITC standard curve.
2. Dilute tissue or cell extracts in suitable amount of assay buffer. Spin down to get a clear fraction and use 50 μl of this per well for a 96-well plate. Assay can also be set up in microfuge tubes. Purified proteases (100 nM final concentrations) can be directly diluted in the assay buffer.
3. Reconstitute FITC-conjugated casein in PBS to make a stock solution of 1 mg/ml (38 μM).
4. Dilute substrate to a desired working concentration using assay buffer. The substrate is activated at 37 °C for 15 min just before use. Add suitable amount of this solution to the control, blank and test wells such that the final substrate concentration in each well is 5 μM. Adjust the final volume is each well to 100 μl.
5. For kinetic readings, record fluorescence at a gap of 5 min over a span of 30 min.
6. For an endpoint reading mix gently and incubate between 30 min and 2 h at room temperature or at the desired temperature and measure the fluorescence intensity.

Precautions

1. Protect fluorogenic reagents from light and store them at –20 °C for long term use.
2. Avoid repeated freeze/thaw of fluorogenic substrate to reduce background.
3. Mix reagents gently and avoid incubation for more than 24 h, to prevent high fluorescence background.

Delineating Mechanism of a Protease Using Fluorescently Quenched Peptide or Substrate: An Example

To understand the role of different domains and critical residues in regulating enzyme activity of HtrA2/Omi mammalian serine protease, the protease activity of Omi and its domain variants was quantitated using substrate FITC β-casein (Sigma) [68]. For each 200 μl of reaction mixture, 0.5 μM protein was incubated with 0.6 μM of FITC-casein substrate in buffer SP. Protease activity was determined over a temperature range of 30–65 °C with 5 °C intervals. Enzyme was pre-incubated at each respective temperature for 15 min, and proteolytic cleavage was monitored using a Fluorolog-3 spectrofluorometer (Horiba Scientific, Edison, NJ, USA) with excitation at 485 nm followed by 535 nm emission. Initial velocities were calculated at each respective temperature using linear regression analysis as mentioned in Sect. 2.2. The kinetic parameters were then determined and compared for HtrA2 and its domain variants, trimeric N-SPD (N-terminal along with the serine protease domain) and the monomeric SPD-PDZ (serine protease domain along with the PDZ domain) was performed. A comparative study showed a three-fold decrease in catalytic efficiency for N-SPD compared to *wt*, whereas, the SPD-PDZ showed

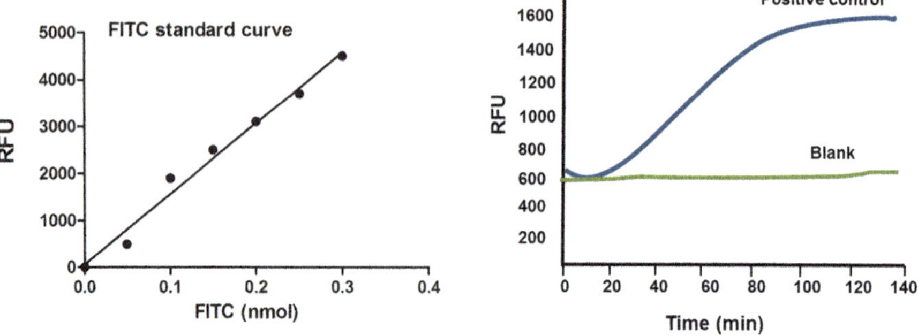

Fig. 5.8 The fluorescence standard curve: The plot of the fluorescence versus dye concentration is drawn and its slope is used to normalize excess background fluorescence

no activity at all [68]. This approach clearly pinpointed the role of trimerization and the PDZ domain toward formation of a catalytically competent HtrA2 molecule.

Data Analysis

Deriving kinetic parameters from experimental fluorescence data

The fluorescence standard curve: The plot of the fluorescence versus dye concentration is drawn as shown in Fig. 5.8 [53, 66, 67]. The standard reading for zero is subtracted from all the readings and then the FITC standard curve is plotted. Normalization of excess background fluorescence is done from the slope of the plot. The fluorescence intensity F1 at time T1 and F2 at T2 at the start and end of incubation is read. The Relative Fluorescence Unit (RFU) is the measure of the fluorescence after proteolytic digestion and is derived by:

$$\Delta RFU = F2 - FI \tag{5.13}$$

It is essential to note that F1 and F2 are chosen such that they fall within the linear reaction range. The ΔRFU value is plotted against time and the slope of this plot yields the initial velocity (v_0). However, the reaction rate needs to be normalized by dividing it with the slope from the standard reaction curve. To derive the kinetic parameters, the normalized reaction rate is plotted against the respective substrate concentrations [43, 53, 66, 67].

2.4.2 FRET (Förster Resonance Energy Transfer) Assays

Background

The principle of Förster energy transfer can be used to model simple, sensitive, continuous fluorescent assays for a variety of proteases [69]. It is defined as the non-radiative transfer of energy from a fluorescent "donor" molecule to a non-identical "acceptor" molecule (not necessarily fluorescent) through long-range dipole-dipole interactions [70, 71]. This is due to energy transfer between the electronic excited states of two molecules having similar resonance frequency. It is one of the few tools available for measuring dynamic information of macromolecular changes in nanometer scale (1–10 nm) with high sensitivity, both *in vitro* and *in vivo* [72, 73]. The energy transfer is typically accompanied by quenching or decrease in donor fluorescence and the concomitant increase in acceptor fluorescence [74]. The fluorescent moieties are chosen such that the emission wavelength of the donor overlaps with the excitation wavelength of the receptor and the degree of overlap is referred to as spectral overlap integral [75]. The rate of this energy transfer is inversely proportional to the sixth power of the distance between the donor and acceptor [76]. This can be summarized by the equation as follows:

$$E = \frac{{R_0}^6}{{R_0}^6 + r^6} \tag{5.14}$$

where, E is the efficiency, R_0 is the Förster distance between the two molecules where FRET has 50 % efficiency and r is the distance between the centers of the two fluorophores. A more simplified and experimentally relevant expression is:

$$E = 1 - \frac{F_{DA}}{F_D} \tag{5.15}$$

with F_{DA} and F_D being the fluorescence intensity of the donor in the presence and absence of the acceptor respectively. The equation suggests that maximum efficiency is attained when the acceptor is close to the donor. The Förster radius can be defined in terms of the fluorescence quantum yield of the donor in the absence of acceptor (f_d), the refractive index of the solution (η), the dipole angular orientation of the fluorophores (K^2) and the spectral overlap integral of the donor-acceptor pair (J) [70, 71]. The quantum yield is the ratio of the number of photons emitted to the number absorbed, which is represented by:

$$R_0 = 9.78 \times 10^3 \left(\eta^{-4}.f_d.J\right)^{1/6} A^0 \tag{5.16}$$

A FRET-based approach to study protease activity essentially has two non-identical fluorescent groups placed on either side of the scissile bond. A number of donor-acceptor pairs are commonly employed for FRET, some of which have been listed below (Table 5.1) [77].

Despite some of its drawbacks such as requirement of multiple labels, sensitiveness to inner-filter effects, error due to auto-fluorescence and its limitation to operate over short distances, it is a popular tool to study structure and dynamics of proteins as well as enzyme kinetics [78, 79].

FRET in Protease Assays

In recent years, FRET-based assays have found broad applications in detecting proteases [80, 81]. The chromophores used in these assays are introduced as extrinsic dye probes attached covalently to particular amino acid residues and placed

Table 5.1 Donor-acceptor pairs commonly used for FRET

Donor	Excitation wavelength (nm)	Acceptor	Emission wavelength (nm)
FITC	520	TRITC	550
Cy3	566	Cy5	649
EGFP	508	Cy3	554
CFP	477	YFP	514
EGFP	508	YFP	514

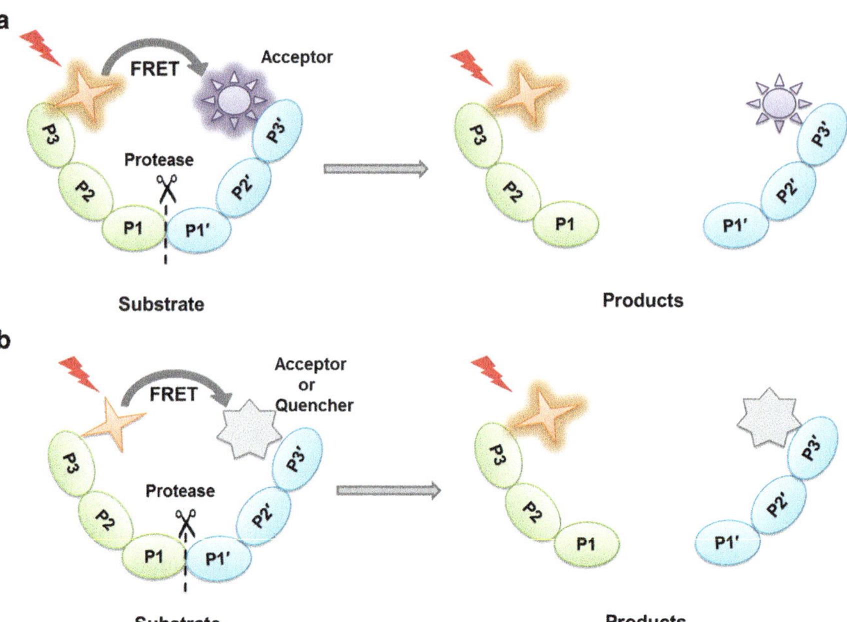

Fig. 5.9 (**a**) Schematic representation of a signal decrease FRET: In a signal-decrease FRET, the proteolytic cleavage leads to the scrambling of the fluorogenic pair, which is detected as a decrease in signal intensity of the acceptor. (**b**) Schematic representation of a signal increase FRET: In the signal increase FRET upon cleavage by a protease, the donor-quencher pair separates which leads to an increase in the fluorescence intensity of the donor

close enough for efficient energy transfer. In a signal-decrease FRET, the donor-receptor pair exhibit maximum overlap in their emission-excitation wavelength. The cleavage of the peptide bond leads to the scrambling of the fluorogenic pair, which is detected as a decrease in signal intensity of the acceptor (Fig. 5.9a) [82, 83]. The limitations of signal decrease FRET such as low signal-to-noise and requirement of a large substrate turnover can be circumvented by adopting a signal increase FRET format with the use of dual-label quenched pairs [84].

In the signal increase FRET, acceptor is a quencher which masks the donor emission by its proximity to it [85]. Upon cleavage by a protease, the pair separates

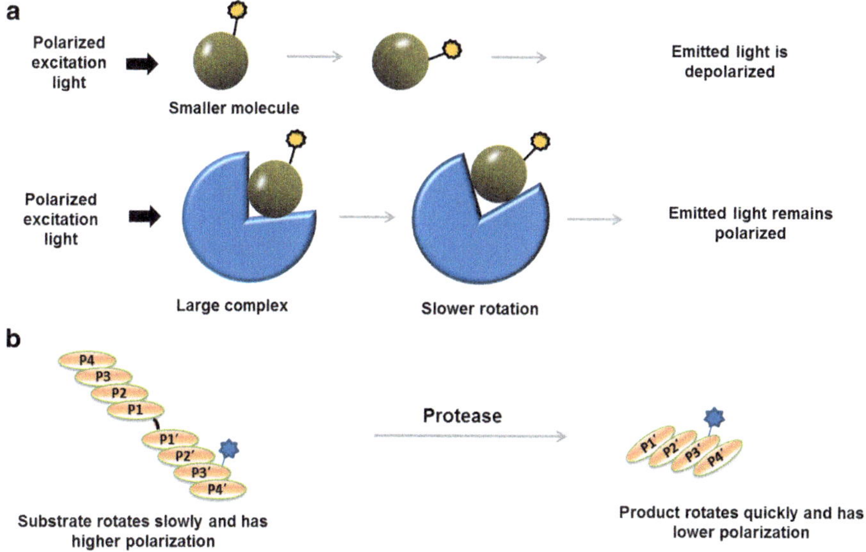

Fig. 5.10 The fluorescence polarization assay: (**a**) A fluorophore covalently attached to a small molecule, on being excited by polarized light will emit largely depolarized light due to the random molecular rotations. In contrast, a larger molecule with lesser rotational speed will allow maximum polarization of emitted light. (**b**) The protease substrate in an FP assay is labeled with one or more fluorescent dyes. Upon cleavage, the smaller substrate fragments lower the polarization signal

which leads to an increase in the fluorescence intensity of the donor (Fig. 5.9b). Fluorescein-Rhodamine (fluorescent reporter-quencher pair) and their derivatives have been commonly used over the years for these assays. The donor assumes the role of the "reporter" for enzymatic activity whereas the quencher can be a fluorescent or a non- fluorescent dye.

Non-fluorescent or "dark quenchers" as they are often popularly known include examples such as Dabcyl (4, (4′- dimethylaminophenylazo) benzyl), Qxl quenchers and IRDye QC-1. These quenchers do not have intrinsic fluorescence and therefore generate very low background noise proving to be an advantage over fluorescent quenchers [86]. The more recent "black hole quenchers" are built on a polyaromatic-azo backbone and have been extensively used to design internally quenched substrates for deubiquitinases. The signal detected is either the reporter fluorescence alone or the ratio of the reporter to quencher fluorescence. The basic protocol for determining the enzyme activity by FRET using coumarin-DABCYL [87, 88] has been summarized as follows.

Materials

1. Substrate: FRET peptide with coumarin (7-hydroxy-4-methyl-3-acetylcoumarinyl; λex 386 nm, λem 448 nm) and DABCYL (emission maxima 454 nm) used at 10 μM.

2. Reagents: Assay buffer (example: 20 mM HEPES, pH 7.0 and 1 mM $CaCl_2$, 0.1 mg/ml BSA and 0.01 % Tween), stopping reagent (example: 1 mM ortho-phenanthroline/10 mM EDTA), and Milli-Q- water.
3. Protease (100 nM).
4. Black 96-well micro plates with flat bottom, pipetors, and tips.
   Working concentrations can be standardized as required.

Instrumentation

1. Plate reader for FRET (excitation-emission wavelength chosen as per substrate).

Method

Set up appropriate positive, negative and substrate controls for the experiment.

1. Prepare working concentrations of substrates, enzymes and assay buffers from stock solutions.
2. Place 25 μl of 10 μM substrate solution in each well.
3. Add 75 μl of protease solution (100 nM) to initiate the reaction.
4. Mix the reagents gently for 30–60 s by shaking the plate.
5. Read the fluorescence emission continuously till the reaction reaches completion.
6. For an endpoint reading, mix gently and incubate 30 min to 2 h at room temperature or desired temperature. Stop the reaction by adding stopping solution. Measure the fluorescence intensity.

Precautions

1. Protect fluorogenic reagents from light.
2. Avoid repeated freeze/thaw of fluorogenic substrate to reduce background.
3. Store reagents at −20 °C.

Data Analysis

The FRET assay is adapted to determine kinetic parameters like K_M, v_{max}, and K_I as described in Sect. 2.2.

Peptide Based FRET Assay to Analyze a Protease – An Example

Bacillus anthracis toxin was proposed to be an attractive pharmacological target, which could be used along with the traditional antibacterial antibiotics to clear the infection. Lethal factor (LF), a zinc-dependent metalloprotease, is a part of the toxin which was found to cleave MAP kinase kinases (MKKs) [88]. MKK cleavage leads to destruction of macrophages and subsequent death of the host. However, the dearth of strong inhibitors and efficient LF activity assays for high throughput screening of potent inhibitors limits LF targeting to prevent cell cytotoxicity. Therefore, a peptide-based FRET assay was designed to study the protease activity of *Bacillus anthracis* LF, which could further be adapted for enzyme inhibition and screening. Modified fluorogenic peptides based on a known substrate MEK1 were synthesized and compared in terms of signal background (S/B) as well as percent cleavage [88]. 10 μM substrate cleavage by 100 nM LF was measured by the increase in fluorescence at λex 355 nm and λem 460 nm after separation of fluorophore and quencher. Out of the four best substrates with K_M > 10 μM, (Cou) consensus (K (QSY-35)GG)-NH2 (S/B

>9, 100 % cleavage), with coumarin as donor and QSY-35[(*N*-({4-[(7-nitro-2,1,3-benzoxadiazol-4-yl)amino]phenyl}-acetyl) as quencher was chosen and used for protease activity studies as well as plate based assays. The assay developed was well suited for studying enzyme kinetics, screening and characterizing inhibitors. Inhibition percentage was directly proportional to the decrease in rate constant for first order reactions or corresponding to the decrease in product formation [88]. The Z' factor for these assays was set at > 0.7.

2.4.3 Fluorescence Polarization (FP) Assays

Digestive fluorescence polarization assay is often used to investigate protease activity due to high sensitivity, accuracy, ease of use, speed and suitability in high throughput screening (HTS) [89–91].

Background

FP makes use of polarization which is a fundamental property of light and is based on the assessment of rotational motions of fluorescent molecules [92]. Fluorophores can be described as electric dipoles with characteristic excitation and emission dipole moments. When a fluorescent molecule is excited with light parallel to its absorption dipole moment, it re-emits in the same parallel plane of polarization. However, if the molecule rotates out of this plane, light is emitted in a plane different from the excitation light (Fig. 5.10). On the other hand, if illuminated with a light perpendicular to its absorption dipole moment, the fluorophore cannot be excited. In other words, a fluorophore covalently attached to a small molecule, say peptide, on being excited by polarized light will emit largely depolarized light due to the random molecular rotations it exhibits during its life time [93]. In contrast, a larger molecule with lesser rotational speed will allow maximum polarization of emitted light. Clearly, the degree of polarization is inversely proportional to the rate of molecular rotation or the molecular volume. The change in molecular volume during a reaction may be due to binding, dissociation, breakdown or conformational changes in the molecular species. This is reflected by the difference in polarization measured indirectly through a change in fluorescence intensity [94].

A linearly polarized light is used to illuminate the fluorophores and a moving polarizer kept in front of the detector is used to detect the intensity of emitted light for a given plane of polarization i.e. vertical or horizontal [95]. The emission in horizontal plane is denoted by $I_{||}$ while that in the horizontal plane is denoted by $I_{\perp}$. A moving polarizer is used to detect the intensity of emitted light and the intensity measurements in the two planes are used to calculate two dimensionless quantities, Polarization Ratio (p), (Eq. 5.17) and the Emission Anisotropy (r), (Eq. 5.18) that are used interchangeably [94].

$$p = \frac{I_{||} - I_{\perp}}{I_{||} + I_{\perp}} \tag{5.17}$$

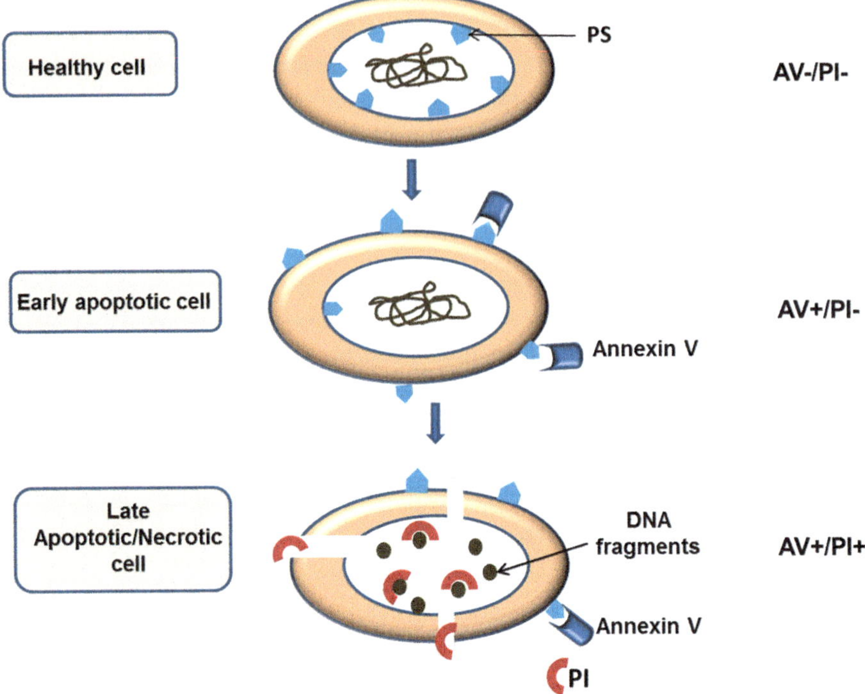

Fig. 5.11 The Annexin V- PI apoptosis assay: Cell death is monitored by the accumulation of phosphatidylserine (PS) on the outer surface of the plasma membrane due to its disruption. PS is detected by Annexin V staining that can be coupled with Propidium Iodide (PI) staining to distinguish between early and late apoptotic cells

$$r = \frac{I_{||} - I_{\perp}}{I_{||} + 2I_{\perp}} \tag{5.18}$$

The relationship between p and r can be shown by

$$r = \frac{2p}{(3 - p)} \tag{5.19}$$

$$\text{or} \quad p = \frac{3r}{(2 - r)} \tag{5.20}$$

p can take values from −0.33 to +0.5 while r can range from −0.25 to +0.4. Emission anisotropy is preferred over p as a way of representing polarization data because of the advantage it offers in terms of mathematical simplicity and ease of interpretation. Polarization measurements are independent of fluorophore concentration which is also a major advantage of this technique.

Fluorescein, rhodamine and boron-dipyrromethene (BODIPY) dyes are commonly used in FP assays [96–98]. The latter have longer lifetimes that add to the sensitivity of the study. Ideally, short linkers that provide rigid binding are used to prevent "propeller effect", a term used to describe a phenomenon where despite binding, the shift in polarization is minimal [89]. Apart from being simple and easily predictable with the involvement of one fluorescent species, FP studies are sensitive even in the subnanomolar range. Moreover, the output is not affected by filter effects since the horizontal and vertical components are controlled internally [99].

Major limitation in an FP set up is that for the change in polarization to be significant, there must be an evident difference in the unprocessed substrate and processed product. Moreover, this procedure is affected by local motion effects and is often prone to autofluorescence [99].

FP in Protease Assays

In these studies, the protease substrate is labeled with one or more fluorescent dyes. Upon cleavage, the large substrate is converted into smaller fragments which lower the polarization signal [100]. Proteolytic activities of many enzymes as well as their inhibition have been regularly studied with enzymes such as trypsin, papain and pronase at different concentrations by FP using FITC labeled substrates [101, 102].

A general protocol for a degradative protease assay has been summarized below:

Materials

1. Fluorescently labeled protease substrate in appropriate buffer: 1 mM Rhodamine-labeled peptides in DMSO, diluted in FP assay buffer to a final concentration of 10 μM.
2. Varying concentrations of purified enzyme in appropriate buffer.
3. FP assay buffer (5X): For example, 5X PBS, 0.025 % Tween 20, pH 7.5, 0.5 % BSA.
4. 384-well black plates (Nalgene Nunc International).

Instrumentation

1. Plate reader for FP measurements (for example, Beacon® 2000 Fluorescence Polarization System or the SpectraMax M5^{e}).
2. Centrifuge.
3. Orbital shaker.

Method

Set up appropriate positive and negative controls.

1. Start by determining the highest concentration of the substrate to be used.
2. Prepare a fresh dilution of it from the stock. Make a series of 12 two-fold dilutions with a final reaction volume of 60 μl in each well.
3. Record the fluorescence polarization and the total fluorescence at room temperature for λex 531 nm and λem 595 nm (Total fluorescence monitors the stability of the reagent and of the lamp).

4. Plot the FP data versus substrate concentration. The lowest concentration of substrate corresponding to a stable FP signal is taken to be the optimal working concentration of the substrate for the polarization assay.
5. Prepare fresh stock enzyme solutions in assay buffer. Thereafter prepare 12 two-fold dilutions in a 384-well plate.
6. Add the pre-determined amount of substrate to each of these wells and incubate at room temperature for different time periods, say, 10 min to overnight.
7. Spin down the plate for 2 min at 200 g then cover the plate, incubate for 30 min and finally measure the fluorescence polarization and total fluorescence in the plate reader.

Precautions

1. Fluorophores are light and pH sensitive so handle them with care.
2. Do not use greater than 10 mM of the peptide stock as it may lead to poor solubility when dissolved in the assay buffer.
3. Centrifuge the plate to remove bubbles.
4. Take the lowest concentration that gives minimal variations of FP signal.
5. Mix the reaction solutions completely and gently using a pipette or orbital shaker at 500–700 rpm.
6. Any interference from the solvent can also be tested by setting up dilutions of the solvent and reading the fluorescence polarization signal in the plate reader.

Data Analysis

Using any curve fitting analysis software, plot the Millipolarization (mP) i.e. the polarization multiplied by 1000 versus the concentrations of the proteins. $I_{||}$ and $I_{\perp}$ are also part of the readout [101]. The proteolytic activity is measured from the initial slope of the graph that is expressed as the change in millipolarization units per min (mP) and the kinetic parameters calculated as described earlier in Sect. 2.2.

Control experiments to test the specificity of the assay can be designed by monitoring unlabeled and labeled substrate along with the enzyme. Experiment is set up as previously described. FP based assays can also be adapted for inhibition assays with great ease [101, 102]. To this end, reactions including inhibitors along with the enzyme and substrate needs to be set up in separate wells. The mP values versus the concentrations of the inhibitor are plotted. Conversion of fluorescence data into percentage inhibition and Z' factor value can be done as described earlier in Sect. 2.2.3.

One such example where fluorescence polarization assay is used to investigate protease activity is as follows:

In a particular study that was aimed at characterizing the protease activity of pepsin in 0.01 N HCl at pH 2.0, 0.5 mg/ml (25 pmol/ml) BODIPY-α-casein conjugate was used as a substrate [102]. Pepsin was used at different concentrations of 0, 0.1, 1, 10, and 100 ng/ml (0, 0.8, 8, 80, and 800 mU/assay respectively). Here, BODIPY-α-casein cleavage was indicated by the drop in mP values with time. An enzyme concentration dependent initial rate of change in millipolarization was observed and the substrate was found to be cleaved in 3 min at sensitivity of 1 ng/ml at pH 2.0 [102].

3 Investigating Proteases in Apoptosis

The general experimental strategy to identify a protease involved in apoptosis follows a four pronged approach, each of which has its own set of pros and cons [103]. These can be grouped under the following heads:

3.1 Studying the effect of over-expressing putative protease genes on cell death [103].
3.2 The use of a protease inhibitor to check whether it can block the normal course of apoptosis [21, 22].
3.3 Studying the apoptotic process on deletion/knockdown of putative protease genes [25] and
3.4 Examining the proteolytic cleavage of specific cellular proteins in apoptotic cells [16–20].

All of these methods have their pros and cons. A few caveats need to be borne in mind before any of the above method(s) is/are employed as standard biophysical/functional assays for a putative apoptotic protease. Inhibition studies to identify proteases are limited by the non-specificity of the inhibitors [24, 103]. Overexpression experiments also have its loopholes where non-apoptotic proteases such as trypsin and chymotrypsin can also cause cell death and hence it becomes difficult to establish a pro-apoptotic role for a putative protease unambiguously [104]. Similarly knockout experiments are problematic due to existence of protease homologues and their redundancy in the apoptotic pathway [105]. Cleavage of polypeptides during apoptosis also needs to be scrutinized carefully. Firstly it is a rare chance event to find a polypeptide getting cleaved during apoptosis and secondly the cleavage products when assayed on a gel cannot distinguish a proteolytic cleavage from formation of intramolecular isopeptide bonds [103, 106]. Hence, it is worthwhile to follow the entire course of proteolysis in cells undergoing apoptosis over time in a context dependent manner. It is also important to distinguish proteolytic events in apoptotic and regular cells. This ensures that the putative protease is indeed implicated in apoptosis.

3.1 Determining the Effect of Overexpressing Putative Protease Genes in Eukaryotic Cells

The depletion or overexpression of a key protease involved in apoptosis is expected to either provoke or abolish apoptotic effects in a cell [103]. In overexpression experiments, cells are transiently transfected with putative protease genes to see whether apoptosis is induced. Common transfection methods use lipids, liposomes, cationic polymers, magnetic beads, DEAE-dextran, calcium phosphate and activated dendrimers as reagents while instrumentation based methods include electroporation, biolistic technology, microinjection and laserfection [107, 108]. A successful transfection is usually indicated by a reporter gene that is included in the plasmid.

A combination of assays is set up for detecting apoptosis in the initial or later stages. Some involve studying the morphological changes, DNA fragmentation, membrane alterations and percentage cell death or cell viability, while others study the effect of inhibitor on apoptotic markers like ced-9/ced-3 or estimate the activity of caspase-3/cytochrome oxidase [109]. Assays designed to gauge the effect of inhibitors on oxidative stress parameters, for example, the generation of reactive oxygen species (ROS) are also employed. A few common assays will be described in depth in the following sections to present a clear picture to the reader.

3.1.1 Fluorescence in Cell Based Studies

Live-cell imaging has become increasingly popular over the years with advances in fluorescent protein and imaging technology [110]. The most important prerequisite for a live-cell imaging experiment is the use of healthy and normally functioning cells. High light intensities and long exposure times should be avoided for live cells even though it may compromise the picture quality. The advent of synthetic fluorescent protein technology has generated new inroads in monitoring cellular processes such as protein localization and dynamics. Moreover, quantitative imaging of fluorescent proteins using widefield fluorescence, confocal microscopy and multiphoton microscopy has facilitated the study of cellular structure and function with high clarity [111].

In particular, enzymatic activity studies in live cells have become routine due to the availability of thiol-reactive, lipophilic and pentafluorobenzoyl (PFB) fluorogenic enzyme substrates [112]. These substrates readily enter the cell and undergo a reactive process which yields a fluorogenic product that is membrane impermeable. The Enzyme-Labeled Fluorescent (ELF) substrates also work on this principle. ELF substrates which are mildly blue fluorescent moieties, undergo enzymatic cleavage to form the intensely fluorescent yellow-green ELF 97 alcohol precipitate at the precise site of reaction [112]. Another fluorescent technology that is popular in cell based studies is the Tyramide Signal Amplification (TSA) which uniquely employs tyramide derivatives labeled with adducts such as biotin or fluorophores that are processed by horseradish peroxidase and are also trapped near the site of its formation. This covalent bond formed results in detection of peroxidase-labeled targets with high spatial resolution.

3.1.2 Experimental Procedures

3.1.2.1 Basic Transfection Protocol

Materials

1. Sterilized tips, centrifuge tubes, tissue culture plates.
2. Reduced Serum media (such as Opti-MEM®).

3. Transfection reagent (such as fugene-6, Lipofectamine LTX™).
4. Mammalian cells.
5. Plasmid DNA.

Instrumentation
1. Sterile hood.
2. CO_2 incubator.

Method
1. Plate 1.25×10^5 cells in 500 μl of complete growth medium in a 24-well plate to achieve 40–70 % confluency on the day of transfection.
2. Prepare transfection complex of DNA and transfection reagent. Dispense 100 μl of Opti-MEM® into the sterile tube and add 0.5 μg plasmid DNA (Optimal reagent to DNA ratio is 2:1). Mix gently.
3. Add 0.75–1.75 μl of transfection reagent (Lipofectamine LTX™) to the above mix. Pipette gently and incubate at room temperature for around 30 min to allow the formation of transfection complexes.
4. Add 100 μl of this mixture dropwise onto each well containing the cells. Gently rock the plate to aid proper mixing.
5. Incubate overnight in a CO_2 incubator at 37 °C.
6. Harvest cells and proceed for apoptotic assays.

Precautions
1. Keep reagents at −20 °C for long term storage.
2. Use healthy and confluent cells for best results.
3. The volume of plating medium, dilution medium, cells per well, DNA and transfection reagent used depends on the culture vessel.
4. Transfection efficiency varies with cell types and densities. Optimize transfection condition as per requirement.

Apoptotic Assays:

3.1.2.2 Annexin V- PI Apoptosis Assay

Background
Loss of membrane integrity is a salient feature of apoptosis. Usually, healthy eukaryotic cells display a characteristic pattern of asymmetry in the distribution of phospholipids between the inner and outer leaflets of the cell membrane [113]. During the early stages of cell death however, phosphatidylserine (PS) which is otherwise located on the inner surface starts accumulating on the outer surface of the plasma membrane due to its disruption. This change in phospholipid translocation is monitored to detect cell death [114]. Vascular anticoagulant α (also called Annexin V), a 36 kDa protein has been shown to have a high affinity for negatively charged phospholipids like phosphatidylserine in the presence of Ca^{2+}. Therefore, fluorochromes conjugated to Annexin V (like FITC) can be used as a sensitive probe to label and detect apoptotic cells by flow cytometry [115]. To further improve the

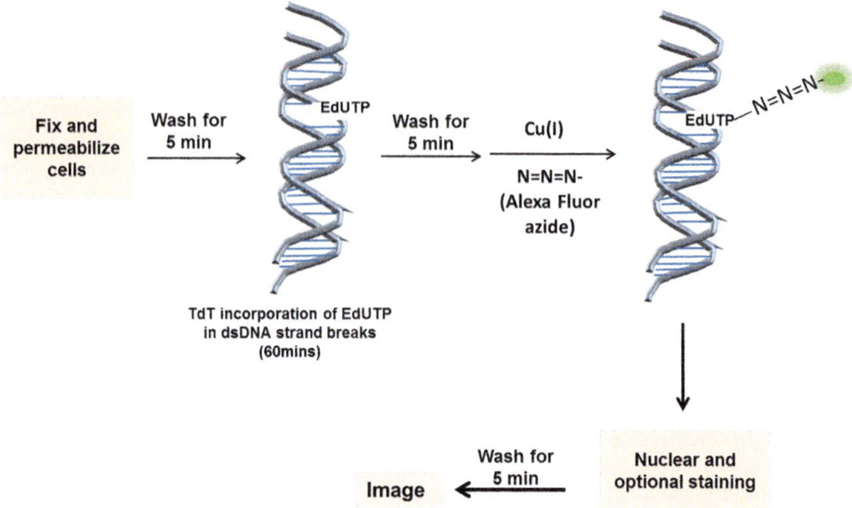

Fig. 5.12 The TUNEL assay: The enzyme terminal deoxynucleotidyl transferase (TdT) labels the free 3′-OH ends of DNA fragments generated during cell death with modified dUTPs (EdUTP). EdUTP is visualised by bond formation with a fluorophore coupled azide group, a reaction catalyzed by a Cu(I)

sensitivity of the assay, Annexin V staining can be coupled with Propidium Iodide (PI) staining to distinguish between early and late apoptotic cells [116]. Propidium iodide is a red-fluorescent dye non-permeant to live cells with intact membranes but is readily taken up by non-viable cells. Hence, Annexin V positive and PI negative cells can be identified as early apoptotic while both Annexin V and PI positive cells are either dead or late apoptotic (Fig. 5.11). Since necrotic and apoptotic cells both stain with Annexin V and PI, tracking cells over time to follow the transition in their staining profile suggests apoptosis without ambiguity. Scraping or tryspinization can grossly affect the binding of Annexin V to PS [117].

Materials

1. Annexin V-FITC.
2. Propidium iodide (PI).
3. 10X Binding buffer: 10 mM HEPES (pH 7.4), 140 mM NaCl, 5 mM KCl, 5 mM $MgCl_2$ and 2.5 mM $CaCl_2$.
4. Mammalian cells.
5. Pipettes, polystyrene/polypropylene tubes.
6. PBS.

Instrumentation

1. Flow Cytometer (such as BD FACSVerse System and BD FACSuite Software).
2. Centrifuge.

Method

1. Prepare a 50 μg/mL working solution of PI in 1X binding buffer.
2. Harvest cells by centrifugation at 300 g for 5 min at RT. Ideally, use a million cells for each sample. Wash cells 2–3 times with either ice cold PBS or binding buffer to completely remove media and other undesired components like EDTA.
3. Add Annexin V-FITC conjugate to the binding buffer in a ratio of 1:100. Resuspend pellet in 500 μl of this buffer.
4. Place cells on ice at RT in the dark for around 30 min or more.
5. Add 5 μg of PI to each sample and mix gently. Transfer sample onto ice until you analyze using the flow cytometer.

Data Analysis

Start the FACS system and run a performance quality control (QC). Set up assay, acquire & analyze data as per instructions in the user guide. The report generated at the end of the analysis includes the fraction of cells corresponding to viable, early and late apoptotic populations [118, 119].

Precautions

1. Care should be taken to prevent damage to cell membrane during sample preparation.
2. Positive and negative controls should be set up along with the test sample.
3. Stained cells should not be stored for a long time before analysis.
4. Contaminated cultures should not be used because it affects the outcome of flow cytometric analysis.

Please refer to Sect. 3.125 for an example on Annexin V-PI apoptotic assay.

3.1.2.3 TUNEL (TdT-Mediated dUTP-Biotin Nick End Labeling) Assay

Background

The TUNEL staining assay was designed for detecting internucleosomal DNA fragmentation which is a hallmark of apoptosis. This method uses the property of the enzyme terminal deoxynucleotidyl transferase (TdT) to label the free 3′-OH ends of double stranded as well as single stranded DNA fragments generated during cell death [120]. TdT mediates incorporation of modified deoxyuridine triphosphate nucleotides (dUTPs) at the free DNA ends (Fig. 5.12). A variety of labels are available for dUTPs namely, digoxigenin, fluorescein, biotin, bromine (BrdUTP) and alkyne (EdUTP) group [121]. The sensitivity of the assay is influenced by the labeling efficiency and nature/size of the modifying group. Smaller labels like bromine and alkyne are suitable for easy labeling. While BrdUTP is easily detected by fluorophore/reporter enzyme conjugated antibody, EdUTP is visualised by a Cu(I) catalyzed bond formation between the alkyne and the fluorophore coupled azide groups. Analysis of the stained cells is usually done through flow cytometry or microscopy. Flow cytometry also gives an idea of the distribution of apoptotic and non-apoptotic cells in the cell cycle in a stage specific manner [122].

The TUNEL assay has its own set of limitations. Firstly, it cannot effectively distinguish between apoptotic and necrotic cell death or cells undergoing DNA repair. Secondly, it is sensitive to fixation and the number of cells under study [123]. Lastly, the cellular protein environment and ongoing events in apoptotic cells like DNA condensation affect the progress of the assay. Despite these limitations, TUNEL assay has been popular due to its simplicity, accuracy and speed. The protocol covered in the following section uses fluorescein-dUTP to mark DNA breaks in cultured cells which can be later visualized by fluorescence microscopy [124, 125].

Materials

1. 1X Phosphate-buffered saline (PBS): 137 mM NaCl, 2.68 mM KCl, 1.47 mM KH_2PO_4, 8.1 mM Na_2HPO_4. pH adjusted to 7.4.
2. Fixation solution: 4 % buffered methanol free formaldehyde or paraformaldehyde made in PBS.
3. Equilibration Buffer: 200 mM potassium cacodylate (pH 6.6 at 25 °C), 25 mM Tris–HCl (pH 6.6 at 25 °C), 0.2 mM DTT, 0.25 mg/ml BSA, 2.5 mM cobalt chloride.
4. Nucleotide Mix: 50 μM fluorescein-12-dUTP, 100 μM dATP, 10 mM Tris–HCl (pH 7.6), 1 mM EDTA.
5. Incubation buffer (freshly made during the equilibration step): For a volume of 100 μl, add 90 μl equilibration buffer +10 μl nucleotide mix+ 2 μl TdT enzyme.
6. Propidium iodide solution (1 mg/ml) or DAPI (1.5 μg/ml). You can also use VECTASHIELD® + DAPI which is DAPI containing mounting medium.
7. 10 % Triton® X-100 solution.
8. TdT Enzyme.
9. Saline Sodium Citrate (SSC) buffer – 20X.
10. 70 % ethanol, DNase I (to be used for positive control), 20 mM EDTA (pH 8.0).
11. Clear nail polish.
12. Glass slides, coverslips, coplin jars, forceps, cell scraper and micropipettors.

Instrumentation

1. Humidified chamber for slides.
2. Incubator (kept at 37 °C).
3. Fluorescence microscope.
4. Tabletop centrifuge.

Method

Include appropriate positive and negative controls.

1. Culture adherent cells or cytospin cells (for cell suspensions) onto poly-L-lysine-coated glass slides. Cytospin preparation can be done by collecting cells on glass slides after centrifugation and washing with PBS. Ideally, $1 \times 10^5 - 5 \times 10^5$ cells are used per slide.
2. Fix cells by immersing in a coplin jar filled with 4 % methanol-free formaldehyde solution for 25 min at 4 °C.

3. Rinse slides with PBS for 5 min at room temperature. At this stage, slides can be stored for future use for up to two weeks in 70 % ethanol at −20 °C or in PBS at 4 °C.
4. Permeabilize cells by treating with PBS containing 0.2 % Triton® X-100 solution for 5 min. A positive control can be prepared at this step by treating cells on a separate slide with DNase I. Repeatedly wash the positive control slide 3–4 times in deionized water before processing it as described for the test sample.
5. Remove excess permeabilization solution and add equilibration buffer to the cells. Incubate at room temperature (RT) for 5–10 min.
6. Remove the equilibration buffer and add 50 μl (optimized for 5 cm^2 surface area of cells) of incubation buffer. For negative control slides, use incubation buffer without TdT enzyme.
7. Place a coverslip on the cells and incubate at 37 °C for 1 h. Avoid direct exposure of slides to light.
8. Prepare 2X SSC and stop the reaction by dipping the slides in it for 15 min at room temperature. Remove coverslip before you do so.
9. Rinse 2–3 times with PBS for 5 min at room temperature to remove unbound fluorescein-dUTP.
10. At this step, cells can be additionally stained with PI or DAPI (4′, 6-diamidino-2-phenylindole) nuclear stain. Depending on the counterstain, the green fluorescence of fluorescein-12-dUTP can be visualized in a red (PI) or blue background (DAPI). Prepare PI solution (1 μg/ml) in PBS and immerse slides for 15 min at room temperature in the dark. Alternatively, add DAPI with mounting medium (VECTASHIELD®) and proceed to step 13.
11. Rinse slides in deionized water 2–3 times for 5 min at room temperature.
12. Remove excess water from the slides.
    Optional: Add a drop of Anti-Fade solution (for sustained fluorescence of dye) to the cells and put a coverslip. Seal edges with nail polish or any other appropriate agent and air-dry for 5–10 min.
13. Use appropriate filters to examine cells by fluorescence or confocal microscopy. Fluorescein emits at 520 ± 20 nm and exhibits green fluorescence; propidium iodide fluoresces red at >620 nm while DAPI is blue at 460 nm.

Precautions

1. Nucleotide mix as well as mixtures containing the same are light sensitive and therefore keep away from direct light exposure.
2. While in use, always keep the nucleotide mix and incubation buffer on ice.
3. Use proper precautionary measures while using highly toxic potassium cacodylate.
4. Take care to use separate coplin jars for positive control slides undergoing DNase I treatment.
5. Avoid precipitation of 20X SSC salts.

Please refer to Sect. 3.125 for an example using TUNEL as an apoptotic assay.

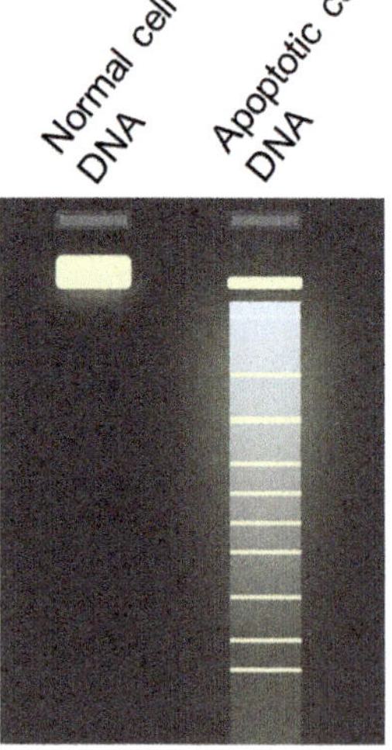

Fig. 5.13 DNA ladder assay. This is a cartoon representation of a DNA ladder pattern generated due to endonucleolytic DNA degradation during apoptosis

3.1.2.4 DNA Ladder Assay

Background

This technique is designed to visualize nucleosomal DNA laddering as a result of apoptotic DNA fragmentation [126]. Caspase activated DNase (CAD), an endogenous endonuclease cleaves genomic DNA at internucleosomal linker regions to generate DNA fragments in multiples of 180–200 bps. This assay involves extraction of oligonucleosomal DNA from lysed cells followed by treatment with RNase A and Proteinase K. The extracted product is then loaded onto an agarose gel to visualize a characteristic 'ladder' pattern by ethidium bromide (EtBr) staining as shown in Fig. 5.13 [127].

DNA laddering is a simple, fast and cheap assay for detecting apoptosis in most cell types. However, necrosis may result in DNA ladders as well, which can be a major drawback [128].

DNA laddering is an early apoptotic marker, and therefore if a ladder is observed at a later time point, it might not represent apoptosis. Moreover, non-specificity of the assay may arise due to damaged, non-apoptotic cells with lower DNA (for example aneuploid) or cells with different chromatin structure [129]. Laddering is also fraught with other disadvantages such as loss of DNA fragments during precipitation, unease of handling large number of samples, requirement of multiple steps and a large starting material. Furthermore, it may fall short in detecting apoptotic cells with 4C amount of DNA, i.e. cells in G2/M phase.

The following protocol describes a typical laddering assay using agarose gel electrophoresis [130, 131].

Materials

1. Mammalian cells.
2. Reagents: EtBr, agarose, PBS, DNase free RNase (50 mg/ml), 10 M ammonium acetate, absolute ethanol, DNA markers, TAE (Tris-acetate-EDTA, 40 mM Tris acetate, pH 8.3, 1 mM EDTA), TE (10 mM Tris–HCl, 1 mM disodium EDTA, pH 8.0.) buffer, 6X gel loading buffer (0.25 % bromophenol blue, 0.25 % xylene cyanol FF, 30 % glycerol).
3. Lysis buffer: 10 mM EDTA, 0.5 % SDS, 50 mM Tris–HCl pH 8.0, 0.5 mg/ml Proteinase K.
4. Pipette tips, microcentrifuge tubes.

Instrumentation

1. Gel electrophoresis equipment.
2. Table top centrifuge.

Method

Include appropriate controls.

1. For adherent cell populations, harvest cells by adding 1 ml of trypsin to each well (100 mm dishes) and scraping out the cells.
2. Pellet cells by centrifugation at 2000 rpm for 5 min at 4 °C. Discard the supernatant. Use around 5×10^6 cells per tube.
3. Resuspend and wash cell pellets with 1X PBS twice.
4. Resuspend cell pellets in about 400 μl of lysis buffer (1 ml for every 10^7 cells) and incubate overnight at 50 °C.
5. Add 10 μl of 50 mg/ml RNase and incubate for 3 h at 50 °C.
6. Add 1/2 volume of 10 M ammonium acetate followed by 2.5 volumes of ice-cold ethanol and mix thoroughly.
7. Keep in −80 °C freezer for at least 30 min for 'ethanol precipitation' of DNA.
8. Centrifuge for 15 min at 14,000 rpm to precipitate DNA. Discard the supernatant. Resuspend and wash the white pellet with ≈200 μl of 80 % ice-cold ethanol. Let it dry for 10 min at room temperature.
9. Dissolve the DNA pellet in about 20–50 μl of TE Buffer and measure its concentration. At this step, the DNA can be stored at 4 °C for future use.
10. Cast a 1.5 % agarose gel in TAE containing 0.5 μg/ml ethidium bromide and run samples at low voltage (≈35 V) using gel loading buffer for proper separation of bands. Visualize under UV light.

Precautions

1. Collect culture media as well as cell monolayer while centrifuging.
2. Avoid using too many cells to prevent incomplete digestion because it results in viscous DNA preparation.
3. If DNA preparations are too viscous, load on dry wells and then transfer to the tank with running buffer.
4. Do not allow vigorous pipetting because it causes shearing of high molecular weight DNA.
5. Autoclave tips, tubes and other consumables to avoid DNase contamination.

DNA Fragmentation Assay Using Flow Cytometry

DNA cleavage can also be analyzed quantitatively by flow cytometry after permeabilization and staining with DAPI [132]. In this case, DNA content analysis is used to measure cells undergoing fragmentation in apoptosis. The cells ($\approx 10^7$) are fixed in 70 % ethanol at −20 °C and extracted with phosphate-citrate buffer (0.05 M Na_2HPO_4 and 25 mM citric acid in 9:1 ratio containing 0.1 % Triton® X-100 at pH 7.8). The cells are stained with staining solution containing 1 μg/ml DAPI and 220 μg/ml of sulforhodamine dissolved in 10 mM PIPES buffer (0.1 N NaC1, 2 mM $MgCl_2$, and 0.1 % Triton® X-100, pH 6.8 at 4 °C). If PI (20 μg/ml) is used to replace DAPI, RNaseA (50 units/ml) is also added to the buffer and cells are incubated at room temperature for 30 min. Since small DNA molecules diffuse out of the cells following permeabilization and washing, staining of such cells with a quantitative DNA binding dye such as DAPI, will be less [118, 133, 134]. These cells with fractional DNA content are referred to as 'Sub-G1'(with less than 2C DNA content) and can be seen left of the G1 peak in a DNA content histogram [133]. The loss in DNA content is concurrent with the loss of S1 phase cells which can be easily detected by the absence of the S peak on the histogram.

Please refer to Sect. 3.125 for an example on DNA ladder assay.

3.1.2.5 Caspase Activity Assay

Background

Caspases are a family of cysteinyl aspartate-specific proteases closely associated with apoptosis as described in the earlier chapters. On induction of apoptosis, initiator caspases such as −8 and −9 use the extrinsic and intrinsic pathways to activate the effector caspases by proteolytic cleavage [135, 136]. The effector caspases (−3 and −9) in turn, cleave downstream targets and commit the cell to its death. A third mechanism is supposed to involve the endoplasmic reticulum and activation of caspase 12. Therefore, caspases have increasingly become targets of detecting apoptosis in cells [137]. Caspase activity assays typically include western blot, immunoprecipitation and immunohistochemistry. Polyclonal and monoclonal antibodies have allowed the recognition of the active form of caspases [138]. Additionally, the availability of substrates for specific caspases has led to the development of numerous luminescent assays for measuring caspase activity. Caspase inhibitors have also been routinely used to label active caspases [139, 140]. Some methods use a combination of the above-mentioned techniques. For e.g., caspase activation is initially detected by lysing the cells and probing the released enzymes with antibodies that are coated onto microwells. The sensitivity is further improved by incorporating fluorescently labeled substrates and detecting their cleavage [141].

This assay format allows rapid, easy and accurate detection as well as quantitation of apoptosis in cells. Moreover, it specifically recognizes the caspases involved and hence the precise mechanism of cell death. The fluorescence based assays are also compatible for HTS [142]. The major disadvantage is that, caspase activation does not always ensure that apoptosis is bound to follow. Moreover, the non-specificity of caspases for their substrates somewhat undermines the sensitivity and accuracy of the assay [138].

Caspase 3 is a target of choice in most caspase activity assays because both the intrinsic and extrinsic pathways converge at this point [143, 144]. Since it is one of the last players in the caspase cascade, measurement of its activity is well suited as a direct read-out of the apoptotic event. Most of the assays described below are aimed at measuring caspase 3 activity unless mentioned otherwise.

Fluorescent Assays

(A) Detecting caspase activity using antibody:

This method uses an antibody against active Caspase 3. Caspase 3 is a proenzyme (procaspase 3) which is activated by its cleavage into a large and a small subunit, called p20 and p12 respectively [145, 146]. The protocol described in this section uses rabbit polyclonal antibody to detect the large fragment (17/19 kDa) of activated caspase 3 resulting from proteolytic processing at conserved aspartic residues. The secondary antibody is fluorescently conjugated to enable easy detection using a fluorescence microscope. There can be various formats of this assay, including the use of cell suspensions or adherent cells. Adherent cells are analyzed using a plate reader. This is referred to as 'In-Cell Western' where microwell plates are scanned with channels of appropriate wavelength [146].

The procedure described here uses a cytospin preparation of cells on glass slides followed by simple detection under a fluorescence microscope.

Materials

1. Mammalian cells.
2. Pipette, slides, coplin jars, ice bucket.
3. Reagents: PBS, 37 % formaldehyde, 10 % Triton® X-100, blocking Buffer (PBS with 0.1 % Tween 20 + 5 % horse serum), mounting medium.
4. Primary antibodies: such as Anti-ACTIVE Caspase 3 pAb.
5. Secondary antibodies: such as donkey anti-rabbit Cy3 conjugate secondary antibody.

Instrumentation

1. Humidified chamber.
2. Fluorescence microscope.

Method

Include appropriate controls. Slide without primary antibody can be used as a negative control.

1. Fix cells using 4 % formaldehyde (by diluting 37 % formaldehyde in 1X PBS) for 20 min at RT.
2. Rinse slides 5 times for 5 min each at RT with 1X PBS containing 0.2 % Triton® X-100 in coplin jars in order to permeabilize the cells.

3. Remove the solution and block cells by adding ≈ 200 μl of blocking buffer.
4. Place slides in the humidified chamber. Incubate for 2 h at RT and follow it up with a single PBS wash to remove blocking buffer.
5. Dilute the primary antibodies in blocking buffer (1:250). Cover cells with 100 μl of the reagent and incubate in a humidified chamber overnight at 4 °C to increase sensitivity of the assay. Alternatively, incubate for 2 h at RT.
6. Rinse 5 times with 1X PBS + 0.1 % Tween 20 for 5 min at RT. Add around 200 μl of the solution each time.
7. Dilute the fluorescently-labeled secondary antibodies in blocking buffer or PBS (1:500). You may add 0.5 % Tween 20 to reduce background. Cover cells with 100 μl of antibody and incubate for 2 h at RT.
8. Wash slides with 1X PBS + 0.1 % Tween 20 for 5 min followed by a single wash with PBS at RT. Avoid direct exposure to light.
9. Remove the washing solution completely. Blot around the area having cells for extra measure. At this step the slides can either be stored for future use at 4 °C in dark for several weeks or immediately used for analysis.
10. Use mounting medium before observing under a fluorescence microscope.

Precautions
1. Take care so that cells do not dry up between washes.
2. Serum used for the blocking buffer should preferably be from a closely related species.
3. Blocking reagent should be standardized for every antigen-antibody pair otherwise it may affect the specificity and sensitivity of the assay.
4. Milk or casein used for blocking and antibody dilution may not be appropriate when using anti-goat antibodies. 0.1 % casein is used in 0.2X PBS buffer for blocking. BSA can also be used but may lead to high background.
5. Do not store or reuse diluted antibodies for long since they tend to deteriorate rapidly even at 4 °C.
6. Protect slides from light exposure after adding secondary antibody. Do not keep antibody vials exposed to light.
7. Test the specificity of the primary antibody prior to the assay.

(B) Detecting caspase activity using fluorogenic substrates:
The executioner caspase 3 cleaves the amino acid sequence Asp-Glu-Val-Asp or DEVD. To study proteolytic cleavage by active caspase 3, a homogenous assay uses fluorogenic derivatives of DEVD such as DEVD-AFC, DEVD AMC, N-Ac-DEVD-N'-MC-R110 (rhodamine110 conjugate) and DEVD-pNA (p-nitroanilide conjugate). Most of these substrates offer high sensitivity for detecting caspase 3 activity in solution and living cells [147]. The fluorogenic substrate $(Ac\text{-}DEVD)_2$-R110 used in this protocol contains two DEVD sequences and follows a two step mechanism for complete hydrolysis by

caspase 3 to release the dye R110 [148]. This intense green fluorescence which is proportional to caspase activity is detected by a Fluorometer with a 470 nm excitation filter and a 520 nm emission filter. Purified enzyme, cell extracts, adherent and suspension cell cultures can be conveniently used for this assay. The procedure described in the next section uses cultured cells in 96-well or 384-well plates [148].

Materials

1. Mammalian cells.
2. Fluorogenic substrate: $(Ac\text{-}DEVD)_2$-R110 (2 mM).
3. Pipette, tips, ice bucket, microwell plates.
4. Reagents: PBS, Cell Lysis/ Assay Buffer (such as Apo-ONE Caspase 3/7 Buffer usually contains 0.1 % Tween 20, NP-40 or Triton® X-100).
5. Caspase 3 detection reagent (such as Apo-ONE Caspase-3/7 Reagent): assay/lysis buffer and substrate mixed in 1:1 ratio.

Instrumentation

1. Centrifuge.
2. Fluorescence plate reader.
3. Plate shaker.
4. Incubator at 37 °C.

Method

Include appropriate positive and negative controls. Untreated cells, cells treated with caspase 3 inhibitor or samples analyzed immediately after apoptotic induction can serve as negative controls.

1. Use 500–50,000 cells per well in culture medium and assay buffer added in equal amounts. If cell extracts are being used, collect cells by centrifugation, wash with PBS and resuspend in ice-cold lysis buffer for 5 min. Thereafter, centrifuge to collect the supernatant and use it for the assay.
2. Thaw all reagents and keep on ice until use.
3. Directly add the caspase detection reagent to equal volume of culture in each well and incubate at 37 °C for upto 1 h. Use a plate shaker to mix the contents of the wells.
4. Read plate continuously at 520 nm emission (470 nm excitation) in a microplate reader.
5. Plot graphs and analyze data to compute kinetic parameters as described earlier in Sect. 2.2.1.

Precautions

1. Storage of fluorogenic reagents should be at −20 °C or below and should be used within 6 months for best results.
2. Frequent freeze-thaw of reagents should be avoided.
3. Use same number of cells and equal reaction volumes to maintain uniformity across test and controls as well as for easy and accurate comparison of results.
4. Keep reagents protected from direct light exposure.

5. Avoid vigorous mixing of reagents in the wells which leads to formation of bubbles that may interfere with fluorescence readings. Gentle mixing using a plate shaker is generally recommended.
6. Optimize the number of cells as well as incubation times to avoid getting misleading results.

(C) Detecting caspase activity using inhibitors:
A number of inhibitors have been used to study caspase (primarily caspase-3) activity within cells [149, 150]. While some of the caspase-3 inhibitors that are used for this study such as DEVD-CHO (Asp-Glu-Val-Asp-aldehyde) and DEVD-fmk (Asp-Glu-Val-Asp-O-methyl-fluoromethylketone) Ac-DEVD-AFC, Ac-DEVD-AMC, Ac-DEVD-CHO and Ac-DEVD-fmk are N-Acetyl-derivatives, others like zDEVD-AFC, zDEVD-fmk are N-benzyloxycarbonyl derivatives of the aforementioned compounds [141]. Cellular extracts have been used for the procedure described below [151–154].

Materials
1. Mammalian cells.
2. Inhibitor: Ac-DEVD-CHO, 0.1 mM (0.05 mg/ml) in DMSO (dimethylsulfoxide).
3. Assay buffer: 50 mM HEPES, 100 mM NaCl, 0.1 % CHAPS, 10 mM DTT, 1 mM EDTA, 10 % glycerol. pH set to 7.4.
4. Substrate and its calibration standard: Ac-DEVD-AMC, 0.3 mM (0.20 mg/ml) diluted in assay buffer, 7-amino-4-methylcoumarin (30 μM) diluted in assay buffer. The calibration standard is used for preparing a standard curve.
5. Lysis buffer: 50 mM HEPES, 0.1 % CHAPS, 5 mM DTT, 0.1 mM EDTA. pH set to 7.4.
6. Caspase-3 enzyme: 10 U/μl in assay buffer.
7. Pipette, ice bucket, tips, microwell plates.

Instrumentation
1. Centrifuge.
2. Fluorescence plate reader.
3. Plate shaker.
4. Incubator at 37 °C.

Method
1. Collect cells by centrifugation, wash with PBS and resuspend in ice-cold lysis buffer for 5 min. Thereafter, centrifuge (10,000 × g, 10 min, 4 °C) to collect the supernatant and keep it on ice until used. At this step, the extract can be stored away at −70 °C for future use. Use $\approx 2 \times 10^7$ cells / ml.
2. Dilute inhibitor to 1:200 in assay buffer (5X stock, 0.5 μM). Use 20 μl of this per well to a final concentration of 1X (0.1 μM) in 100 μl of reaction volume. Dilute the substrates to make a 2X stock using assay buffer. Use 50 μl of this per well to a final concentration of 1X (30 μM for Ac- DEVD-AMC) in 100 μl of reaction volume. Dilute enzyme to 1:5 in assay buffer. Use 15 μl of this per well (30 U).

3. Set up assay in microwell plates as indicated below. Make total reaction volume up to 100 μl. Add assay buffer first and substrate last to start the reaction. Maintain reaction temperature at 37 °C.

Components	Test1	Test2	Positive control	Negative control	Blank
Buffer	40 μl	20 μl	35 μl	40 μl	50 μl
Extract	10 μl (treated)	10 μl (treated)	–	10 μl (untreated)	–
Enzyme	–	–	15 μl	–	–
Inhibitor	–	20 μl	–	–	–
Substrate	50 μl	50 μl	50 μl	50 μl	50 μl

Read plate at appropriate wavelength and compute the kinetic parameters as shown in Sect. 2.2

Precautions

1. Store all reagents at −70 °C if not in continuous use.
2. Thaw by bringing reagents to RT and mix thoroughly before use.

One such study on caspase activity is described below:

The expression and activation of caspase-3 (CPP32) was studied in a reversible *mouse Ischemia model* [155]. Ischemia was induced by insertion of a filament in the left common carotid artery and was later removed for reperfusion. Mice were sacrificed at different time points after reperfusion for various assays. Caspase like DEVDase activity assay was performed with homogenates from ischemic brain using fluorogenic substrate *N*-benzyloxycarbonyl-Asp-Glu-Val-Asp-7-amino-4-trifluoromethyl coumarin or zDEVD-afc. There was an increase in enzyme activity over time, the highest being at 30 min post reperfusion. Preincubation with 100 μM of an irreversible caspase inhibitor, *N*-benzyloxycarbonyl-Asp(OMe)-Glu(OMe)-Val-Asp(OMe)-fluoromethyl ketone or zDEVD-fmk caused a drop in enzyme activity by 64 %. Caspase 3 and its cleavage products were probed by immunoblotting of lysates with caspase-3p32 (CPP32) or caspase-3p20 (cleaved active 20 kDa fragment) antiserum. A time dependent increase in p20 was observed, which was within a 1–12 h time frame after reperfusion whereas p32 was prominent in control as well as ischemic brain and its level did not change over time. Caspase-3p20 positive neurons were then subjected to TUNEL assay to observe DNA fragmentation. A majority of p20 positive cells were found to be TUNEL positive 12 and 24 h post reperfusion. This observation was confirmed by a DNA ladder assay for ischemic brain homogenates. DNA ladder was observed within a time frame of 6–24 h [155]. These results show that appearance of active caspase 3 and caspase like enzyme activity at or post-reperfusion is a characteristic event in ischemic injury and that there is a direct correlation between the two. It also indicates that this phenomenon eventually leads to the morphological and biochemical changes that happen during ischemic cell death as is evident by the DNA ladder and TUNEL assays.

3.2 Use of Inhibitors to Identify Proteases in Apoptosis

These experiments are carried out in two phases. The first phase of the experiment involves treating cells with an inhibitor specific to protease being tested. The second phase is aimed at investigating the ability of the inhibitor to block induced programmed cell death [156]. The inhibition studies can be done in multiple ways. In one, cell cultures are induced with apoptotic stimuli following which the inhibitor is directly added to the culture wells [34, 157]. In another method, cells are transiently transfected with the inhibitor-expression vector for constitutive expression in the cells [158]. The positive clones are selected by limiting dilution, treated with apoptotic inducers and then used for the apoptotic assays. A third assay system can be used to test the specificity of the inhibitor towards a particular apoptotic protease [159]. This requires the overexpression of both the protease and the inhibitor expression constructs in cell lines followed by cell death assays to compute cell viability/cytotoxicity. These assays can be modified to test dose dependent response of the inhibitor to apoptosis.

The blocking percentage can be determined from the percentage of cell death calculated through a number of apoptotic assays most of which are described elaborately in Sect. 3.1 [22]. A popular technique in measuring cell viability/cytotoxicity is the Trypan blue dye exclusion assay [160]. It is based on the principle of dye uptake by non-viable cells and its exclusion by viable ones. The 960 Da dye molecule enters dead cells and binds to intracellular proteins because of which the cells appear blue. Cell death can be measured by counting stained cells under the microscope or any other suitable instrument [160, 161]. In case percent viability needs to be reported, it can be done by counting the unstained cells. Trypan blue has been shown to fluoresce at 660 nm when bound to proteins. At 0.002 % (w/v) trypan blue can also be easily detected by a flow cytometer and is stable for 30 min (Fig. 5.14).

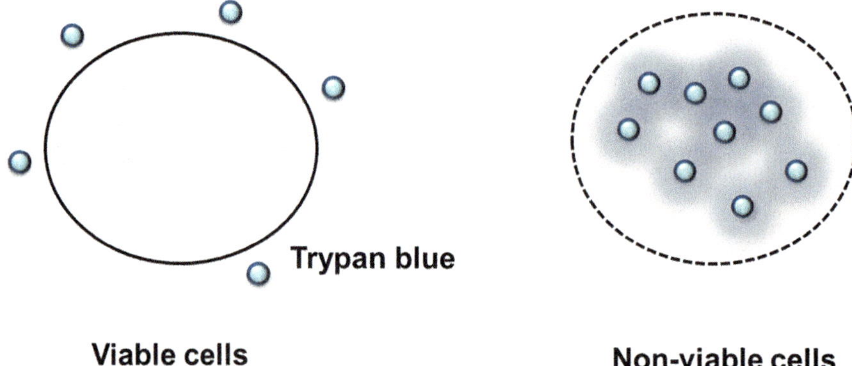

Fig. 5.14 Trypan blue cell viability assay: Trypan blue dye is taken by non-viable cells while it is excluded by viable cells

Despite being a fairly easy-to-perform, fast, cheap and widely used technique, its major disadvantage lies in the fact that it cannot distinguish between healthy cells and the cells that are alive but losing cell function [160].

A protocol to study proteases in cells undergoing activation induced cell death (AI-PCD), using exogenous inhibitors has been described below [22]:

Materials

1. Mammalian cells ($\approx 2.5 \times 10^6$cells/ml).
2. Cell culture medium, Trypan blue (0.4 %), PBS.
3. Protease inhibitors (10 mM stock in DMSO).
4. Apoptosis inducing drugs (for example pokeweed mitogen or staphylococcal entorotoxin B, 1 μg/ml to 10 μg/ml)
5. 96-well plates, tips, pipettes

Instrumentation

1. CO_2 incubator.
2. Hematocytometer/cell counter and microscope.
3. Flow cytometer or microplate reader.
4. Centrifuge.

Method

Keep appropriate positive and negative controls.

1. Culture 200 μl of cells in 96-well plates using suitable medium.
2. Directly add apoptotic inducer in the required amount to the culture wells.
3. Directly add inhibitor in the required amount to the culture wells.
4. Incubate for 48–60 h in a CO_2 incubator at 37 °C.
5. Aspirate 0.1 ml of supernatant from each well and discard.
6. Add 0.4 % trypan blue solution to dilute cells in each well in a 1:5 (cell: stock dilution) ratio. Use PBS to dilute.
7. Analyze cells using the instrument of your choice.

For cell suspensions: add inhibitor, apoptotic inducer and dye directly to the culture vessel in proper order and in the required amount.

Precautions

1. Sterile filter Trypan Blue before using it to get rid of particles that might interfere with cell counting.
2. Do not treat cells with Trypan blue for extended periods of time. It may lead to non-specificity in the assay because of dye uptake by both viable and non-viable cells.
3. Serum proteins in medium or salt solution may interfere with the assay because Trypan blue has a lower affinity for cellular proteins comparatively. This gives a dark background on visualization. To circumvent this problem, collect cells by centrifugation and reconstitute in fresh protein-free medium or salt solution before counting.
4. If a hematocytometer is used, take care so that chambers are not over or under-filled with samples.

5. If more than 10 % cells appear clustered, repeat the entire process after homogenously dispersing the cells by vigorous pipetting.

Data Analysis
Percentage death is initially measured by cell counting in the control and test reactions [22]. A positive AI-PCD response is one where the percentage of induced death is greater than percentage death in control cells by 20 ($\%D_{in} > 20 + \%D_c$),

The percentage blocking by an inhibitor is then calculated by the following:

$$100 \times \left[1 - \left(\frac{\%D_{in,I} - \%D_I}{\%D_{in} - \%D_c}\right)\right]$$

where, $\%D_{in}$ = percentage death due to inducer alone; $\%D_I$ = percentage death due to inhibitor alone; $\%D_{in,\,I}$ = percentage death due to inducer and inhibitor combined; $\%D_c$ = percentage death in control cells.

Blocking by the inhibitor is considered positive if the score is above 50 %.

Inhibition of Protease Activity in Apoptotic Pathway: An Example
An inhibition study was designed to check whether a calpain dependent programmed cell death pathway (PCD) leads to the decline of T lymphocytes from HIV^+ donors [22]. Pokeweed mitogen (PWM) or staphylococcal enterotoxin B (SEB) antigen was used to induce AICD in these cells. Following this, calpain inhibitors E-64 (50 μM), leupeptin (25 μM) and calpain inhibitor II were added to the cultures and cell death was analyzed through trypan blue staining. The cell death percentage was used to compute blocking percentage of the inhibitor. The E-64 inhibitor showed 40–60 % inhibition of AI-PCD responses, while leupeptin and calpain inhibitor II demonstrated 60–67 % inhibition. This suggested that a calpain-dependent PCD pathway plays a role in HIV-associated immunodeficiency [22].

3.3 Determining the Effect of Knockout/Knockdown of Putative Protease Genes in Eukaryotic Cells

Apoptosis regulation by a putative protease can be assessed by targeted gene depletion *in vitro* or *in vivo* [25]. While gene knockout ensures that the gene is not transcribed at all and usually occurs through a process of homologous recombination/cre-lox excision, knockdown allows sub-normal expression of the target gene [162–164]. Knockdown is achieved by introducing short RNA sequences homologous to the target gene RNA and depleting it through specific binding [165–167]. A knockdown gives an approximation of the knockout phenotype and is achievable in a week; whereas a knockout may take a couple of months.

Gene knockout can be used to determine how apoptosis gets affected and how/where the gene product may appear in the apoptotic pathway [168]. This method can further be improved by allowing the rescue of the knockout phenotype by using suitable vectors or simply transfecting proteins downstream of the protease target. A general *in vitro* knockout experiment starts with the design of insertion or replacement constructs used to disrupt the target gene. This is followed by the growth, transfection, and selection of the knockout cells [169, 170]. Finally, these cells are induced by a variety of cell death agents and evaluated with any of the apoptotic assays mentioned in Sect. 3.1. Moreover, cDNA constructs can be transfected back to verify the function of the protease.

A knockdown experiment on the other hand begins with the generation of siRNA *in vitro* and its transfection into the target cells using lipid based delivery methods [171]. Real-time PCR and functional assays are performed to verify the knockdown and analyze its consequences.

Though knockout studies are clearly a foolproof method of establishing the role of a protease in apoptosis, it is extremely expensive and time consuming. The low frequency (10^{-2} to 10^{-5}) of homologous recombination in mouse ES cells for generating knockouts makes the procedure even more challenging. Other setbacks include developmental lethality ($\approx$15 %), altered gene function in adult mice, no observable change in phenotype and appearance of a completely different phenotype. Knockdown studies on the other hand, may cause little reduction of the mRNA level even after successful transfection. Additionally, a lot of optimization may be required in some cell types that are not easily transfected.

A general protocol for generation of knockout cells *in vitro* using the recently developed and highly popular CRISPR/Cas9 system is given below [172–175]. Cas 9 RNA guided nuclease generates double strand breaks (DSBs) at a specific genomic locus using $\approx$20 nucleotides (nt) targeting sequence cloned in the guide RNA (gRNA). The cellular repair system prompts an error prone non-homologous end joining (NHEJ) response leading to mutations in the coding region, which finally leads to genome editing by frameshifts or premature stop codons. However, if a homology-directed repair (HDR) response is initiated, precise knockouts can be generated by using donor vectors having the left and right homologous arms as well as the replacement cassette [172–175] as shown in Fig. 5.15. The protocol below uses co-transfection of the guide and donor vectors to achieve the knockout through HDR.

Materials

1. gRNA expression construct (3–5 μg), dissolved in TE buffer. Vector has cas9 gene as well as the $\approx$20-nt targeting sequence.
2. Donor vector (3–5 μg), dissolved in TE buffer. Vector has left and right homologous arms and the replacement cassette.
3. Negative control scramble vector. Dilute to a final concentration of 100 ng/ μL.
4. Culture medium, LB broth, LB agar plates with appropriate antibiotic, 6-well plates, micropipettors, micropipettor tips, microcentrifuge tubes, 50 ml conical tubes, transfection reagent (Turbofectin used here).
5. Mammalian cells.

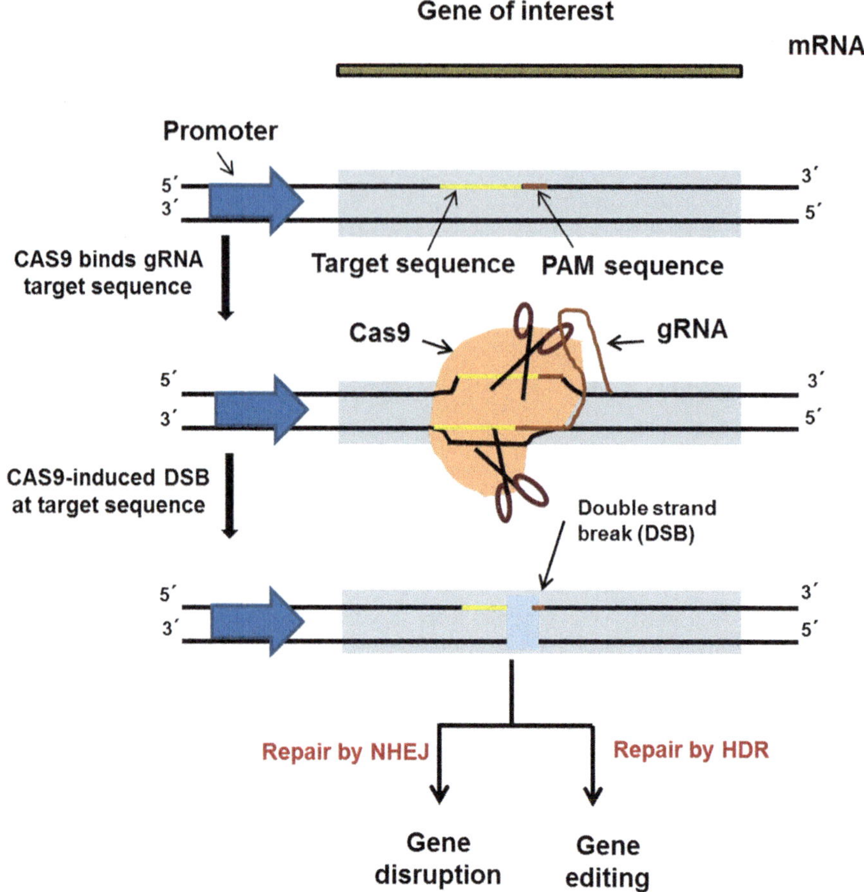

Fig. 5.15 Schematic representation of gene knockout using the CRISPR/Cas9 system: Cas9 nuclease introduces double strand breaks (DSBs) at the target using the sequence on the guide RNA (gRNA). This is either repaired by an error prone non-homologous end joining (NHEJ) or a homology-directed repair (HDR), generating knockouts by gene disruption or gene editing respectively

Instrumentation

1. Centrifuge.
2. CO_2 incubator.

Method

Include appropriate controls

1. Culture $\approx 3 \times 10^5$ adherent cells in each well using 2 ml of media 18–24 h before transfection. Allow cells to reach $\approx 70\ \%$ confluency. For cell suspensions, use 5×10^5 cells per well.

2. In a separate microfuge tube, add 250 μl of reduced serum media (Opti-MEM®). Reconstitute 1 μg of the gRNA vector. Add 1 μg of donor vector in it and mix gently.
3. Add 6 μl turbofectin in this mixture to obtain turbofectin-DNA in 3:1 ratio. Mix gently and incubate for 15 min at RT.
4. Add the mixture directly and dropwise onto each well containing the cells. Gently rock the plate to aid proper mixing.
5. Incubate in a CO_2 incubator at 37 °C for 48 h.
6. Split cells in 1:10 ratio seven times (G1-G7) with 3 days of growth in between. At G2, a genomic PCR can be done to check the integrity of the inserted cassette.
7. Use appropriate antibiotic to select G8 cells by growing in 10 cm dishes containing complete medium.
8. Confirm knockout by western blot or genome PCR followed by sequencing. Scramble control and donor vector may give colonies even on selection, however the number of colonies will be significantly less compared to test control.
9. Isolate single cell colonies and proceed for apoptotic assays.

Precautions
1. Design at least two sgRNAs for each locus in case one does not work.
2. Store cDNA clones at −20 °C.
3. For genomic PCR at G2, the sequence outside the homology region on the donor vector should be included in one primer while the other primer should have a region of the functional cassette.
4. During selection, use the lowest concentration of antibiotic that can completely kill non-transfected cells.
5. Separate a batch of cells and maintain them without subjecting to antibiotic selection. These cells can be used later on in case the selected cells need to go through a second round of antibiotic selection.
6. Single cell colonies may contain mono-allelic knockout. In order to obtain a bi-allelic knockout, a second donor vector with a different selection marker can be used to repeat the whole procedure.

Described below is a study on how knockout of a protease gene affects cell death:
The role of caspases in mitochondrial dysfunction-induced neuronal apoptosis was explored in a recent experimental study [176]. A neurotoxin, rotenone was used to induce cell death in cultured mice neurons. Caspase-2 was found to be activated during this process by an *in situ* trapping approach. To confirm whether loss of caspase-2 inhibits Rotenone-induced apoptosis, neurons were cultured from knockout mice (*casp2−/−*). In wild type (*wt*) versus *casp2−/−* cells, rotenone treatment showed no significant difference in basal level of cell death. However, after 36 h it caused around 55 % cell death in *wt* cells whereas no significant change was observed in *casp2−/−* cells [176]. Viability assays indicated that *casp2−/−* cells offered a survival advantage compared to *wt* cells apart from confirming the apoptotic mode of cell death. This clearly suggests that the initiator caspase-2 plays a key role in mitochondrial dysfunction-induced neuronal apoptosis and its loss inhibits apoptosis.

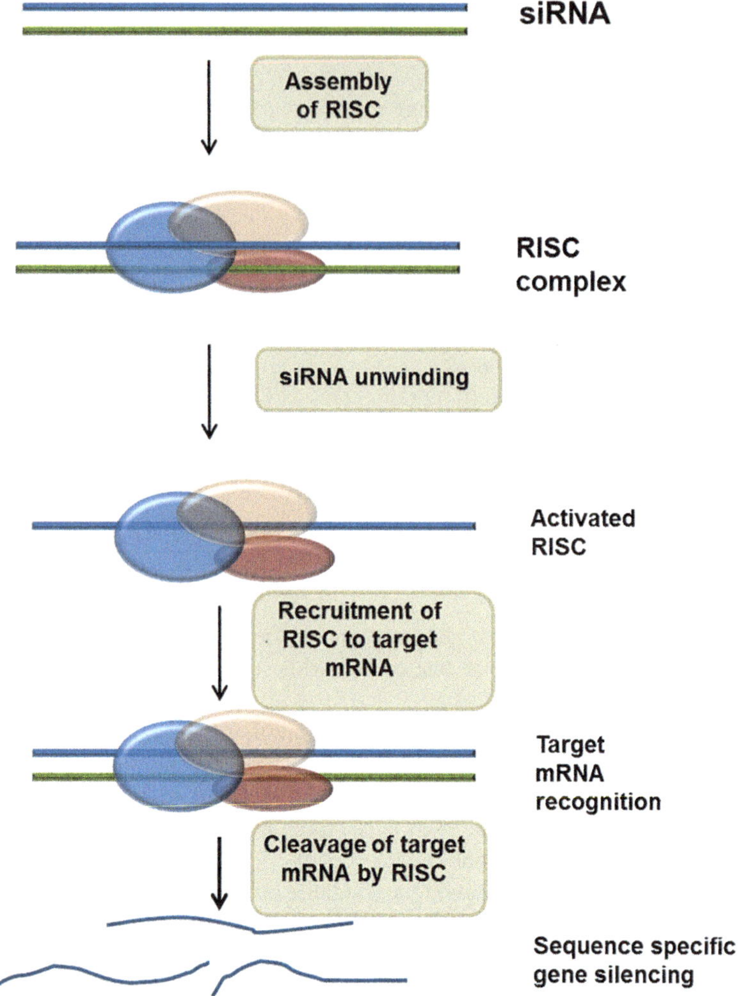

Fig. 5.16 Schematic representation of gene knockdown using siRNA: Transfection of siRNA into the target cells causes the assembly of the RNA-induced silencing complex (RISC). RISC is activated by siRNA unwinding and consequently recruited to the target mRNA followed by degradation of the mRNA leading to sequence specific gene silencing

A general protocol for generation of knockdown cells *in vitro* using siRNA transfection is given below [177] as shown in Fig. 5.16:

Materials

1. siRNA (0.25–1 μg).
2. Culture medium without antibiotic and supplemented with FBS, 6-well plates, micropipettors, micropipettor tips, 1.5-ml microcentrifuge tubes.

3. siRNA Transfection Medium (reduced-serum medium, modification of Eagle's Minimal Essential Medium).
4. Control siRNAs (scrambled sequence).
5. siRNA transfection reagent (suitable for both serum containing and serum free medium).

Instrumentation

1. Centrifuge.
2. CO_2 incubator.

Method

Include appropriate controls.

1. Culture ≈3×10^5 adherent cells in each well using 2 ml of media 18–24 h before transfection. Allow cells to reach ≈70 % confluency.
2. Add 0.25–1 μg of siRNA duplex in 100 μl siRNA transfection medium. Simultaneously prepare a solution of 2–8 μl of siRNA transfection reagent in 100 μl siRNA transfection medium. Mix the solutions gently and keep at RT for 15–45 min.
3. Use 2 ml of siRNA transfection medium to wash cells. Remove the medium completely.
4. Add 0.8 ml of siRNA transfection medium to the mixture prepared in step 2. Mix gently and completely. Add this to the cells and keep in a CO_2 incubator for 5–7 h at 37 °C
5. Now add 1 ml of 2X normal growth medium (2X serum and antibiotic containing) and keep in a CO_2 incubator for 18–24 h at 37 °C.
6. Remove the medium completely and replace with fresh 1X growth medium. Incubate for 24–72 h.
7. Use these cells for the apoptotic assay.

Precautions

1. Use healthy and sub-confluent cells for efficient transfection.
2. Optimize siRNA amount for target protein if required.
3. Check for siRNA transfection reagent compatibility while using different cell lines.
4. Optimize transfection times if required. However, bear in mind that longer periods may lead to serum starvation and cell detachment.
5. If cell toxicity is observed at step 5, remove the transfection mixture completely before adding growth medium.

Described below is a study on how knockdown of a protease gene affects the cell death phenomena:

In a study focused on delineating the role of uPAR and cathepsin B in radiation-induced apoptosis in gliomainitiating cells (GICs), it was found that radiation led to an increase in the levels of these molecules 24 h post treatment [178]. A bicistronic shRNA construct was used to target these molecules in irradiated and nonirradiated cells to understand the mechanism involved. It was observed that knockdown resulted in an increase in apoptosis in both non-GICs and GICs as seen in a TUNEL assay. While 58 % of U87 non-GICs and 42 % U87 GICs were found to be TUNEL positive in non-irradiated cells, the count increased to 71 % and 69 % respectively on radiation. It was also observed that uPAR and cathepsin B downregulation enhanced the expression of γH2AX which suggested that the disruption of transcription leads to sensitization of these cells to apoptosis [178].

3.4 Studying Cleavage of Specific Polypeptides During Apoptosis

Proteolytic cleavage of specific cellular proteins is one of the salient features in cellular apoptosis and the 70 kDa polypeptide of U1 RNP particles are some of the primary polypeptide substrates that undergo proteolytic cleavage during early apoptosis [16, 103, 179]. The degradation of lamins brings about disassembly of the nuclear membrane while PARP cleavage results in the formation of ADP-ribose polymers that bind to proteins such as topoisomerases, histones, and PARP itself. This leads to inhibition of their cellular functions. The 70 kDa polypeptide of U1 snRNP on the other hand, is an important component of the splicing machinery and its cleavage disrupts RNA processing [17]. It has been shown that in apoptotic cells, there is >80 % decrease in intact lamin levels within 1–2 h with the simultaneous increase of smaller lamin fragments when assayed on an SDS-polyacrylamide gel. On the other hand, 116 kDa PARP is cleaved into 25 kDa and 85 kDa fragments with concurrent DNA fragmentation in apoptotic cells. In a cell free model system PARP was shown to be >90 % cleaved within 3 min [16]. The U1 RNP polypeptide, on apoptotic induction gets cleaved to a 40 kDa fragment and is believed to occur in the Fas/TNF-mediated apoptosis as well.

The role of a protease in apoptosis can therefore be partially proved by investigating whether specific proteins get cleaved during the time course of apoptosis when expressed. For example, active caspase 3 cleaves PARP, in many cells undergoing apoptosis and is an early *in situ* apoptotic marker. Caspase 3 is also known to process prolamin to active lamin, thereby suggesting its role in PCD [16].

PARP cleavage can be detected by immunohistochemistry. Proteins from the test cells can be isolated and run on an 8 % (w/v) SDS-polyacrylamide gel before transferring on a nitrocellulose membrane and probing with specific antibodies [180]. Alternatively, an in-cell western can be done directly in the culture wells, or a cytospin sample can be prepared and the PARP fragment can be visualized by a method suitable for the antibodies being used. In the protocol below, the p85 fragment of PARP is specifically recognized by a rabbit polyclonal antibody and visualized using a fluorescently conjugated secondary antibody [179].

Materials

1. Mammalian cells.
2. Primary antibody: Anti-PARP p85 Fragment pAb.

3. Reagents: PBS, 0.2 % Triton® X-100/PBS, blocking buffer (PBS/0.1 % Tween 20 + 5 % horse serum), ultrapure water.
4. Secondary antibody: donkey anti-rabbit Cy3 conjugate.
5. Coplin jars, pipettes, tips, culture plates, slides.

Instrumentation

1. Humidified chamber.
2. Fluorescence microscope.

Method

Please follow the protocol provided in Sect. 3.125 under detection of caspase activity and include appropriate controls.

Precautions

Please follow the precautions provided in section 3.125.

One such study to assign the role of a protease in apoptosis by investigating the cleavage of specific proteins is described below:

In a study aimed at understanding the role of PARP in chemotherapy-induced apoptosis, HL-60 cells were initially induced with 68 μM etoposide for a period of 0–4 h [16]. The proteins from these cells were then separated by SDS-PAGE and subjected to western blotting. Monoclonal antibody C-2-10 specific to the p85 fragment, a polyclonal anti-F2 (D) that recognizes the 25 kDa fragment of PARP, and an antibody against the nucleolar protein B2 were used. It was shown that PARP was cleaved into two smaller fragments while the nucleolar polypeptide B23/nucleophosmin, with a nuclear localization similar to poly-(ADP-ribose) polymerase (pADPRp) remained uncleaved. Activity blot was performed to check whether the 85 kDa fragment retains its enzyme activity. Polypeptides were subjected to SDS-PAGE, transferred to nitrocellulose, renatured *in situ* and incubated with 2 μCi/ml ^{32}P-labeled NAD and finally washed with isotope-free buffer F. In catalytically active fragments, the radiolabel remains bound even after treatment with 3 % SDS but loses it when incubated with purified glycohydrolase. Extracts from control and etoposide treated cells analyzed thus indicated that the 116 kDa intact PARP and the 85 kDa cleaved fragment from test cells both showed activity [16]. The cleavage of pADPRp was also studied in human KG1a acute myelogenous leukemia cells, Molt 3 acute lymphocytic leukemia cells, MDA-MB-468 human breast cancer cells and rat thymocytes [16]. Therefore from the observations of this study we can infer that etoposide-induced apoptosis led to PARP cleavage that was concomitant with endonucleolytic DNA degradation.

4 Conclusions

Proteolysis is undoubtedly an integral part of many biological processes; one of the most fundamental being apoptosis. Deregulation of the apoptotic pathway leads to various pathologies such as heart and liver failure, stroke, sepsis, neurodegenerative syndromes, autoimmune diseases and cancers, adequate treatment for many of which is still elusive [5–11]. Fortunately, the identification of key players in this pathway including the proteases, has boosted the development of diagnostic/prognostic markers and therapeutics in this direction [181–183]. In this context, it is important to mention that this has been possible mainly due to the technical advancements achieved over the years in identifying and studying the major apoptotic proteases using different assays; most of which form the base of protease-dependent therapies [184]. For example, caspase knockout and transgenic mice experiments have highlighted that the inhibition of caspases can be used for therapeutic purposes in a variety of disorders. Eventually, caspase-8, caspase-3 and PARP have evolved as potential therapeutic targets in regulating sepsis-induced cell death through use of inhibitors [181]. Moreover, caspase inhibitors (z-VAD-fmk, IAP family of proteins) have also been shown to be effective in treating liver destruction by blocking Fas-mediated apoptosis. Other apoptotic proteases such as cathepsin B, serine proteases in addition to caspases play key roles in cancer progression as well as neurodegenerative, pulmonary and cardiovascular diseases [185, 186].

More recently, protease activatable prodrugs (PAP) have been developed as therapeutic agents [187, 188]. These are molecules where a therapeutic drug is linked to a specific peptide substrate (promoiety) and/or targeting ligand(s). Cleavage of the substrate by the protease leads to release of the active parent drug at a specific cellular location [189]. A case in point is the cathepsin activatable prodrug, poly (L-glutamic acid) conjugated with paclitaxel (PTX) that has been successfully used in ovarian cancer models [190]. Another strategy uses direct selective activation of caspases. One such activator used in antithrombotic therapeutics is RGD peptide which directly binds to and activates caspase-3 [181, 191]. In a different approach, screening of compounds capable of inducing/inhibiting caspase activity in living cells or *in vitro* was carried out to identify a potent indolone compound. It was found to be cytostatic on 40 and cytotoxic on 8 tumor cell lines [192, 193]. Another novel strategy uses the specificity of HIV protease to induce apoptosis in infected cells by replacing the maturation sites of procaspase-3 with the protease recognition motif [194]. Apart from the above therapeutic approaches, proteases have been used as diagnostic/prognostic markers of pathological conditions [195–198]. In this context, protease activatable probes are used which typically act as molecular beacons by emitting fluorescence on being processed by the protease.

This chapter discussed several popular proteolytic activity assays using site-specific fluorescent probes, fluorogenic assays, bioluminescent imaging, or immunochemistry among others. Some variants of these basic techniques include nanoparticle based substrates, rare metal complexes, bioluminescence peptide and molecular probes. Quantum dots (QD) which are essentially nano-metal fluorophore conjugates have been used for caspase and thrombin assays in a FRET based format [199–202]. One such study utilizes gold nanostructures as quenchers in conjunction with QD or other fluorophores as donors in a caspase-3 assay [203]. Bioluminescence based peptide substrates that release chemical moieties and emit light when processed, are also used [204, 205]. For example, D-aminoluciferin, a luciferase substrate conjugated to a furin consensus sequence was used as a substrate in presence of luciferase to assay live breast cancer cells *in vitro* and *in vivo*. Molecular probes having a reactive group, an inert linker and a fluorescent tag are used to effectively quantitate protease activity within live cells or in whole organism due to their specificity for only active forms of the protease [206, 207]. Therefore, apart from its localization and cellular distribution, it also gives a measure of the reduction in protease activity.

Newer assays are constantly replacing the existing ones by improving signal intensity and sensitivity thus lowering noise and developing better reporter molecules. The nanomaterials based substrates are undergoing changes to develop materials with better spectral properties while nanoarrays, graphene-based platforms are already laying down the future of improved proteinase assays [208]. Protease engineering is an emerging and rapidly evolving field now being used to improve protease properties such as substrate specificity, selectivity, stability and solubility, thus making them more suitable for industrial and therapeutic applications [208, 209]. This has led to the development of therapeutic proteases with improved efficacy, reduced immunogenicity and extended serum half-lives. Many factors are responsible for breakthroughs achieved in the field of protease engineering; the most important factors are increase in understanding of the biology of proteases, their catalytic characterization, understanding the protease-inhibitor interactions and functional diversity, all of which have been achieved through a variety of robust assay platforms [208, 209]. Therefore, with the advent of even better and more advanced techniques as well as methodologies, the future of proteases in drug development and commercial applications appears very promising indeed.

References

1. Bond JS, Butler PE (1987) Intracellular proteases. Annu Rev Biochem 56:333–364
2. Dixon M, Webb E (1979) Enzymes. Academic Press, New York
3. Thompson CB (1995) Apoptosis in the pathogenesis and treatment of disease. Science 267:1456–1462
4. Turk B (2006) Targeting proteases: successes, failures and future prospects. Nat Rev Drug Discov 5:785–799
5. Elmore S (2007) Apoptosis: A review of programmed cell death. Toxicol Pathol 35:495–516
6. Favaloro B, Allocati N, Graziano V, Di Ilio C, De Laurenzi V (2012) Role of apoptosis in disease. Aging (Albany NY) 4:330–349
7. Mattson MP (2000) Apoptosis in neurodegenarative disorders. Nat Rev 1:120–129
8. Wyllie AH (1997) Apoptosis and carcinogenesis. Eur J Cell Biol 73:189–197
9. Eguchi K (2001) Apoptosis in autoimmune diseases. Intern Med 40:275–284
10. Hayashi T, Faustman DL (2003) Role of defective apoptosis in type 1 diabetes and other autoimmune diseases. Recent Prog Horm Res 58:131–153
11. Ameisen JC, Capron A (1991) Cell dysfunction and depletion in AIDS: the programmed cell death hypothesis. Immunol Today 12:102–105
12. Lockshin RA, Zakeri Z (2001) Programmed cell death and apoptosis: origins of the theory. Nat Rev Mol Cell Biol 2:545–550

13. Alberts B, Johnson A, Lewis J et al (2002) Molecular biology of the cell, 4th edn. Garland Science, New York
14. Kaufmann SH, Mesner PW Jr, Samejima K, Tone S, Earnshaw WC (2000) Detection of DNA cleavage in apoptotic cells. Methods Enzymol 322:3–15
15. Steller H (1995) Mechanisms and genes of cellular suicide. Science 267:1445–1449
16. Kaufmann SH, Desnoyers S, Ottaviano Y, Davidson NE, Poirier GG (1993) Specific proteolytic cleavage of poly(ADP-ribose) polymerase: an early marker of chemotherapy-induced apoptosis. Cancer Res 53:3976–3985
17. Casciola-Rosen LA, Miller DK, Anhalt GJ, Rosen A (1994) Specific cleavage of the 70-kDa protein component of the U1 small nuclear ribonucleoprotein is a characteristic biochemical feature of apoptotic cell death. J Biol Chem 269:30757–30760
18. Hebert L, Pandey S, Wang E (1994) Commitment to cell death is signaled by the appearance of a terminin protein of 30 kDa. Exp Cell Res 210:10–18
19. Oberhammer FA, Hochegger K, Froschl G, Tiefenbacher R, Pavelka M (1994) Chromatin condensation during apoptosis is accompanied by degradation of lamin A+B, without enhanced activation of cdc2 kinase. J Cell Biol 126:827–837
20. Martin S, O'Brien CA, Nishioka W et al (1995) Proteolysis of fodrin (non-erythroid spectrin) during apoptosis. J Biol Chem 270:6425–6428
21. Bruno S, Del Bino G, Lassota P, Giaretti W, Darzynkiewicz Z (1992) Inhibitors of proteases prevent endonucleolysis accompanying apoptotic death of HL-60 leukemic cells and normal thymocytes. Leukemia 6:1113–1120
22. Sarin A, Adams DH, Henkart PA (1993) Protease inhibitors selectively block T cell receptor-triggered programmed cell death in a murine T cell hybridoma and activated peripheral T cells. J Exp Med 178:1693–1700
23. Ray CA, Black RA, Kronheim SR, Greenstreet TA, Sleath PR, Salvesen GS, Pickup DJ (1992) Viral inhibition of inflammation: cowpox virus encodes an inhibitor of the interleukin-1 beta converting enzyme. Cell 69:597–604
24. Komiyama T, Ray CA, Pickup DJ, Howard AD, Thornberry NA, Peterson EP, Salvesen G (1994) Inhibition of interleukin-1 beta converting enzyme by the cowpox virus serpin CrmA. An example of cross-class inhibition. J Biol Chem 269:19331–19337
25. Kuida K, Lippke JA, Ku G, Harding MW, Livingston DJ, Su MS, Flavell RA (1995) Altered cytokine export and apoptosis in mice deficient in interleukin-1 beta converting enzyme. Science 267:2000–2003
26. Izzo JL Jr, Weir MR (2011) Angiotensin-converting enzyme inhibitors. J Clin Hypertens (Greenwich) 13:667–675
27. Thompson MA, Aberg JA, Hoy JF, Telenti A, Benson C, Cahn P, Eron JJ, Gunthard HF, Hammer SM, Reiss P, Richman DD, Rizzardini G, Thomas DL, Jacobsen DM, Volberding PA (2012) Antiretroviral treatment of adult HIV infection: 2012 recommendations of the International Antiviral Society-USA panel. JAMA 308:387–402
28. Nachman S, Zheng N, Acosta EP, Teppler H, Homony B, Graham B, Fenton T, Xu X, Wenning L, Spector SA, Frenkel LM, Alvero C, Worrell C, Handelsman E, Wiznia A (2014) Pharmacokinetics, safety, and 48-week efficacy of oral raltegravir in HIV-1-infected children aged 2 through 18 years. Clin Infect Dis 58:413–422
29. He Y, King MS, Kempf DJ, Lu L, Lim HB, Krishnan P, Kati W, Middleton T, Molla A (2008) Relative replication capacity and selective advantage profiles of protease inhibitor-resistant hepatitis C virus (HCV) NS3 protease mutants in the HCV genotype 1b replicon system. Antimicrob Agents Chemother 52:1101–1110
30. Kawada T, Okada Y, Hoson M, Endo S, Yokoyama M, Kitanaka Y, Kimura K, Abe H, Yamate N (1999) Argatroban, an attractive anticoagulant, for left heart bypass with centrifugal pump for repair of traumatic aortic rupture. Jpn J Thorac Cardiovasc Surg 47:104–109
31. Matsuo T, Kario K, Matsuda S, Yamaguchi N, Kakishita E (1995) Effect of thrombin inhibition on patients with peripheral arterial obstructive disease: A multicenter clinical trial of argatroban. J Thromb Thrombolysis 2:131–136

32. Herman GA, Stevens C, Van Dyck K, Bergman A, Yi B, De Smet M, Snyder K, Hilliard D, Tanen M, Tanaka W, Wang AQ, Zeng W, Musson D, Winchell G, Davies MJ, Ramael S, Gottesdiener KM, Wagner JA (2005) Pharmacokinetics and pharmacodynamics of sitagliptin, an inhibitor of dipeptidyl peptidase IV, in healthy subjects: results from two randomized, double-blind, placebo-controlled studies with single oral doses. Clin Pharmacol Ther 78:675–688
33. Karasik A, Aschner P, Katzeff H, Davies MJ, Stein PP (2008) Sitagliptin, a DPP-4 inhibitor for the treatment of patients with type 2 diabetes: a review of recent clinical trials. Curr Med Res Opin 24:489–496
34. Wadhawan M, Singh N, Rathaur S (2014) Inhibition of cathepsin B by E-64 induces oxidative stress and apoptosis in filarial parasite. PLoS One 9, e93161
35. Saeki Y, Fukunaga K, Tanaka K (2010) Proteasome inhibitors. Nihon Rinsho 68:1818–1822
36. Marchi E, Paoluzzi L, Scotto L, Seshan VE, Zain JM, Zinzani PL, O'Connor OA (2010) Pralatrexate is synergistic with the proteasome inhibitor bortezomib in in vitro and in vivo models of T-cell lymphoid malignancies. Clin Cancer Res 16:3648–3658
37. von Schwarzenberg K, Held SA, Schaub A, Brauer KM, Bringmann A, Brossart P (2009) Proteasome inhibition overcomes the resistance of renal cell carcinoma cells against the PPARgamma ligand troglitazone. Cell Mol Life Sci 66:1295–1308
38. Carreno FLG (1992) Protease inhibition in theory and practice. Biotechnol Educ 3:145–150
39. Sharma R (2012) Enzyme inhibition: mechanisms and scope, enzyme inhibition and bioapplications. In: Sharma R (ed) InTech, pp 1–35. ISBN: 978-953-51-0585-5. doi:10.5772/39273. Available from: http://www.intechopen.com/books/enzyme-inhibition-and-bioapplications/enzyme-inhibition-mechanisms-and-scope
40. Berg JM, Tymoczko J, Stryer L (2002) Biochemistry, 5th edn. W H Freeman, New York
41. Cleland WW (1967) Enzyme kinetics. Annu Rev Biochem 36:77–112
42. Briggs G, Haldane J (1925) A note on the kinetics of enzyme action. Biochem J 19:338–339
43. Rogers A, Gibon Y (2009) Chapter 4: Enzyme kinetics: theory & practice. In: Schwender J (ed) Plant metabolic networks. Springer, Berlin/Heidelberg/New York, pp 71–103. ISBN 978-0-38-778744-2
44. Gubler H, Schopfer U, Jacoby E (2013) Theoretical and experimental relationships between percent inhibition and IC50 data observed in high-throughput screening. J Biomol Screen 18:1–13
45. Krishna PN (2011) Enzyme technology : pacemaker of biotechnology. PHI Learning Pvt. Ltd., New Delhi
46. Shen P, Larter R (1994) Role of substrate inhibition kinetics in enzymatic chemical oscillations. Biophys J 67:1414–1428
47. Holt A (1999) On the measurement of enzymes and their inhibitors. In: Cell neurobiology techniques. Humana Press, Totowa, pp 131–194
48. Waley SG (1993) The kinetics of slow-binding and slow, tight-binding inhibition: the effects of substrate depletion. Biochem J 294(Pt 1):195–200
49. Brooks HB, Geeganage S, Kahl SD et al (2012) Basics of enzymatic assays for HTS. In: Sittampalam GS, C N, Nelson H et al (ed). Eli Lilly & Company/National Center for Advancing Translational Sciences, Bethesda
50. Kraut J (1977) Serine proteases: structure and mechanism of catalysis. Annu Rev Biochem 46:331–358
51. Baruch A, Jeffery DA, Bogyo M (2004) Enzyme activity–it's all about image. Trends Cell Biol 14:29–35
52. Funovics M, Weissleder R, Tung CH (2003) Protease sensors for bioimaging. Anal Bioanal Chem 377:956–963
53. Lakowicz JR (1983) Principles of fluorescence spectroscopy. Springer Science+Business Media, New York
54. Royer CA, Scarlata SF (2008) Fluorescence approaches to quantifying biomolecular interactions. Methods Enzymol 450:79–106

55. Burns B, Mendz G, Hazell S (1998) Methods for the measurement of a bacterial enzyme activity in cell lysates and extracts. Biol Proced Online 1:17–26
56. Shaner NC, Steinbach PA, Tsien RY (2005) A guide to choosing fluorescent proteins. Nat Method 2:905–909
57. Combs CA (2010) Fluorescence microscopy: a concise guide to current imaging methods. Curr Protoc Neurosci, Chapter 2, Unit 2 1
58. Delgadillo RF, Parkhurst LJ (2010) Spectroscopic properties of fluorescein and rhodamine dyes attached to DNA. Photochem Photobiol 86:261–272
59. Sjoback R, Nygren J, Kubista M (1995) Absorption and fluorescence properties of fluorescein. Spectrochim Acta A Mol Biomol Spectrosc 51:L7–L21
60. Batchelor RH, Zhou M (2002 Jun 1) A resorufin-based fluorescent assay for quantifying NADH. Anal Biochem 305(1):118–9
61. Benson JR, Hare PE (1975) O-phthalaldehyde: fluorogenic detection of primary amines in the picomole range. Comparison with fluorescamine and ninhydrin. Proc Natl Acad Sci U S A 72:619–622
62. Kiernan JA (2007) Indigogenic substrates for detection and localization of enzymes. Biotechnic Histochem 82:73–103
63. Grant SK, Sklar JG, Cummings RT (2002) Development of novel assays for proteolytic enzymes using rhodamine-based fluorogenic substrates. J Biomol Screen 7:531–540
64. Cilenti L, Lee Y, Hess S, Srinivasula S, Park KM, Junqueira D, Davis H, Bonventre JV, Alnemri ES, Zervos AS (2003) Characterization of a novel and specific inhibitor for the pro-apoptotic protease Omi/HtrA2. J Biol Chem 278:11489–11494
65. Sittampalam GS (2012) Assay guidance manual [Internet]. In: GS Sittampalam, C N, Nelson H et al (ed) Protease assays. Eli Lilly & Company/National Center for Advancing Translational Sciences, Bethesda
66. Farmer WH, Yuan ZY (1991) A continuous fluorescent assay for measuring protease activity using natural protein substrate. Anal Biochem 197:347–352
67. Wickstrom C, Herzberg MC, Beighton D, Svensater G (2009) Proteolytic degradation of human salivary MUC5B by dental biofilms. Microbiology 155:2866–2872
68. Chaganti LK, Kuppili RR, Bose K (2013) Intricate structural coordination and domain plasticity regulate activity of serine protease HtrA2. FASEB J 27:3054–3066
69. Shi J, Dertouzos J, Gafni A, Steel D (2008) Application of single-molecule spectroscopy in studying enzyme kinetics and mechanism. Methods Enzymol 450:129–157
70. Selvin PR (2000) The renaissance of fluorescence resonance energy transfer. Nat Struct Biol 7:730–734
71. Wu P, Brand L (1994) Resonance energy transfer: methods and applications. Anal Biochem 218:1–13
72. Packard BZ, Komoriya A (2008) A method in enzymology for measuring hydrolytic activities in live cell environments. Methods Enzymol 450:1–19
73. Jares-Erijman EA, Jovin TM (2003) FRET imaging. Nat Biotechnol 21:1387–1395
74. Packard BZ, Komoriya A (2008) Chapter 1: A method in enzymology for measuring hydrolytic activities in live cell environments. In: Ludwig B, Michael LJ (ed) Methods in enzymology. Academic Press, New York, pp 1–19
75. Berney C, Danuser G (2003) FRET or no FRET: a quantitative comparison. Biophys J 84:3992–4010
76. Didenko VV (2001) DNA probes using fluorescence resonance energy transfer (FRET): designs and applications. Biotechniques 31:1106–1116, 1118, 1120–1101
77. Sapsford KE, Berti L, Medintz IL (2006) Materials for fluorescence resonance energy transfer analysis: beyond traditional donorБ—ſacceptor combinations. Angew Chem Int Ed 45:4562–4589
78. Piston DW, Kremers GJ (2007) Fluorescent protein FRET: the good, the bad and the ugly. Trends Biochem Sci 32:407–414
79. Szollosi J, Damjanovich S, Matyus L (1998) Application of fluorescence resonance energy transfer in the clinical laboratory: routine and research. Cytometry 34:159–179

80. Tian H, Ip L, Luo H, Chang DC, Luo KQ (2007) A high throughput drug screen based on fluorescence resonance energy transfer (FRET) for anticancer activity of compounds from herbal medicine. Br J Pharmacol 150:321–334
81. Stryer L (1978) Fluorescence energy transfer as a spectroscopic ruler. Annu Rev Biochem 47:819–846
82. Stojanovic MN, de Prada P, Landry DW (2000) Homogeneous assays based on deoxyribozyme catalysis. Nucleic Acids Res 28:2915–2918
83. van der Krogt GNM, Ogink J, Ponsioen B, Jalink K (2008) A comparison of donor-acceptor pairs for genetically encoded FRET sensors: application to the epac cAMP sensor as an example. PLoS One 3, e1916
84. Woehler A, Wlodarczyk J, Neher E (2010) Signal/noise analysis of FRET-based sensors. Biophys J 99:2344–2354
85. Okamura Y, Kondo S, Sase I, Suga T, Mise K, Furusawa I, Kawakami S, Watanabe Y (2000) Double-labeled donor probe can enhance the signal of fluorescence resonance energy transfer (FRET) in detection of nucleic acid hybridization. Nucleic Acids Res 28, E107
86. Le Reste L, Hohlbein J, Gryte K, Kapanidis AN (2012) Characterization of dark quencher chromophores as nonfluorescent acceptors for single-molecule FRET. Biophys J 102:2658–2668
87. Tatham MH, Hay RT (2009) FRET-based in vitro assays for the analysis of SUMO protease activities. Methods Mol Biol 497:253–268
88. Cummings RT, Salowe SP, Cunningham BR, Wiltsie J, Park YW, Sonatore LM, Wisniewski D, Douglas CM, Hermes JD, Scolnick EM (2002) A peptide-based fluorescence resonance energy transfer assay for Bacillus anthracis lethal factor protease. Proc Natl Acad Sci U S A 99:6603–6606
89. Lea WA, Simeonov A (2011) Fluorescence polarization assays in small molecule screening. Expert Opin Drug Discov 6:17–32
90. Inglese J, Shamu CE, Guy RK (2007) Reporting data from high-throughput screening of small-molecule libraries. Nat Chem Biol 3:438–441
91. Owicki JC (2000) Fluorescence polarization and anisotropy in high throughput screening: perspectives and primer. J Biomol Screen 5:297–306
92. Pu Y, Wang W, Dorshow RB, Das BB, Alfano RR (2013) Review of ultrafast fluorescence polarization spectroscopy [Invited]. Appl Opt 52:917–929
93. Jameson DM, Croney JC (2003) Fluorescence polarization: past, present and future. Comb Chem High Throughput Screen 6:167–176
94. Halfman CJ, Schneider AS (1982) Direct measurement of fluorescence polarization or anisotropy. Anal Chem 54:2009–2011
95. Popelka SR, Miller DM, Holen JT, Kelso DM (1981) Fluorescence polarization immunoassay. II. Analyzer for rapid, precise measurement of fluorescence polarization with use of disposable cuvettes. Clin Chem 27:1198–1201
96. Brinkley M (1992) A brief survey of methods for preparing protein conjugates with dyes, haptens, and cross-linking reagents. Bioconjug Chem 3:2–13
97. Wessendorf MW, Brelje TC (1992) Which fluorophore is brightest? A comparison of the staining obtained using fluorescein, tetramethylrhodamine, lissamine rhodamine, Texas red, and cyanine 3.18. Histochemistry 98:81–85
98. Wood EJ (1994) Molecular probes: handbook of fluorescent probes and research chemicals: by R P Haugland, pp 390. Interchim (Molecular Probes Inc, PO Box 22010 Eugene, OR 97402–0414, USA, or 15 rue des Champs, 92600 Asnieres, Paris). 1992–1994. $15. Biochem Educ 22:83–83
99. Nasir MS, Jolley ME (1999) Fluorescence polarization: an analytical tool for immunoassay and drug discovery. Comb Chem High Throughput Screen 2:177–190
100. Levine LM, Michener ML, Toth MV, Holwerda BC (1997) Measurement of specific protease activity utilizing fluorescence polarization. Anal Biochem 247:83–88
101. Roehrl MH, Wang JY, Wagner G (2004) A general framework for development and data analysis of competitive high-throughput screens for small-molecule inhibitors of protein-

protein interactions by fluorescence polarization. Biochemistry 43:16056–16066
102. Schade SZ, Jolley ME, Sarauer BJ, Simonson LG (1996) BODIPY-alpha-casein, a pH-independent protein substrate for protease assays using fluorescence polarization. Anal Biochem 243:1–7
103. Patel T, Gores GJ, Kaufmann SH (1996) The role of proteases during apoptosis. FASEB J 10:587–597
104. Williams M, Henkart P (1994) Apoptotic cell death induced by intracellular proteolysis. J Irnmunol 153:4247–4255
105. Fernandes-Alnemri T, Litwack G, Alnemri ES (1995) Mch2, a new member of the apoptotic Ced-3/Ice cysteine protease gene family. Cancer Res 55:2737–2742
106. Fesus L, Davies PJ, Piacentini M (1991) Apoptosis: molecular mechanisms in programmed cell death. Eur J Cell Biol 56:170–177
107. Kim TK, Eberwine JH (2010) Mammalian cell transfection: the present and the future. Anal Bioanal Chem 397:3173–3178
108. Jiang JK, Ma X, Wu QY, Qian WB, Wang N, Shi SS, Han JL, Zhao JY, Jiang SY, Wan CH (2014) Upregulation of mitochondrial protease HtrA2/Omi contributes to manganese-induced neuronal apoptosis in rat brain striatum. Neuroscience 268:169–179
109. Saraste A, Pulkki K (2000) Morphologic and biochemical hallmarks of apoptosis. Cardiovasc Res 45:528–537
110. Michalet X, Kapanidis AN, Laurence T, Pinaud F, Doose S, Pflughoefft M, Weiss S (2003) The power and prospects of fluorescence microscopies and spectroscopies. Annu Rev Biophys Biomol Struct 32:161–182
111. Telford WG, King LE, Fraker PJ (1992) Comparative evaluation of several DNA binding dyes in the detection of apoptosis-associated chromatin degradation by flow cytometry. Cytometry 13:137–143
112. Whitaker JE, Haugland RP, Moore PL, Hewitt PC, Reese M (1991) Cascade blue derivatives: water soluble, reactive, blue emission dyes evaluated as fluorescent labels and tracers. Anal Biochem 198:119–130
113. Andree HA, Reutelingsperger CP, Hauptmann R, Hemker HC, Hermens WT, Willems GM (1990) Binding of vascular anticoagulant alpha (VAC alpha) to planar phospholipid bilayers. J Biol Chem 265:4923–4928
114. Fadok VA, Voelker DR, Campbell PA, Cohen JJ, Bratton DL, Henson PM (1992) Exposure of phosphatidylserine on the surface of apoptotic lymphocytes triggers specific recognition and removal by macrophages. J Immunol 148:2207–2216
115. Koopman G, Reutelingsperger CP, Kuijten GA, Keehnen RM, Pals ST, van Oers MH (1994) Annexin V for flow cytometric detection of phosphatidylserine expression on B cells undergoing apoptosis. Blood 84:1415–1420
116. Riccardi C, Nicoletti I (2006) Analysis of apoptosis by propidium iodide staining and flow cytometry. Nat Protoc 1:1458–1461
117. Rieger AM, Nelson KL, Konowalchuk JD, Barreda DR (2011) Modified annexin V/propidium iodide apoptosis assay for accurate assessment of cell death. J Vis Exp (50). pii: 2597. doi:10.3791/2597
118. Shapiro HM (2005) Practical flow cytometry. Wiley, New York
119. Hingorani R, Deng J, Elia J, McIntyre C, Mittar D (2011) Detection of apoptosis using the BD annexin V FITC assay on the BD FACSVerse™ system. BD Biosciences, San Jose, pp 1–12
120. Gavrieli Y, Sherman Y, Ben-Sasson SA (1992) Identification of programmed cell death in situ via specific labeling of nuclear DNA fragmentation. J Cell Biol 119:493–501
121. Ben-Sasson SA, Sherman Y, Gavrieli Y (1995) Identification of dying cells–in situ staining. Methods Cell Biol 46:29–39
122. Darzynkiewicz Z (1994) Flow cytometry. Methods Cell Biol 41:27–442
123. Li X, Traganos F, Melamed MR, Darzynkiewicz Z (1995) Single-step procedure for labeling DNA strand breaks with fluorescein- or BODIPY-conjugated deoxynucleotides: detection of apoptosis and bromodeoxyuridine incorporation. Cytometry 20:172–180

124. Loo DT (2002) TUNEL assay. An overview of techniques. Methods Mol Biol 203:21–30
125. Walker PR, Carson C, Leblanc J, Sikorska M (2002) Labeling DNA damage with terminal transferase. Applicability, specificity, and limitations. Methods Mol Biol 203:3–19
126. Suman S, Pandey A, Chandna S (2012) An improved non-enzymatic "DNA ladder assay" for more sensitive and early detection of apoptosis. Cytotechnology 64:9–14
127. Zhivotosky B, Orrenius S (2001) Assessment of apoptosis and necrosis by DNA fragmentation and morphological criteria. Curr Protoc Cell Biol, Chapter 18, Unit 18 13
128. Archana M, Yogesh TL, Kumaraswamy KL (2013) Various methods available for detection of apoptotic cells–a review. Indian J Cancer 50:274–283
129. Matassov D, Kagan T, Leblanc J, Sikorska M, Zakeri Z (2004) Measurement of apoptosis by DNA fragmentation. Methods Mol Biol 282:1–17
130. Kasibhatla S, Amarante-Mendes GP, Finucane D, Brunner T, Bossy-Wetzel E, Green DR (2006) Analysis of DNA fragmentation using agarose gel electrophoresis. CSH Protoc 2006
131. Georgiou CD, Papapostolou I, Grintzalis K (2009) Protocol for the quantitative assessment of DNA concentration and damage (fragmentation and nicks). Nat Protocols 4:125–131
132. Belfield H, Chikh A, Ramadan S (2005) Apoptosis methods and protocols. Cell Death Differ 12:834–834
133. Darzynkiewicz Z, Huang X (2004) Analysis of cellular DNA content by flow cytometry. Curr Protoc Immunol, Chapter 5, Unit 5 7
134. Otto F (1990) DAPI staining of fixed cells for high-resolution flow cytometry of nuclear DNA. Methods Cell Biol 33:105–110
135. Wang ZB, Liu YQ, Cui YF (2005) Pathways to caspase activation. Cell Biol Int 29:489–496
136. Lu CX, Fan TJ, Hu GB, Cong RS (2003) Apoptosis-inducing factor and apoptosis. Sheng Wu Hua Xue Yu Sheng Wu Wu Li Xue Bao (Shanghai) 35:881–885
137. Fan TJ, Han LH, Cong RS, Liang J (2005) Caspase family proteases and apoptosis. Acta Biochim Biophys Sin (Shanghai) 37:719–727
138. Niles AL, Moravec RA, Riss TL (2008) Caspase activity assays. Methods Mol Biol 414:137–150
139. Scholz J, Broom DC, Youn DH, Mills CD, Kohno T, Suter MR, Moore KA, Decosterd I, Coggeshall RE, Woolf CJ (2005) Blocking caspase activity prevents transsynaptic neuronal apoptosis and the loss of inhibition in lamina II of the dorsal horn after peripheral nerve injury. J Neurosci 25:7317–7323
140. Callus BA, Vaux DL (2006) Caspase inhibitors: viral, cellular and chemical. Cell Death Differ 14:73–78
141. Amstad PA, Yu G, Johnson GL, Lee BW, Dhawan S, Phelps DJ (2001) Detection of caspase activation in situ by fluorochrome-labeled caspase inhibitors. Biotechniques 31:608–610, 612, 614, passim
142. Preaudat M, Ouled-Diaf J, Alpha-Bazin B, Mathis G, Mitsugi T, Aono Y, Takahashi K, Takemoto H (2002) A homogeneous caspase-3 activity assay using HTRF technology. J Biomol Screen 7:267–274
143. Butterick TA, Duffy CM, Lee RE, Billington CJ, Kotz CM, Nixon JP. Use of a caspase multiplexing assay to determine apoptosis in a hypothalamic cell model. J Vis Exp (86). doi:10.3791/51305
144. Porter AG, Janicke RU (1999) Emerging roles of caspase-3 in apoptosis. Cell Death Differ 6:99–104
145. Nestal de Moraes G, Carvalho E, Maia RC, Sternberg C (2011) Immunodetection of caspase-3 by Western blot using glutaraldehyde. Anal Biochem 415:203–205
146. Mandal D, Mazumder A, Das P, Kundu M, Basu J (2005) Fas-, caspase 8-, and caspase 3-dependent signaling regulates the activity of the aminophospholipid translocase and phosphatidylserine externalization in human erythrocytes. J Biol Chem 280:39460–39467
147. Bardet P-L, Kolahgar G, Mynett A, Miguel-Aliaga I, Briscoe J, Meier P, Vincent J-P (2008) A fluorescent reporter of caspase activity for live imaging. Proc Natl Acad Sci 105:13901–13905

148. Hug H, Los M, Hirt W, Debatin KM (1999) Rhodamine 110-linked amino acids and peptides as substrates to measure caspase activity upon apoptosis induction in intact cells. Biochemistry 38:13906–13911
149. Darzynkiewicz Z, Bedner E, Smolewski P, Lee BW, Johnson GL (2002) Detection of caspases activation in situ by fluorochrome-labeled inhibitors of caspases (FLICA). Methods Mol Biol 203:289–99
150. Nicholson DW, Ali A, Thornberry NA, Vaillancourt JP, Ding CK, Gallant M, Gareau Y, Griffin PR, Labelle M, Lazebnik YA et al (1995) Identification and inhibition of the ICE/CED-3 protease necessary for mammalian apoptosis. Nature 376:37–43
151. Reyland ME, Anderson SM, Matassa AA, Barzen KA, Quissell DO (1999) Protein kinase C delta is essential for etoposide-induced apoptosis in salivary gland acinar cells. J Biol Chem 274:19115–19123
152. Li L, Lorenzo PS, Bogi K, Blumberg PM, Yuspa SH (1999) Protein kinase Cdelta targets mitochondria, alters mitochondrial membrane potential, and induces apoptosis in normal and neoplastic keratinocytes when overexpressed by an adenoviral vector. Mol Cell Biol 19:8547–8558
153. Bellido T, Huening M, Raval-Pandya M, Manolagas SC, Christakos S (2000) Calbindin-D28k is expressed in osteoblastic cells and suppresses their apoptosis by inhibiting caspase-3 activity. J Biol Chem 275:26328–26332
154. Han H, Wang H, Long H, Nattel S, Wang Z (2001) Oxidative preconditioning and apoptosis in L-cells. Roles of protein kinase B and mitogen-activated protein kinases. J Biol Chem 276:26357–26364
155. Namura S, Zhu J, Fink K, Endres M, Srinivasan A, Tomaselli KJ, Yuan J, Moskowitz MA (1998) Activation and cleavage of caspase-3 in apoptosis induced by experimental cerebral ischemia. J Neurosci 18:3659–3668
156. Porn-Ares MI, Samali A, Orrenius S (1998) Cleavage of the calpain inhibitor, calpastatin, during apoptosis. Cell Death Differ 5:1028–1033
157. Barreiro-Iglesias A, Shifman MI (2012) Use of fluorochrome-labeled inhibitors of caspases to detect neuronal apoptosis in the whole-mounted lamprey brain after spinal cord injury. Enzyme Res 2012:7
158. Kumar S, Baglioni C (1991) Protection from tumor necrosis factor-mediated cytolysis by overexpression of plasminogen activator inhibitor type-2. J Biol Chem 266:20960–20964
159. Miura M, Friedlander RM, Yuan J (1995) Tumor necrosis factor-induced apoptosis is mediated by a CrmA-sensitive cell death pathway. Proc Natl Acad Sci U S A 92:8318–8322
160. Strober W (2001) Trypan blue exclusion test of cell viability. Curr Protoc Immunol Appendix 3, Appendix 3B
161. Avelar-Freitas BA, Almeida VG, Pinto MC, Mourao FA, Massensini AR, Martins-Filho OA, Rocha-Vieira E, Brito-Melo GE (2014) Trypan blue exclusion assay by flow cytometry. Braz J Med Biol Res 47:307–315
162. Santiago Y, Chan E, Liu P-Q, Orlando S, Zhang L, Urnov FD, Holmes MC, Guschin D, Waite A, Miller JC, Rebar EJ, Gregory PD, Klug A, Collingwood TN (2008) Targeted gene knockout in mammalian cells by using engineered zinc-finger nucleases. Proc Natl Acad Sci 105:5809–5814
163. Hertzog PJ, Kola I (2001) Overview. Gene knockouts. Methods Mol Biol 158:1–10
164. Wilson TJ, Kola I (2001) The LoxP/CRE system and genome modification. Methods Mol Biol 158:83–94
165. Szulc J, Aebischer P (2008) Conditional gene expression and knockdown using lentivirus vectors encoding shRNA. Methods Mol Biol 434:291–309
166. Hobel S, Aigner A (2010) Polyethylenimine (PEI)/siRNA-mediated gene knockdown in vitro and in vivo. Methods Mol Biol 623:283–297
167. Mocellin S, Provenzano M (2004) RNA interference: learning gene knock-down from cell physiology. J Transl Med 2:39
168. Hertzog PJ (2001) Isolation of embryonic fibroblasts and their use in the in vitro characterization of gene function. Methods Mol Biol 158:205–15

169. DeChiara TM (2001) Gene targeting in ES cells. Methods Mol Biol 158:19–45
170. Tessarollo L (2001) Manipulating mouse embryonic stem cells, pp 47–63
171. Moore CB, Guthrie EH, Huang MT, Taxman DJ (2010) Short hairpin RNA (shRNA): design, delivery, and assessment of gene knockdown. Methods Mol Biol 629:141–158
172. Ran FA, Hsu PD, Wright J, Agarwala V, Scott DA, Zhang F (2013) Genome engineering using the CRISPR-Cas9 system. Nat Protocols 8:2281–2308
173. Deveau H, Garneau JE, Moineau S (2010) CRISPR/Cas system and its role in phage-bacteria interactions. Annu Rev Microbiol 64:475–493
174. Horvath P, Barrangou R (2010) CRISPR/Cas, the immune system of bacteria and archaea. Science 327:167–170
175. Makarova KS, Haft DH, Barrangou R, Brouns SJ, Charpentier E, Horvath P, Moineau S, Mojica FJ, Wolf YI, Yakunin AF, van der Oost J, Koonin EV (2011) Evolution and classification of the CRISPR-Cas systems. Nat Rev Microbiol 9:467–477
176. Tiwari M, Lopez-Cruzan M, Morgan WW, Herman B (2011) Loss of caspase-2-dependent apoptosis induces autophagy after mitochondrial oxidative stress in primary cultures of young adult cortical neurons. J Biol Chem 286:8493–8506
177. Sawitzke JA, Thomason LC, Bubunenko M, Li X, Costantino N, Court DL (2013) Recombineering: using drug cassettes to knock out genes in vivo. Methods Enzymol 533:79–102
178. Malla RR, Gopinath S, Alapati K, Gorantla B, Gondi CS, Rao JS (2012) uPAR and cathepsin B inhibition enhanced radiation-induced apoptosis in gliomainitiating cells. Neuro Oncol 14:745–760
179. Zhou Y, Liang S, Williams LR (2002) Markers of poly (ADP-ribose) polymerase activity as correlates of DNA damage. Methods Mol Biol 203:247–55
180. Whitacre CM, Zborowska E, Willson JK, Berger NA (1999) Detection of poly(ADP-ribose) polymerase cleavage in response to treatment with topoisomerase I inhibitors: a potential surrogate end point to assess treatment effectiveness. Clin Cancer Res 5:665–672
181. Fischer U, Schulze-Osthoff K (2005) New approaches and therapeutics targeting apoptosis in disease. Pharmacol Rev 57:187–215
182. Oberholzer C, Oberholzer A, Clare-Salzler M, Moldawer LL (2001) Apoptosis in sepsis: a new target for therapeutic exploration. FASEB J 15:879–892
183. Cao Y, Mohamedali KA, Marks JW, Cheung LH, Hittelman WN, Rosenblum MG (2013) Construction and characterization of novel, completely human serine protease therapeutics targeting Her2/neu. Mol Cancer Ther 12:979–991
184. Drag M, Salvesen GS (2010) Emerging principles in protease-based drug discovery. Nat Rev Drug Discov 9:690–701
185. Lopez-Otin C, Bond JS (2008) Proteases: multifunctional enzymes in life and disease. J Biol Chem 283:30433–30437
186. Karikari CA, Roy I, Tryggestad E, Feldmann G, Pinilla C, Welsh K, Reed JC, Armour EP, Wong J, Herman J, Rakheja D, Maitra A (2007) Targeting the apoptotic machinery in pancreatic cancers using small-molecule antagonists of the X-linked inhibitor of apoptosis protein. Mol Cancer Ther 6:957–966
187. Mahato R, Tai W, Cheng K (2011) Prodrugs for improving tumor targetability and efficiency. Adv Drug Deliv Rev 63:659–670
188. Stella VJ (2004) Prodrugs as therapeutics. Expert Opin Ther Pat 14:277–280
189. Rautio J, Kumpulainen H, Heimbach T, Oliyai R, Oh D, Jarvinen T, Savolainen J (2008) Prodrugs: design and clinical applications. Nat Rev Drug Discov 7:255–270
190. Li C, Yu DF, Newman RA, Cabral F, Stephens LC, Hunter N, Milas L, Wallace S (1998) Complete regression of well-established tumors using a novel water-soluble poly(L-glutamic acid)-paclitaxel conjugate. Cancer Res 58:2404–2409
191. Roy S, Bayly CI, Gareau Y, Houtzager VM, Kargman S, Keen SL, Rowland K, Seiden IM, Thornberry NA, Nicholson DW (2001) Maintenance of caspase-3 proenzyme dormancy by an intrinsic "safety catch" regulatory tripeptide. Proc Natl Acad Sci U S A 98:6132–6137
192. Nguyen JT, Wells JA (2003) Direct activation of the apoptosis machinery as a mechanism to target cancer cells. Proc Natl Acad Sci U S A 100:7533–7538

193. Jiang X, Kim HE, Shu H, Zhao Y, Zhang H, Kofron J, Donnelly J, Burns D, Ng SC, Rosenberg S, Wang X (2003) Distinctive roles of PHAP proteins and prothymosin-alpha in a death regulatory pathway. Science 299:223–226
194. Vocero-Akbani AM, Heyden NV, Lissy NA, Ratner L, Dowdy SF (1999) Killing HIV-infected cells by transduction with an HIV protease-activated caspase-3 protein. Nat Med 5:29–33
195. Jang B, Choi Y (2012) Photosensitizer-conjugated gold nanorods for enzyme-activatable fluorescence imaging and photodynamic therapy. Theranostics 2:190–197
196. Kim GB, Kim YP (2012) Analysis of protease activity using quantum dots and resonance energy transfer. Theranostics 2:127–138
197. Yhee JY, Kim SA, Koo H, Son S, Ryu JH, Youn IC, Choi K, Kwon IC, Kim K (2012) Optical imaging of cancer-related proteases using near-infrared fluorescence matrix metalloproteinase-sensitive and cathepsin B-sensitive probes. Theranostics 2:179–189
198. Lee S, Kim K (2012) Protease activity: meeting its theranostic potential. Theranostics 2:125–126
199. Jokerst JV, Raamanathan A, Christodoulides N, Floriano PN, Pollard AA, Simmons GW, Wong J, Gage C, Furmaga WB, Redding SW, McDevitt JT (2009) Nano-bio-chips for high performance multiplexed protein detection: determinations of cancer biomarkers in serum and saliva using quantum dot bioconjugate labels. Biosens Bioelectron 24:3622–3629
200. Hu M, Yan J, He Y, Lu H, Weng L, Song S, Fan C, Wang L (2010) Ultrasensitive, multiplexed detection of cancer biomarkers directly in serum by using a quantum dot-based microfluidic protein chip. ACS Nano 4:488–494
201. Zajac A, Song D, Qian W, Zhukov T (2007) Protein microarrays and quantum dot probes for early cancer detection. Colloids Surf B: Biointerfaces 58:309–314
202. He H, Xie C, Ren J (2008) Nonbleaching fluorescence of gold nanoparticles and its applications in cancer cell imaging. Anal Chem 80:5951–5957
203. Swierczewska M, Lee S, Chen X (2011) The design and application of fluorophore-gold nanoparticle activatable probes. Phys Chem Chem Phys 13:9929–9941
204. Richard JA, Jean L, Schenkels C, Massonneau M, Romieu A, Renard PY (2009) Self-cleavable chemiluminescent probes suitable for protease sensing. Org Biomol Chem 7:2941–2957
205. Giron P, Dayon L, Turck N, Hoogland C, Sanchez JC (2011) Quantitative analysis of human cerebrospinal fluid proteins using a combination of cysteine tagging and amine-reactive isobaric labeling. J Proteome Res 10:249–258
206. Simon GM, Cravatt BF (2010) Activity-based proteomics of enzyme superfamilies: serine hydrolases as a case study. J Biol Chem 285:11051–11055
207. Berger AB, Vitorino PM, Bogyo M (2004) Activity-based protein profiling: applications to biomarker discovery, in vivo imaging and drug discovery. Am J Pharmacogenomics 4:371–381
208. Li Q, Yi L, Marek P, Iverson BL (2013) Commercial proteases: present and future. FEBS Lett 587:1155–1163
209. Pogson M, Georgiou G, Iverson BL (2009) Engineering next generation proteases. Curr Opin Biotechnol 20:390–397

Chapter 6
Preclinical Animal Model and Non-invasive Imaging in Apoptosis

Pradip Chaudhari

Abstract This chapter deals with preclinical studies involving animal model systems and non-invasive methodologies toward understanding the role of proteases in apoptotic pathways and their potential as therapeutic targets. Several critical diseases are associated with imbalance in the apoptotic machinery including neurodegeneration and cancer, and hence an effective strategy to target these molecules might be a tractable solution for combating these ailments. Proapoptotic proteases and their binding partners therefore have always been of special interest for designing and evaluating the efficacy of several drug-like molecules (both activators and inhibitors). This chapter is focused on discussing the development of few such molecules with specific examples. It also vividly describes the different *in vivo* model systems that are essential for rigorous screening of these molecules at different stages of drug development. Current role and future prospects of preclinical translational imaging platforms (PET, SPECT, CT and MRI) and their utility in clinical trials are also outlined in this chapter.

Keywords Animal model • Xenograft • Knockout • Non invasive imaging • Preclinical • Drug-like molecules • Proteases • Apoptosis

1 Introduction

Basic research has played a significant role in delineating the pathophysiological processes underlying various diseases as well as in determining potential leads for translational applications. The understanding of intricate pathways provides clue towards identifying the key players or potential molecular targets for therapeutic intervention. However, it takes a series of rigorous steps for a molecule of interest to emerge as a drug. This long process initiates with identification of a macromolecule, followed by designing small molecules for targeting the macromolecule with an aim at modulating its activity with desired characteristics. A multidisciplinary

P. Chaudhari (✉)
Small Animal Imaging Facility, Advanced Centre for Treatment, Research and Education in Cancer (ACTREC), Tata Memorial Centre, Navi Mumbai, Maharashtra 410210, India
e-mail: pchaudhari@actrec.gov.in

K. Bose (ed.), *Proteases in Apoptosis: Pathways, Protocols and Translational Advances*, DOI 10.1007/978-3-319-19497-4_6

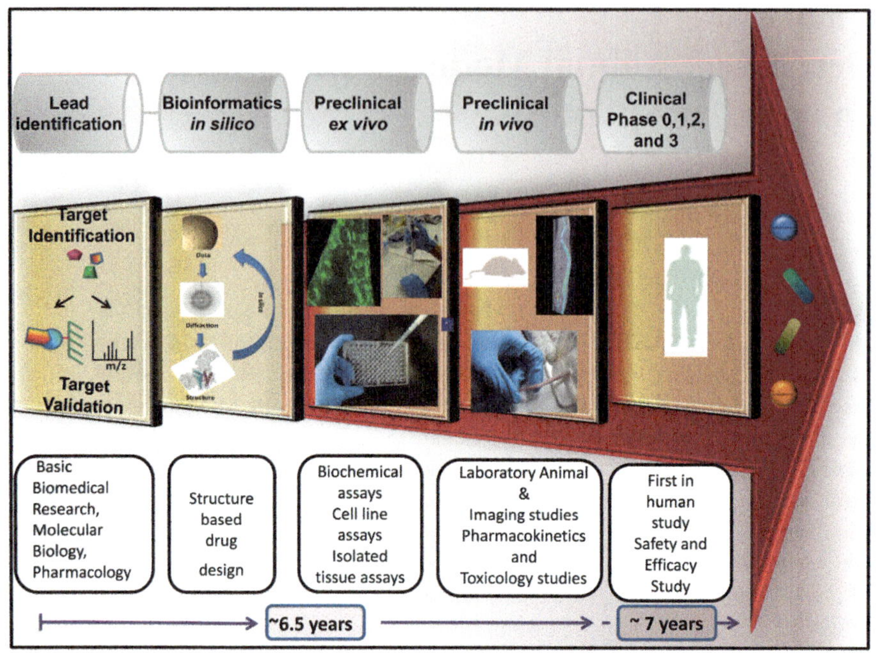

Fig. 6.1 Schematic representation of steps towards drug discovery with an average time line

approach including *in silico*, biochemical, pharmacological, *ex vivo*, *in vivo* imaging and animal model studies are required for target identification i.e. finding potential '*leads*' from innumerable '*hits*'. These '*leads*' are further taken forward through several steps of preclinical trials prior to successful clinical development (Fig. 6.1).

The strategies to cure and control disease progression due to impaired apoptosis led to targeting of proteases involved in this pathway. Less effective conventional methods that include activating proapoptotic signals by administration of cytotoxic and chemotherapeutic agents to induce cell damage are making way for more targeted therapy. Specific targeting of diseased cells can be achieved in multiple ways such as through activating or inhibiting proapoptotic proteins or designing antagonists of antiapoptotic molecules. Several such molecules (antisense oligonucleotides, peptidomimetics and synthetic compounds) are being currently explored as drug targets. The previous chapter vividly describes different laboratory protocols with relevant examples to understand and characterize the role of proapoptotic proteases in cell death. Here, we continue this effort with a step further toward discovering their potential in treatment of different diseases such as cancer and neurodegenerative disorders using *in vivo* animal model system and non-invasive preclinical imaging.

Although *in vitro* and *ex vivo* studies, for example, experiments with tumor cell lines, are essential in lead identification and optimization, they are not sufficient to mimic the complex disease environment in human systems [1, 2]. Moreover, this system fails to answer questions related to drug distribution, drug uptake, and pharmacokinetics and so on, which are essential steps for clinical development. To fill in this huge gap between fundamental studies and clinical applications, preclinical *in vivo* model systems have played a pivotal role over the last few decades. Different animal model systems have been used for understanding the complexities of a disease, its progression and effects of drugs in a three-dimensional scaffold. Both nude and SCID (severe combined immunodeficiency) mouse xenograft model systems are routinely used to study initiation and progression of different diseases. Advancement in the field of genetic engineering has also led to the use of transgenic mouse models in elucidating pathophysiological conditions [3–5]. A major shift in the focus of drug discovery in the modern era has identified various limitations that are associated with murine models, which consequently led way to exploring alternative models such as zebrafish and canines with spontaneous cancer [6–8]. Thus, a requirement for establishing a perfect model system becomes imperative where molecular characterization of the disease along with effect of the candidate drugs and drug-like molecules can be tested effectively. However, in this chapter, the most widely used mouse model systems have been described in context of apoptotic molecules and their associated diseases.

Non-invasive imaging in small animal model system is extremely useful toward elucidation of molecular events in disease progression, recurrent disease management and monitoring of response to drugs [9]. With increasing interest in harnessing apoptotic molecules for understanding and treatment of various diseases, an upsurge in advancement of imaging techniques has been observed in the past decade to monitor cell death. Various fluorescent and radiolabeled agents for PET/SPECT imaging have been specifically designed to study activation of proteases (mainly caspases) for direct visualization and quantitative evaluation of apoptosis *in vivo*. Development of several studies in animal model systems combined with tremendous advancement in the field of small animal imaging therefore showcase very high potential to translate basic research into clinical medicine.

2 Animal Model

In vitro and *ex vivo* studies have provided a wealth of information in biomedical research with critical leads in drug designing and development [10, 11] in the last few decades. However, they are limited by their inability to simulate the cellular milieu during a physiological process. This led to extensive research toward development of an appropriate animal model system so as to understand the intricacies of a complex cellular mechanism or pathophysiology of a disease.

Studies on animals led to significant advancement in translational research. Understanding of a human disease at the molecular level has been made possible either by studying spontaneous diseases in animal counterparts or by inducing different factors in normal animals for development of a disease of interest [12].

A varied array of species ranging from bacteria, protozoa, insects, zebra fish, guinea pigs, rabbits, felines, dogs and non-human primates [6, 7, 13–17] have been used in studying molecular mechanisms of different human ailments. Mouse has so far been the most popular and widely used model system for *in vivo* studies of human diseases [12].

2.1 Mouse as a Laboratory Model System

Quite intriguingly, mouse was used as a tool to study human anatomy way back in the second century AD by Galen, a Greek physician [18]. After a long recess, mouse regained its prominence as a popular animal model in experiments ranging from properties of air by Robert Hooke in the mid seventeenth century, to studies on inheritance in early 1900s by William Castle [19]. Owing to several advantages such as small size, easy and inexpensive maintenance, abundant breeding, optimal gestation period and most importantly ability to be genetically manipulated, mouse has become the most popular ($\sim$70 %) laboratory animal with nearly 450 inbred varieties for studying human physiology, cellular pathways and effects of drugs on diseases [20, 21]. Here we will look into various applications of mouse model system in understanding the role of proapoptotic proteases in modulating neurodegenerative diseases and cancer.

Although the laboratory mouse bears several similarities with domestic mouse, artificial breeding has led to marked differences in certain characteristics that are required for study of specific diseases or macromolecules in cellular pathways. Currently, there are different types of genetic modifications available in mouse models. Therefore, strategically devising proper sets of experiments is the key toward understanding the regulatory steps in a cellular process or pathophysiology of a disease. The different types of laboratory mice that have been bred, maintained and used for biomedical research are discussed as follows:

2.1.1 Nude Mice

The pioneer in breeding nude mice was Miroslav Holub (1923–1998), an eminent Czech immunologist [22]. The basis for generating nude mice is to study tumor biology and obtain xenografts. They are made immunodeficient by genetic removal of FOXN1 gene [23] leading to absence of thymus. This results in hairless phenotype from which they derive their name. These mice do not produce T lymphocytes, therefore do not exhibit major types of immune responses that include

(a) Generation of antibody due to lack of $CD4^+$ helper T cells
(b) Cell-mediated immune responses due to absence of one or both $CD4^+$ and/or $CD8^+$ T cells
(c) Rejection of grafts that require both $CD4^+$ and $CD8^+$ T cells

(d) Elimination of virus-infected or cells exhibiting malignancy due to lack of $CD8^+$ cytotoxic T cells

Overall, the nude mice do not exhibit immune response toward foreign tissues or tumor grafts and have been a popular tool to be used in imaging and transplantation.

Nude mice served a popular experimental model for study of genesis, progression and treatment of cancer in the past few decades. However, precise experimental control is required for obtaining reproducible data. Several parameters need to be taken into careful consideration including tumor type, origin, kind of mouse host and its age for propagation of different human malignancies. Studies in nude mice have provided interesting findings regarding metastatic processes as well. Attempts have also been made to study spontaneous tumors in these animals but their frequency of occurrence compared to the thymus-bearing counterparts remains questionable. Nevertheless, for over 25 years, the nude mice system has provided a wealth of information on tumor biology and progression of cancer and other diseases [24]. The nude mice, although widely used, sometimes produce leaky phenotypes with presence of a few residual T lymphocytes. As a consequence, they are losing their prominence in translational research and making way for SCID mice.

2.1.2 Severe Combined Immunodeficient (SCID) Mice

Scid mice are homozygous for an autosomal recessive mutation (scid) on chromosome 16 and exhibit an apparent detention in the initial developmental stages of B and T cells. The pro-B cells that are formed in the bone marrow, do not mature into pre-B and B cells. Similarly, although early T-lineage cells (Thy-1 + IL-2R +) are found in thymus, absence of functional $CD3^+$ T cells are not observed. These conditions arise due to loss of function of PRKDC gene [25] which leads to impairment in resolving DNA strand breaks in developing T and B lymphocytes by the DNA repair enzyme (DNA-dependent protein kinase catalytic subunit) that the gene encodes [26]. All other nonlymphoid blood cells exhibit normal differentiation. SCID mice are being routinely used for transplantation and xenografting tumors as well as normal and malignant tissues. In addition, they are also used to test vaccines [27, 28].

The major application of SCID mice lies in engraftment of xenogeneic tumors. Although the quality of xenografts are similar to that of nude mice, in some conditions such as retinoblastoma, acute lymphoblastic leukemia, lung metastasis etc [29, 30], SCID show improved engraftment. Thus for several human tumors SCID mouse has established itself as a model with high efficacy to study initiation, progression metastasis and effect of drugs. However, as observed in case of nude mice, SCID unfortunately is also not completely devoid of leaky phenotypes as 15–25 % of young adult mice and almost all old mice develop a few clones of B and T cells.

A variant of SCID called NSG, or NOD scid gamma (NOD.Cg-$Prkdc^{scid}$ $Il2rg^{tm1Wjl}$/SzJ), where the genetic background is derived from inbred NOD (non obese diabetic) mouse strain, is among the most immunodeficient inbred mice varieties described till date [31]. First developed in the Jackson laboratory [32], these mice do not contain mature T, B and natural killer (NK) cells. NSGs have impaired cytokine signaling pathways as well as defective innate immune system [33]. These multiple immunodeficiencies allow them to engraft a variety of primary human cells as well as to mimic complex cellular networks and diseases.

2.1.3 Transgenic Mice

The idea of introducing a foreign gene material into mouse germ line revolutionized the field of genetic engineering and translational research. This discovery has given a new dimension to the study of specific genes as well as human diseases in a living system. Apoptosis research as well as preclinical studies targeting diseases such as cancer and neurodegeneration have benefited tremendously with the advent of this technology [34, 35]. This advancement came hand-in-hand with emergence of technologies such as gene cloning, chromosomal mapping and DNA sequencing [36].

Transgenic mice can be generated mainly via three possible ways:

(a) Delivery of foreign DNA by retroviral infection of mouse embryos at various developmental stages. However, due to technical complications, this technique has not become very popular [36].
(b) The second method is very commonly used where foreign genetic material is injected into the pronuclei of fertilized one-cell mouse embryos leading to random integration of multiple copies of the transgene. This is followed by transfer of the embryos in oviducts of foster mothers that eventually produce the transgene [37, 38].
(c) Finally, another way of procuring transgenic mice is through introduction of 'loss or gain of function' mutations of varied sizes at loci of interest in mouse embryonic stem cell (ES) [37, 38].

Detailed protocol for developing transgenic mice is provided at the end of this section.

2.1.4 Knockout Mice

These '*designer mice*' have become an integral part of preclinical and clinical research where a particular gene of interest is made non-functional by knocking it out. This simple method aids in understanding the importance of a gene in an intricate cellular process and complexities of a disease. This is achieved using the homologous recombination technology [39] where mutations are introduced in pluripotent mouse stem cell lines (ES) that subsequently get transmitted to the progeny [40–42]. Knockout technology has been tremendously used in the last few decades for understanding apoptotic pathway and the diseases it is associated with [43]. *Oncomice* is such a variant of transgenic mice that has been genetically designed to develop cancer by expressing prominent oncogenes, way back in the 1980s.

Oncomice has been phenomenal in providing in-depth understanding of different stages of cancer as well as to devise strategies for therapeutic intervention [44].

3 Proteases in Preclinical and Clinical Research

Proteases are tightly regulated during tissue development to maintain homeostasis and proper cellular functions. However, genetic aberrations, exo- as well as endogenous factors, abnormal (both less and excessive) proteolytic activity that occurs in the cellular milieu, often lead to several diseases including cancer, inflammation, neurodegeneration as well as microbial infections [45]. Modulating activity of specific proteases so as to control or eliminate a disease condition has been one of the major foci of pharmaceutical research [46]. History of drug design involving proteases goes way back to the mid twentieth century when two drugs against thrombosis (heparin and warfarin-a vitamin K analog) became available that indirectly regulate thrombin activity. Two major breakthroughs in targeting proteases is development of ACE (peptidyl dipeptidase A) inhibitors against cardiovascular diseases [47] and HIV protease inhibitors [10]. Although, designing a protease modulator (mostly inhibitors) seems straightforward, it requires an in-depth understanding of the structure, dynamics and activation mechanism of the protease. Inhibitors can either target the protease directly, allosterically or indirectly through inactivation of downstream protease molecules [48–52]. Moreover, absence of any direct correlation between expression levels and altered activity of an enzyme adds to the complexity. Recently, tremendous efforts are being made to circumvent these barriers through an interdisciplinary holistic approach including *in silico* drug design, structural probes, proteomics and other high throughput approaches that act as an excellent support system for *in vivo* studies. This technical advancement has led to a huge upsurge in the overall clinical and pharmaceutical research involving proteases with more than 50 proteases being characterized as potential targets. A majority of these targets were identified in knockout studies in mouse model systems [47, 53–59] reiterating the indispensability of this modality in preclinical and clinical research.

3.1 Different Strategies to Modulate Protease Activities

Targeting a protease though challenging can be achieved through multiple ways. Based on the requirement for a particular disease, an inhibitor or an activator may be designed. Moreover, allosteric regulators also play important roles in modulating a protease activity. Inhibitors in turn, can be reversible or irreversible. Studies have shown that reversible non-covalent inhibitors have better effect in the intracellular milieu compared to their irreversible counterparts as the former provide more specificity and lesser side effects. For designing reversible, non covalent inhibitors, transition state analogs are used as templates [60, 61]. On the other hand, allosteric modulators are considered better than the orthosteric ones due to greater sub-site specificity and lower-dose requirement. In a unique approach called '*tethering*', a cysteine residue in the vicinity of the active site of caspases-3 and -7 was reversibly attached via a disulfide bond to a thiol group of the potential small molecule inhibitor(s) [62, 63]. This led to allosteric inhibition of the executioner caspases and freezing them in a proenzyme conformation. Similarly, enhancing the activity of a protease is possible indirectly through inhibiting its inhibitor. One such brilliant example is SMAC mimetics or small molecule inhibitors of IAP proteins, which in turn release inhibition on caspases thus facilitating apoptosis [34, 48]. These molecules have shown cancer specific apoptosis both in mammalian cell culture as well as in xenograft tumor model system in mice and are currently in clinical trials.

3.2 Mouse Models Involving Proteases of the Apoptotic Pathway

The proteases in the apoptotic pathway have been extensively explored in biomedical research. A few successful stories emanating from these efforts are substrates of caspases and granzyme B [64, 65]. In addition, SMAC mimetics and inhibitors of cathepsin K are under clinical trials. While SMAC analogs are devised to combat cancer, few inhibitors of cathepsin K are currently being tested against osteoporosis, osteoarthritis and bone metastasis [66]. Understanding the intricacies of a cellular pathway as well as effects of drugs or drug-like molecules on a pathological condition requires a living model system. As mentioned earlier, knockout studies have been an extremely important tool for understanding the role of proteases in a particular disease and devising strategies for therapeutic intervention. A slightly modified model system where an animal disease model is combined with a rodent knockout model has led to specifically understand the role of the target protease, for example, cathepsins in pancreatic cancer [67]. Knockouts have also been used as a model system to understand role of proteases in neurodegeneration [35]. Apart from these, tumor xenograft models in rodents, mainly mice have given interesting insights into initiation, progression and metastasis of cancer.

Here we would discuss general strategies involving *in vivo* tumor xenograft and knockout model systems for targeting proapoptotic proteases or their interacting partners in diseases such as cancer or neurodegeneration using two specific examples.

***Case I: Designing mimetics of a proapoptotic molecule to relieve inhibitory effect of its antiapoptotic binding partner on caspases*:**

An enormous amount of research endeavors to understand the protein-protein interactions involving apoptotic proteases resulted in a few molecules to enter clinical trials. One such molecule is a mimetic of proapoptotic SMAC/DIABLO (second mitochondria-derived activator of caspase/direct IAP binding protein with low pI) that relieves inhibition of inhibitor of apoptosis proteins (IAPs) on upstream caspases [68, 69]. It has been observed that IAPs have been associated with different cancers such as breast, prostrate, lung, renal and bladder carcinomas [70–73]. Moreover, this antiapoptotic family of proteins including XIAP and cIAPs facilitate metastatic progression in several cancers [74, 75]. At the cellular level, antiapoptotic property of IAPs is manifested through their ability to interact with upstream caspases which make them potential therapeutic targets against cancer. SMAC/DIABLO is a mitochondrial protein that gets released into the cytosol upon apoptotic trigger and relieves the inhibition of IAPs on caspases by interacting with IAPs through their N-terminal tetrapeptide (AVPI) motif. While the BIR3 (Baculovirus Inhibitor of apoptosis protein Repeat domain 3) of IAPs are known to bind caspase-9, BIR2 domain inhibits caspase-3 and -7 [76]. Therefore, agonists of SMACs (peptidomimetics and small molecule analogs) can be tested as therapeutic targets against several types of cancers where over-expression of IAPs has been observed. It has been proven that SMAC analogs induce apoptosis in tumor cells and reduce tumor growth in mice [77–82].

Overall Strategy The mimetics of 'AVPI' have been developed using structure guided design either by, 'shape-based screening' of a virtual drug library or by designing small molecule analogs of 'AVPI' by molecular docking using SMAC-XIAP complex structure as a template (PDB IDs: 1G73 and 1TW6) [83, 84]. A few best molecules were selected based upon binding score and were further tested for their drug-like properties. A 100 total '*leads*' were shortlisted which were combined to get a final 71 '*hits*' to be tested in *ex vivo* model systems. In the other work, structure-guided chemical synthesis of a series of small molecule analogs was performed directly. In both the cases, a battery of *ex vivo* studies were performed such as cytotoxicity, caspase activation, cell phase distribution and IAP inhibition assays to screen these molecules for *in vivo* testing of their efficacy. Xenografts were generated in nude mice for studying the effect of the synthesized SMAC analog on prostrate and breast tumor xenograft models. The mice were subjected to treatment intraperitoneally with different doses of the '*hit*' compound. The animals were sacrificed post experiment, and the tumors were removed, a series of experiments that followed provided vital information on the ability of the compounds to reduce the weight and volume of the tumors. Elaborate preclinical pharmacokinetic studies and ADME (absorption, distribution, metabolism, and excretion) analyses were performed in various animal models to identify the best molecule to take forward toward clinical trials.

Case II: Knock out of a caspase gene: Implications in neurodegenerative diseases [35]

Caspases are the most prominent proteases involved in programmed cell death and therefore their deregulation leads to several diseases that include neurodegeneration. Caspase 6, like any other protease of the family exists as a proenzyme which gets processed into active dimeric protease by upstream caspases -7, -8 and -10 post apoptotic signal. In addition, self-activation in caspase 6 has also been observed

[85]. Although predominantly an executioner protease, caspase 6 has also been found as an initiator of the caspase cascade. This statement is supported by the observation that caspase 6 is present in brains of Huntington and Alzheimer patients long before apparent cell death and in turn has the ability to activate caspase-2, -3 and -8 [86–89]. In addition, inhibition of β-amyloid precursor protein at caspase 6 recognition site (residue 664), rescues Alzheimer-like phenotype [90–97]. Caspase 6 activation is also associated with axonal degeneration in an Alzheimer's Disease (AD) mouse model [98]. Similarly, cleavage of a mutant Huntington (mhtt) at a particular residue (AA 586) leads to pathogenesis of Huntington's disease (HD). It has also been reported in the literature that prevention of this cleavage in mice prevents neurotoxicity and aberrant neuronal behaviour [99–103]. These observations strongly implicate caspase 6 in progression of neurological diseases and therefore designing caspase 6 inhibitors could be a successful strategy to combat those ailments. Since understanding the role of caspase 6 in normal brain development is the first step in this process, this study performs behavioural and neuropathological characterization of mouse brain lacking caspase 6 (caspase 6-/-) [35].

The first step in this process was to precisely design, develop and validate a *caspase 6-/-* mouse model system (refer to step-by-step protocol below). These knockout mice were then used for several experiments pertaining to caspase 6 mediated axonal degeneration and excitotoxicity, determination of Mendelian ratio and size of the cerebellar cortex, so as to understand the effect of caspase 6 on normal functions of the brain, determination of body weight as well as motor coordination. All these studies in caspase (6-/-) knockouts confirm the role of caspase 6 activation primarily in AD and/or HD. This suggests designing caspase 6 inhibitors would be a promising therapeutic strategy against Alzheimer's and other neurodegenerative diseases.

4 General Protocol for Generation of Tumor Xenograft and Knockout Mice

4.1 *Tumor Xenograft in Nude Mice: A Step-by-Step Protocol*

A flowchart providing a pictorial representation of xenograft generation is shown in Fig. 6.2a.

Step 1. Preparation of donor tumor-bearing mouse:

(a) Thaw the cryo-preserved tumor sample and transplant it subcutaneously using sterile trocar with proper aseptic care
(b) Observe the tumor growth daily
(c) As the tumor diameter reaches ~1.5 cm, the tumor can be used for transplantation in other mice. This tumor bearing mouse is known as *donor mouse*

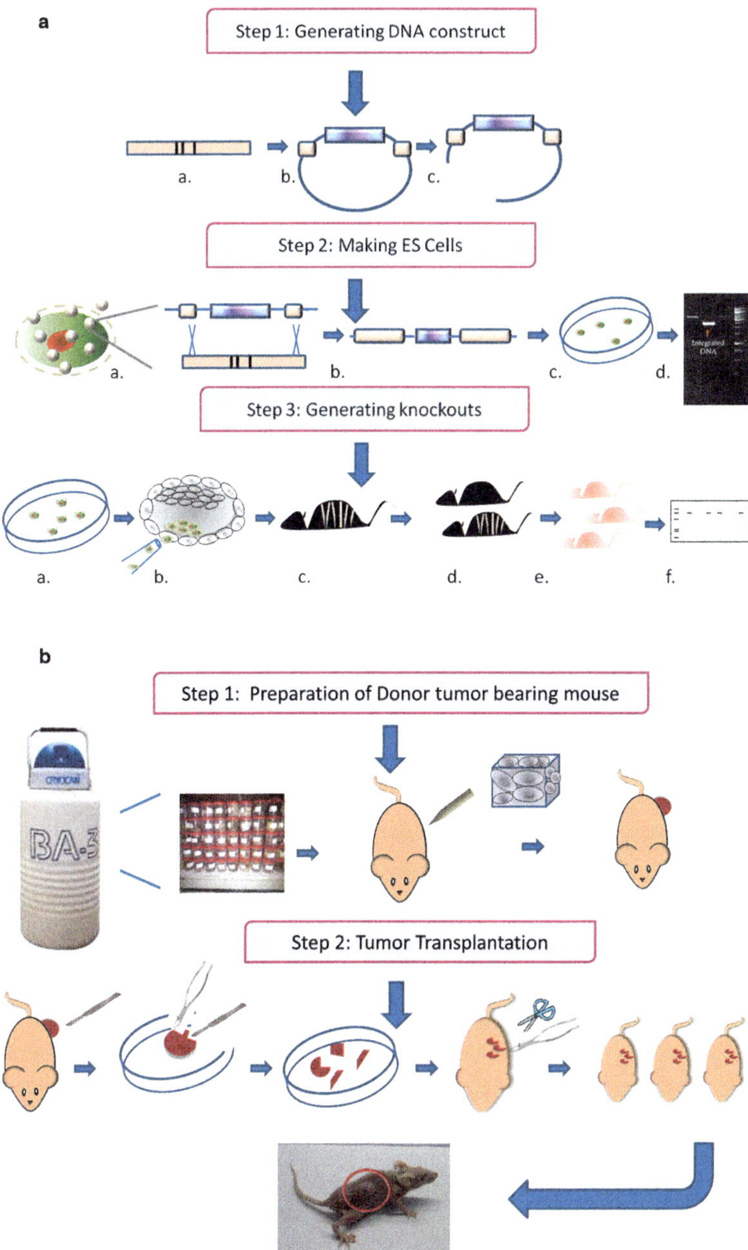

Fig. 6.2 Flowcharts illustrating generation of tumor xenograft and knockout mice. (**a**) Generation of knock-out mice. The steps (designated as *a*, *b*, *c* etc.) in the figure are provided in detail in the text. (**b**) Generation of xenograft bearing mice. Step 1 in the figure represents the following (*from left to right*): cryo-preserved tumor samples, subcutaneous transplantation of the tumor in donor mouse with a trocar and follow the growth of the tumor to an optimum size. Step 2 (*from left to right*): Select the tumor bearing mouse and remove the tumor with a scalpel blade, transfer tumor mass in a medium with suitable antibiotics, cut the tumor into several small pieces, place a small piece under the skin of *recipient* mouse using forceps and scissors, growth of tumor in *recipient* mouse (tumor is encircled in *red*)

Step 2. Tumor transplantation:

Precautions: The person carrying out tumor transplantation must wear sterile protective gears i.e. long surgical lab-coat, cap, mask, shoe covers, and surgical latex gloves. This procedure should be conducted in laminar hood. The ultraviolet lamp in laminar hood should be turned on 15–30 min prior to initiation of the procedure.

(a) Select the *donor* tumor-bearing mouse
(b) Sacrifice the donor tumor-bearing mouse by cervical dislocation and dip it in a jar containing absolute alcohol for approximately 3 min
(c) Remove the mouse from jar and wipe the excess alcohol using a paper towel
(d) Pull the skin over the tumor and detach it from the tumor with the help of scalpel blade, if required
(e) Remove the tumor mass completely and put it into suitable medium with broad spectrum antibiotics (streptomycin and penicillin) in a petri-plate
(f) Transfer the donor mouse into the discard-bag and dispose it following proper procedure of biological waste disposal
(g) Cut the tumor into small pieces of approximately 2 cubic mm with the help of sterile forceps and scalpel blade. Make as many pieces as number of mice needed for the study along with an additional piece for histological examination of the tumor
(h) Anaesthetize the *recipient* mice using isoflurane (2-chloro-2-(difluoromethoxy)-1, 1, 1-trifluoro-ethane) gaseous anesthesia
(i) Pick up one mouse and place it on a sterile paper towel in a position so that its dorsal side faces the sky
(j) Dip a small piece of cotton gauze in absolute alcohol (ethanol) and wipe the skin on the posterior side of the back
(k) Lift this skin with blunt bent sterile forceps and make a small incision (approximately 3 mm) with 1.5 cm blunt-end scissors. Insert the scissors under the skin through this incision towards the right hind limb area. Gently open the scissors to widen the skin pocket and then quickly bring it back to closed position
(l) Remove the scissors from the skin pocket still holding the skin at the incision using forceps with left hand. With the right hand, pick up one piece of tumor from the petri-plate using blunt bent forceps and insert it deep into the skin pocket that has already been made. After releasing the tumor into the pocket, remove the forceps and roll it over the pocket skin in a sweeping movement so as to push the tumor piece to the interior of the pocket
(m) Release the skin at the incision and put 1 drop of 100× antibiotic cocktail on the incision followed by one or two drops of tissue adhesive
(n) Transfer the mouse to the recovery cage. Maximum recovery-time required with isoflurane is 2–3 min
(o) Repeat the procedure from step 11–14 for transplant in the next mouse. This procedure takes 3–5 min of time

Carry out tumor transplants in animals required for one study in single sitting, using the tumor from the same *donor* mouse.

4.2 Protocol for Generating Knockout Mouse

An overview of the experimental procedure is shown in Fig. 6.2b

Step 1. Acquiring DNA of Interest:

(a) Screen genomic library that is generated preferably from 129Sv-derived embryonic stem (ES) cell lines. The advantage of 129Sv over ES cell lines derived from other strains is that they are more reliable at forming germ cell line colonies [104]. The 129Sv cells that dramatically changed the field of transgenic mouse technology was derived from a cross breed of 129 substrain and another inbred strain [105]. Alternatively, use the conventional method of PCR amplification of the gene of interest from genomic databases
(b) Insert the PCR amplified region of genomic DNA that encompasses the DNA of interest into a bacterial plasmid with compatible restriction sites. This DNA of interest should also have a selectable drug resistance marker
(c) Linearize the DNA using suitable restriction enzymes

Step 2. Development of ES cells:

(a) Insert the DNA of interest into embryonic stem cells via electroporation.
(b) This piece of DNA replaces a part of the normal gene in the mouse through homologous recombination.
(c) Grow the cells in suitable medium with proper antibiotics. Cells that uptake the plasmid with the antibiotic resistant insert will grow and the rest will die. Hand-pick ~500–600 colonies and expand them in plates
(d) Perform DNA extraction (using standard molecular biology protocols) from individual colonies Perform Southern Blot experiment or PCR to identify the colonies with integrated DNA at the proper loci

These clones will be further used to generate the knockout mouse.

Step 3. Generation of Knockout mouse:

(a) Grow the ES cells harboring the gene of interest in a suitable culture medium
(b) Inject ES cells into mouse blastocysts (having brown and black phenotypes due to different strains of donor and recipient mouse respectively) so that it becomes an integral part of the embryonic tissue. Transfer these blastocysts into the uterus of pseudo-pregnant recipient female mouse
(c) Pups that are born with integrated ES cells are chimeras with black and brown stripes owing to the respective coat colors of the donor and recipient. This will help in selecting the chimeras easily
(d) Breed female chimeras with black male for approximately 8 weeks
(e) Select the brown offspring's (suggests modified ES cells have contributed to the germ line)
(f) Perform genetic tests to identify mice with knockout gene. Breed desired colonies and initiate phenotypic analysis

5 Current Scenario and Future Perspectives

The ever expanding biomedical literature and successes in preclinical and clinical trials underscore the importance of animal models in basic as well as translational research. Techniques to utilize this unique research tool has been extensively modified and fine-tuned to cater to existing as well as novel research ideas. However, although mouse models have been successful in aiding the basic biomedical research, it requires more fidelity and reproducibility for using it in preclinical settings. For example, a rigorous screening procedure should precede preclinical studies with any mouse model. The mouse model should undergo a quality control check to ensure its ability to mimic genetics of a human system or pathophysiology of a disease. Secondly, the histopathology of the model should precisely match that of human tumors and it needs to be evaluated by a scientist with expertise both in human and animal pathology. Another interesting way of establishing credibility of mouse models is through back-validation i.e. studying the effect of drugs in mice whose effects have already been established in humans. This global and transparent approach would definitely lead toward development of better mouse models that would efficiently simulate human physiology as well as diseased conditions.

Although transgenic rodent models have proved their mettle in biomedical and cancer research, another less explored yet excellent alternative is use of spontaneous tumor models that provide perfect environment to test novel therapeutics. Some of these promising immune competent and syngeneic models are genetically engineered mice (GEM) and companion (pet) animals that naturally develop cancers [7]. An appropriate use of both xenograft and spontaneous models, taking into consideration their *pros* and *cons*, might help understand complex physiological processes as well as combat different diseases with higher precision and efficacy.

6 Noninvasive Apoptosis Imaging

6.1 Introduction

In vivo noninvasive imaging being a potential tool for visualizing and understanding pathophysiological processes, plays an important role in development of novel diagnostic and therapeutic molecules. Apoptosis is an evolutionary conserved and tightly regulated biological phenomenon that is crucial for maintenance of cellular homeostasis [106]. It is a fundamental process observed in normal as well as in cells exposed to cytotoxic agents used in anticancer therapy regime. Apoptosis induced by anticancer drug is often seen as an indicator of therapy response. Hence *in vivo* apoptosis imaging is an active area of research since last decade, and efforts are being directed towards making use of it as a prognostic marker [107]. Since, apoptotic switch is highly regulated by a subset of proteases that are activated by exogenous and endogenous factors, these proteins have become an integral part of apoptosis imaging.

6.2 Imaging Techniques

The field of diagnostic imaging includes a variety of non-invasive modalities that are pivotal for visualization and assessment of molecular events in several diseases such as neoplastic and neurodegenerative processes. Imaging modalities including x-ray computed tomography (CT), positron emission tomography (PET), single photon emission computed tomography (SPECT) and magnetic resonance imaging (MRI) have routine application in imaging laboratory animals as well as in clinics [108–111]. However modalities such as optical (bioluminescence and fluorescence) and cerenkov imaging are still at preclinical stage and gradually leaping towards clinical applications [112, 113]. Most of these tools are complementary to each other and help visualization and precise measurement of structural changes, biochemical, physiological and molecular processes. Currently dual or hybrid imaging (HDI) is preferred over single imaging modality as it provides an advantage to visualize molecular/biochemical/functional events along with structural information. However, multi-modality requires precise image co-registration algorithm besides complementarity of the technologies. The two most commonly used imaging combinations or 'fusion imaging' systems are PET-CT and SPECT-CT where molecular/functional/biochemical data can be co-registered on anatomical platform and read together to enhance our understanding of the underline molecular processes in both preclinical and clinical settings [114].

The above mentioned imaging techniques provide a large amount of information without using any invasive tool. Moreover, these procedures do not alter the disease process or cause unacceptable discomfort to the animal. Although these procedures are painless, sedation is often desirable to reduce associated anxiety and stress so as to enable acquisition of good diagnostic data with minimal repeats. The use of anesthetic agents also controls stress or pain associated with handling of tumor bearing animals or other disease models. However, the most challenging aspect of these preclinical systems is to achieve sub millimeter spatial resolution, which is plausible with the use of high-end detector materials, efficient signal transmission electronics, appropriate data acquisition and processing algorithms.

In vivo preclinical imaging has high impact in drug development as it can precisely monitor disease progression and therapeutic response longitudinally in the same subject. *In vivo* experiments are typical longitudinal studies where each individual animal serves as its own control thus enhancing the reproducibility and accuracy of the data. PET Tracers (e.g. Fluorine-18 based), SPECT tracers (e.g. ^{99m}Technetium and ^{125}Iodine), contrast agents, bioluminescent markers/reporters (luciferins, proluciferins etc.) are the most often used reagents in these imaging studies [115–118].

The selection of an imaging technique mainly depends upon the disease to be monitored and its pathophysiological impact. Translation from preclinical evaluation to clinical studies has been highly facilitated by these present generation imaging modalities with the imaging protocols being easily applicable in clinical setup. They are routinely used in characterizing newer and more realistic models of human diseases such as invasive disease in the tissue of origin as well as transgenic mouse models. Another important reason for present interest in preclinical imaging is making the process of drug discovery and development more dependable. The process of drug discovery is both time consuming and resource intensive mostly with uncertain outcomes. Therefore to reduce the failure rate of *drug-like* molecules in the later stage of clinical trials, incorporation of advanced translational imaging technologies in preclinical studies is essential. These cutting-edge translational imaging platforms help predict and understand the importance and limitations of the animal model system in a particular clinical evaluation [119]. The different imaging modalities are described below in detail.

6.2.1 Positron Emission Tomography

Small animal PET imaging has become an indispensable preclinical tool in the last decade due to advanced technological development in instrumentation and detector technology. It has tremendous potential to translate basic understanding derived from animal imaging to clinical medicine. PET is being used increasingly to advance the understanding of cellular and molecular processes that are altered in cancer initiation and progression [109]. Compared with other molecular imaging technologies, PET enables higher sensitive and quantitative measurements of biological and biochemical processes *in vivo* through specific labeling of organic compounds (or close analogs) with positron emitters, such as ^{18}F [120, 121]. Human PET scanner has revolutionized biomedical research since its introduction in mid-1970s. Simultaneously, there was a strong interest in developing preclinical scanners due to inability of the human system to image a mouse, which is $\sim$2000 times lighter. Over a period of time, scanners for imaging animals have been developed from analog to today's completely digital systems with advanced solid state detector materials [122].

Instrumentation Advanced microPET scanner has one-to-one coupled Phoswich Avalanche Photodiode (APD) detector. This detector is optically coupled with a pair of scintillation crystals that are made up of materials with nanosecond decay time. This configuration provides high spatial resolution and advanced coupling of scintillator to APD enhances energy resolution. This combination of scintillation detectors efficiently attenuates high energy (511 keV) gamma ray photons. However, the short half-life of PET radioisotope is a challenging factor in terms of synthesis of molecules and further execution of the animal imaging studies. Hence availability of a medical cyclotron onsite is highly desirable. Among all PET radionuclides, ^{18}F has sufficiently long half-life, hence there is a huge interest in developing ^{18}F-based molecules [121]. There are also several long-lived radioisotopes such as ^{64}Cu and ^{124}I having half-life of 12.7 h and 4.2 days respectively [123]. These isotopes can be shipped from production site to various preclinical labs. Usefulness of these radionuclides is in metabolic/accumulation/clearance studies where tracer accumulation over a longer period of time is of interest. Alternatively, there are also generator systems consisting of a long-lived parent radionuclide ($^{62}Zn/^{62}Cu$ and $^{68}Ge/^{68}Ga$ generator) that are supplied directly to the preclinical labs as per the research requirements. The parent radioisotope continuously decays into a short-lived daughter radionuclide (^{62}Cu, $T_{1/2} = 9.7$ min; ^{68}Ga, $T_{1/2} = 68$ min) [124]. The use of long-lived radiotracers or availability of generator systems facilitates the preclinical imaging without having an on-site cyclotron. Typical injected tracer dosages for mice are in the range of 50- to 350-μCi. The limiting factor in selection of dose in human studies is a radiation dose, but in case of animals it is count rate capability of the scanner, specific activity and volume of injected dose [125].

6.2.2 Single Photon Computed Tomography (SPECT)

There has been a huge advancement and application of micro-SPECT systems in preclinical research in the last decade. Most important advantage of SPECT over PET is easy availability of ^{99m}Tc (Technetium), which is the most common SPECT radioisotope. The half-life logistics and labeling versatility of ^{99m}Tc adds to the utility of SPECT technology in research setup. Besides, multiple radioisotope imaging can be done simultaneously using SPECT, which enables performing studies with 2 to 3 different radioisotopes having distinguishable gamma emissions [110].

Instrumentation SPECT imaging device is basically a gamma ray detector, which can localize the distribution pattern of systemically administered radiotracer noninvasively. The SPECT instrumentation has significantly improved in the last decade in terms of performance characteristics and its diagnostic/research analytical qualities. Initial gamma cameras had smaller field of view, low sensitivity and resolution as compared to present generation systems. Advanced digital μSPECT scanners have solid-state CZT (Cadmium Zink Telluride) detector due to its superior energy resolution. The detector is mounted on a rotating gantry, which is common for SPECT and CT scanner. These preclinical scanners possess detachable SPECT and CT assembly, which significantly improve the scanner utility. Availability of common animal imaging beds allows its use with multiple scanners of different modalities. The collimator present in this system assists gamma cameras in locating the site of emission and also plays significant role in improving image resolution. Similar to clinical scanners, preclinical scanners are also well equipped with high resolution parallel hole, single and multiple pinhole low energy collimators, which are useful for imaging animals using different research protocols. The achievable resolution using this scanner is ~0.5 mm after data reconstruction. The extraordinary energy resolution of this preclinical system allows multi isotope studies simultaneously.

6.2.3 microCT

microCT owes its popularity as an indispensable imaging tool in preclinical studies to high-quality spatial and temporal resolution. This significant technological advancement has made capture of detailed anatomical images possible so as to precisely monitor the progress of a disease condition in small animals [108].

Earlier its application was limited to high-contrast structures such as bones. Currently, technological advancements and use of better contrast agents made *in vivo* application of microCT possible such as studying soft tissue structures as well as vasculature. This anatomical imaging modality in preclinical research is required for distinguishing structural abnormalities and evaluating the location and extent of disease. Precise location of any lesion can be achieved by co-registration with PET and SPECT image data. The other applications of microCT include mapping of tumor vascularity, visualization of bone metastasis and evaluating novel contrast agents. microCT scanner provides images with ~50 μm resolution and allows faster whole body image acquisition. High-resolution (HIRES) scanning acquisition mode is used for specimen imaging with a resolution of 15–30 μm.

Although, the modality provides images with excellent resolution, its major setback is the high radiation dose that is prescribed for the animals under investigation. This might lead to disruption of biological networks, immune system and functioning of major cellular pathways thus interfering with the results of the study such as tumor size. Proper control studies need to be designed to circumvent this problem and maximal utilization of this technique.

Instrumentation X-ray is the major source for Computed Tomography (CT) imaging. Unlike PET and SPECT, where the subject under probe emits radiation (source), microCT requires external X-ray for imaging. Here, the animal under test is placed in the centre of the scanner and X-rays from a focused radiation source is rotated around the animal under investigation [126]. The two parameters viz. current and voltage that are measured in milliamperes (mA) and kilovolts (kV) respectively determine the strength and number of x-rays produced by the X-ray tube. Increasing the current on the machine increases the number of x-rays produced and hence enhances image contrast. The strength of the X-ray is varied at different rates depending on the density of tissue it is focusing on. Tissue density that influences the absorption of x-rays results in contrast differences in the image. Since the x-ray source is rotated around the test animal, an array of two-dimensional images are acquired, which are later combined to produce 3D images using suitable computer software.

The different major imaging modalities are shown in Fig. 6.3.

6.3 Other Imaging Modalities

6.3.1 Micro Magnetic Resonance Imaging (microMRI)

Based on the principles of nuclear magnetic resonance (NMR) spectroscopy [127, 128], MRI is used in preclinical setting for generating images of different soft tissues. The major difference between microMRI and MRI is the strength of the magnet that is used to generate the magnetic field, which is higher in the former. With its excellent spatial and contrast resolutions as well as better safety

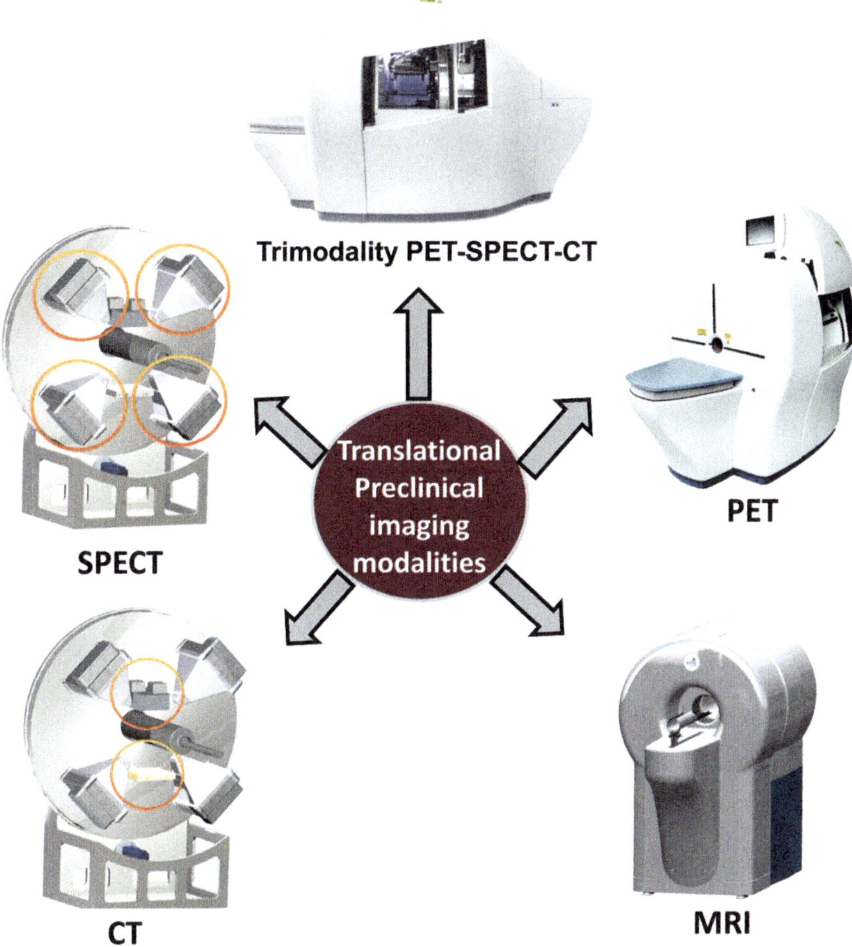

Fig. 6.3 Cartoon depicting different translational preclinical imaging modalities (*Courtesy:* Tri-Foil Imaging, CA, USA)

features (radiation-free), the application of microMRI is somewhat limited due to its astronomical cost, longer scanning time (often hours) and inability to study real time processes and fluids. Currently, it is used for imaging of brain and small tumors in combination with other imaging modalities such as microPET/SPECT.

6.3.2 Cerenkov Luminescence Imaging (CLI)

Almost known for a century, CLI has recently been adopted as an imaging modality for biomedical research. This developing optical imaging tool uses several common medical isotopes for its functionality [112]. Charged particles passing through a dielectric medium at a speed greater than light emit Cherenkov radiation. This modality combines the principles of luminescence with applications of PET radioisotopes and several therapeutic radionuclides for molecular imaging. High signal to noise ratio, ability to perform imaging on multiple subjects parallely makes it a promising tool in small animal imaging. However, several technical challenges need to be overcome such as low signal intensity, before it can be widely used in preclinical and clinical settings.

6.3.3 Optical Imaging

As its name suggests, optical imaging is based on two major principles: fluorescence and bioluminescence [118, 129]. While fluorescence imaging requires a fluorophore such as GFP and YFP (green and yellow fluorescent protein), bioluminescence utilizes enzymatic reactions based on chemiluminescence. However, due to autofluorescence of tissues below 700 nm, near infra-red probes are being used to minimize this interference in fluorescence based assays. Although both these tools are fast, simple and extremely sensitive, their application in preclinical research is limited due to extremely poor penetration ability and hence they find their primary application in studies of biological molecules.

A list of specific features and applications of different imaging modalities is provided in Table 6.1.

7 Protocol for Animal Handling and Imaging

Studies with mouse will be described here with a general protocol applicable for various imaging modalities. The schematic flow of small animal imaging is illustrated in Fig. 6.4.

7.1 Immobilization of the Animal

For imaging studies, it is important for the animal to be immobile during the entire process and therefore anesthesia is the first and foremost step. However, the body temperature of the animals needs to be maintained using heated pads, beds and heating lamps. Extreme care need to be taken for recovery of animals post imaging procedure. Scientists conducting studies that require animals injected with radioactive contrast materials should ensure special handling of the animals after the experiment.

Table 6.1 List of features and applications of different imaging modalities

Imaging modality	Resolution (spatial)	Sensitivity	Penetration distance	Type of emission	Application type	Strengths	Limitations
PET	1–2 mm (pre-clinical)	10^{-11}–10^{-12} M	No limit	Ionizing radiation	Clinical preclinical	User-friendly algorithms	Short-lived radiotracers
							Additional cyclotron required
						High affinity of binding the targets	Not economical
	5–7 mm (clinical)					High sensitivity	Complex radiochemistry
SPECT	0.5–2 mm (pre-clinical)	10^{-10}–10^{-11} M	No limit	Ionizing radiation	Clinical preclinical	Highly sensitive	Safety issues
							Expensive
	8–10 mm (clinical)					Multiple probes	Limited accuracy
						Emission is proportional to probe concentration	
CT	50–200 μm (pre-clinical)	Data not available	No limit	Ionizing radiation	Clinical preclinical	Provides anatomic details	Higher radiation exposure
	0.5–1 mm (clinical)					Very high resolution	Precise control studies need to be performed
MRI	25–100 μm (pre-clinical)	10^{-3}–10^{-5} M	No limit	Non Ionizing radiation but high magnetic field	Clinical preclinical	High resolution	Low sensitivity

(continued)

Table 6.1 (continued)

Imaging modality	Resolution (spatial)	Sensitivity	Penetration distance	Type of emission	Application type	Strengths	Limitations
							Noise arising due to motion
						Equip-ment easily available	
							Expensive
						Compara-tively safer than PET, SPECT and CT	Long scanning time
	mm (clinical)						Claustrop-hobic in clinical setup
CLI	$\sim$220 μm	Data not available	No limit	Ionizing radiation	Preclinical	High signal to noise ratio	Low signal intensity
						Choice of several probes	No preclinical and clinical application so far
						Simulta-neous imaging on multiple subjects	
Optical fluorescence	2–3 mm	10^{-9}–10^{-12} M	$\sim$1 cm or less	Non-ionizing emission	Preclinical	Extrem-ely high ex vivo resolution	Application limited to *ex vivo* studies so far
						Availa-bility of a variety of probes	
Optical biolu-minescence	3–5 mm	10^{-15}–10^{-17} M	1–2 cm	Non-ionizing emission	Preclinical	Extrem-ely high *ex vivo* resolution	Not suitable for *in vivo* application
						Good safety feature	

PET positron emission tomography, *SPECT* single-photon emission computed tomography, *CT* computed tomography, *MRI* magnetic resonance imaging, *CLI* Cerenkov luminescence imaging

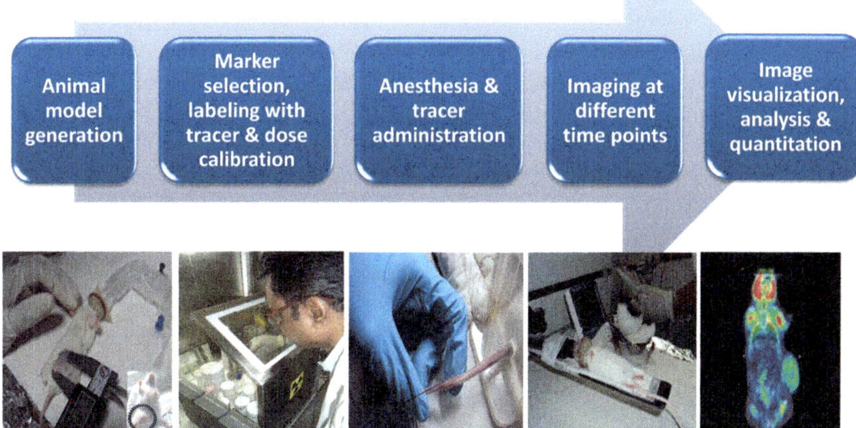

Fig. 6.4 Stepwise self-explanatory illustration of small animal imaging procedure

Anesthesia The mice under investigation will be exposed to an inhalational anesthetic, isoflurane which will be given at an approximate dosage of 2–5 % in an induction chamber. During the entire procedure, this anesthetic state will be maintained in the imaging instrument by supplying 1–2 % isoflurane through a vaporizer-connected nose cone. For studies requiring imaging for a short period of time (30 min or less), injectable anesthetics are used such as Ketamine/Xylazine which are usually administered intraperitoneally at the rate of 100–200 mg or 5–16 mg per kg body weight respectively [130].

Following anesthesia, certain physiological parameters such as electrocardiogram (ECG) and respiratory rate need to be measured using ECG and respiratory sensor pads for PET, CT, SPECT and MRI imaging following proper protocols [111, 131].

7.2 mPET, mCT, mSPECT Imaging

Prior to image acquisition, animals are given injections of contrast agents which might either be iodine-based tracers (CT) or radio tracers (mPET, mSPECT). The role of appropriate tracers is crucial for obtaining high resolution images of target organ/system. Based on the requirement for some radiotracers, overnight fasting is required for optimal tracer distribution. Minimal amount of tracers (in micro- or nanograms) will be administered to the animal so that it does not evoke any pharmacological effect. The tracer materials are either injected directly or introduced with the aid of a catheter in the tail vein of the conscious or anesthetized mouse. In case there is no anesthesia, restraining is required for the procedure. Sometimes, to increase dilation of the tail vein, it is immersed for a minute in warm water or wiped with absolute alcohol prior to administration of the tracer material. As a precautionary measure, the syringe containing the radiotracer is properly shielded to minimize exposure of the administrator. The entire procedure of the tracer uptake and completion of the experiment takes approximately 30 min to 2 h.

Prior to imaging by any of the above-mentioned modalities, the mouse is placed on an absorbent paper towel that is plastic-coated at the bottom to contain any excreted material, Although the approximate time for imaging varies between few minutes to a couple of hours, some studies might require repetition of the same experiment with the same animal several times over a period of several weeks.

For fusion imaging using PET/SPECT and CT, sometimes fiducial markers with low radioactivity are placed on the skin (after hair removal at that region) of the animal so as to mark the anatomic site of interest while acquisition of both PET/SPECT and CT images. This is important in overlaying PET/SPECT and CT images for analysis of the data with higher precision.

Some protocols also demand blood sampling at various time points to study kinetics of novel tracers in the animal system.

7.3 Imaging with Other Modalities (MRI and Optical)

Although the overall procedure is very much similar for all the modalities, there are certain specific requirements for MRI and optical imaging. Since MRI is associated with high magnetic field, care should be taken to use compatible tools and accessories during the entire process. Special training is required for personnel involved in these imaging studies.

In case of optical imaging (both fluorescence and bioluminescence), hair is removed from where signal is expected for optimum light transmission and image acquisition. The use of nude mice is highly recommended in this case to reduce the signal interference. For both bioluminescence as well as fluorescence imaging, anesthetized animals are placed in a light-tight box and imaged with a CCD camera for the time period required for the experiment (typically between few seconds to an hour). Suitable imaging materials are used based on the requirement of the study. For example, in bioluminescence, the animal is injected with a reporter gene with luciferin whereas, a fluorescent-tagged red/near-infrared emitting optical contrast agent or mouse antibodies are used for fluorescence imaging. Present systems are now well equipped with anatomic imaging modality for improved understanding of target sites.

7.4 Post-imaging Care and Record-Keeping

After the experiment, proper care should be taken to house the animals in the vivarium. This is important for regular monitoring and follow-up experiments. Prior to returning the animals, their cages should be treated with disinfectants to eliminate any scope for infection. The cages housing animals with radioactive tracers should contain a proper radioactive label that will provide the details such as the date, name of the isotope, dose, and the approximate date when there will permissible radioactive burden. Regular monitoring of the radioactivity level will be performed by a survey meter till their safe release in general housing areas.

8 *In vivo* Apoptosis Imaging with Specific Apoptosis Targets

Non-invasive imaging of apoptosis is of great interest both in biomedical research as well as in clinical settings where real-time monitoring of several markers of apoptosis and effects of drugs or radiation is possible.

In programmed cell death, both the receptor mediated extrinsic as well as the mitochondrial pathways converge downstream with respect to activation of caspase-3. Different apoptotic markers can therefore be labeled to follow the pathways both in normal cells as well as in tumors. The common tracers and probes used in different apoptotic imaging studies are shown in Fig. 6.5. Moreover, effect of chemotherapy and other drugs can also be monitored in small animal model systems. Different steps in this pathway can be studied using individual or fusion imaging techniques that include microPET, microSPECT, CT, microMRI, and optical imaging [111].

8.1 Phosphatidylserine Imaging

The most popular method of visualizing apoptosis in small animals is through imaging of radiolabeled annexin V [132]. Annexin V is a protein belonging to the annexin family that has high affinity for phosphatidylserine or PS (an anionic phospholipid). Although, it is found on the inner side of plasma membrane in viable cells, apoptotic induction leads to activation of γ-scramblase which flips PS to the outer side of plasma membrane allowing it to interact with annexin V. In addition, drop in membrane potential enhances annexin V binding to PS in a dose-dependent manner [133]. Therefore, an abundance of PS on the cell surface of apoptotic cells post caspase-3 activation, makes annexin V an excellent probe for *in vivo* detection of programmed cell death. ^{99m}Tc-labelled Annexin V derivatives have been routinely used in monitoring apoptosis and necrosis in cell death related disorders [134, 135]. ^{99m}Tc with its optimal radionuclidic properties,

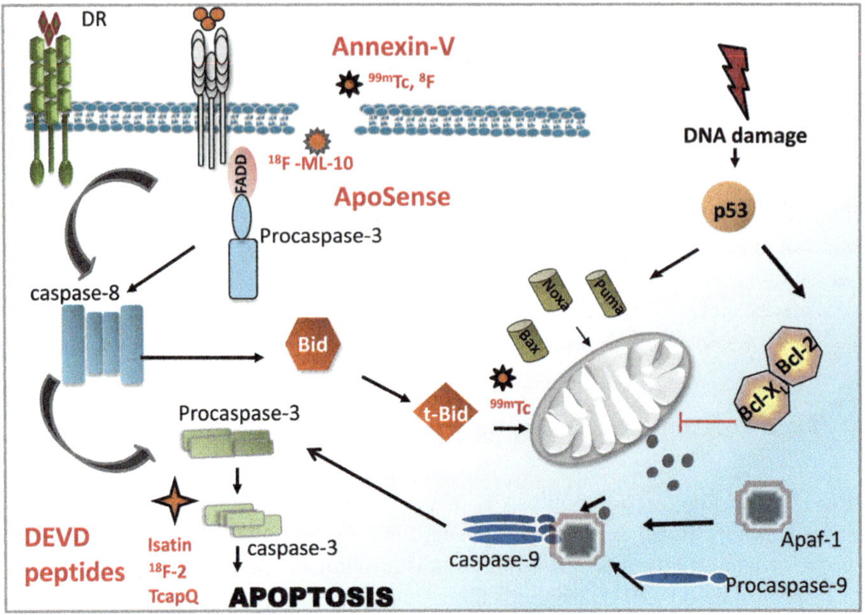

Fig. 6.5 Cartoon representing various probes and tracers for apoptosis imaging. The tracers are shown as *orange symbols* of different shapes

easy availability and low cost is one of the most popular labels for non invasive imaging such as SPECT [136].

^{99m}Tc-annexin V was first used in a mouse lymphoma model that was treated with cyclophosphamide. Post chemotherapy, these animals showed a significant increase in annexin V uptake (300 %) after 20 h compared to control mice [137]. In another more recent study it was used for monitoring apoptosis in a hereditary breast cancer mouse model after docetaxel treatment [138]. Immunohistochemical analysis of the sensitive tumors provided important clues on apoptotic changes that occurred due to the treatment. Moreover, in the early 2000s, ^{99m}Tc-annexin V has been used in clinical trials involving cancer patients with leukemia, small cell lung cancer, lymphoma and so on [139]. Although extensively used, several setbacks such as uptake variability in different subjects, slow pharmacokinetics and low signal to noise ratio limit its use in clinical studies. Furthermore, annexin V is incapable of distinguishing apoptosis from necrosis since phosphatidylserine is also exposed in the latter.

Due to its several drawbacks, annexin V has given way to many other radiolabeled molecules. For example, ^{99m}Tc-labeled C2A domain of synaptotagmin I has been used to study apoptosis in small lung cell carcinoma [140]. More specific binding to target cells and better contrast has made it a more promising candidate than annexin V in microPET and SPECT imaging. Further research has led to development of several peptides and small molecule based probes to target the anionic phosphatidylserine. These small molecules and peptides have the advantages of better clearance from circulation, enhanced target specificity and tumor homing. One such study describes monitoring of xenograft tumor model in nude mice by optical imaging after administration of a fluorescent-based 9-mer peptide with promising results [141]. In addition, ApopSense molecule that harbours a fluorine atom is an excellent probe to study the apoptotic process [142, 143] due to its preferential accumulation in the cytoplasm of apoptotic cells. The fluorine atom makes it most suitable for radiolabeling with its radioisotope (^{18}F) to be monitored by PET imaging.

8.1.1 Morphology of Plasma Membrane

Apart from phosphatidylserine imaging, changes in the morphology of plasma membrane due to membrane acidification, loss of membrane potential and activation of γ-secretase during early apoptotic process have been monitored using small molecule 'Aposense' probes (in several cancer models [142, 144, 145]). One such novel probe, ^{18}F-ML-10 that is selectively taken up by radio or chemotherapy treated apoptotic cells in a tumor, suggests loss of mitochondrial membrane potential, caspase activation and degradation of DNA [143]. One severe drawback however is lack of understanding in the mechanism of uptake of these compounds.

Studying the fate of caspases (primarily the executioner caspase-3) is another possibility in understanding the dynamism of apoptotic process. ^{18}F radiolabeled sulfonamide derivatives of Isatin (^{1}H-indole-2,3-dione) that have high metabolic stability and considerable lipophilicity are the best molecules to trace executioner caspases by microPET [146–148] as they bind to caspases-3/-7 with nanomolar affinity. However, due to the stringent requirement of higher lipophilicity or a cell penetration moiety, these molecules are limited only to preclinical studies. Lack of selectivity of several caspase inhibitors prevents their effective use in the clinical and preclinical studies. Recently, this problem has been circumvented with the development of irreversible active site inhibitor probes of caspases that show negligible or no activity against other cysteine proteases such as cathepsins. These molecules were utilized in optical imaging of caspases with a near infrared fluorescent probe and peptide transduction domain [149]. Design of probes such as TcapQ(647) comprising activatable caspase recognition (DEVD tetrapeptide) sequences have taken this technique one step ahead where the probe is attached with a fluorophore- quencher pair that gets activated only post recognition by caspases. This molecule successfully probed the fate of caspases in different xenograft murine models [150, 151]. Other techniques that have been explored include fluorescent nanoparticles alone or in combination with activatable caspases [152, 153]. However, further research to reduce non-specific recognition of proteases will lead to their wider application in biomedical research as well as in clinics. Imaging of caspase activation with tagged reporter genes that have also been extensively experimented with a few successful endeavors, led their way to the market. One such example is Caspase-Glo 3/7 that has been developed by Promega [154, 155], where a luciferase acts as a reporter.

9 Future Perspectives

Non-invasive imaging has revolutionized biomedical and translational research in the current decade. With introduction of fusion imaging modalities and significant technological advancements, it has transcended itself from being a mere visualization tool to an indispensable preclinical and clinical aide in designing, quantifying and testing preventive interventions. With its ability to build three dimensional images of different forms of tissues, organs, bones and vasculature, it has taken up a more responsible task of understanding the intricacies of cellular networks and pathophysiological processes. Along with other biological pathways, apoptosis imaging has become a popular tool for studying progress and treatment of associated diseases in preclinical settings. However, better comprehension of various molecular features of cell death is required for development of better imaging agents with enhanced specificity and optimal pharmacokinetic properties. With all-round advancement including technology and probe efficacy, apoptosis imaging can be one of the most advanced tools to study the molecules involved in this pathway as well as its association with several disease conditions. Effective amalgamation of apoptosis imaging with other important biological processes will provide a global picture of pathophysiological conditions which will certainly improve clinical decision making in apoptosis-related diseases and interventions.

References

1. Call JA, Eckhardt SG, Camidge DR (2008) Targeted manipulation of apoptosis in cancer treatment. Lancet Oncol 9:1002–1011
2. Levi M, Dorffle-Melly J, Johnson GJ, Drouet L, Badimon L (2001) Usefulness and limitations of animal models of venous thrombosis. Thromb Haemost 86:1331–1333
3. Rygaard J, Povlsen CO (1969) Heterotransplantation of a human malignant tumour to "Nude" mice. Acta Pathol Microbiol Scand 77:758–760
4. Abdulkadir SA, Kim J (2005) Genetically engineered murine models of prostate cancer: insights into mechanisms of tumorigenesis and potential utility. Future Oncol 1:351–360. doi:10.1517/14796694.1.3.351
5. Talmadge JE, Singh RK, Fidler IJ, Raz A (2007) Murine models to evaluate novel and conventional therapeutic strategies for cancer. Am J Pathol 170:793–804. doi:10.2353/ajpath.2007.060929 [pii]
6. Feitsma H, Cuppen E (2008) Zebrafish as a cancer model. Mol Cancer Res 6:685–694. doi:10.1158/1541-7786.MCR-07-2167
7. Hansen K, Khanna C (2004) Spontaneous and genetically engineered animal models; use in preclinical cancer drug development. Eur J Cancer 40:858–880. doi:10.1016/j.ejca.2003.11.031
8. Chakraborty S et al (2015) Evaluation of 177Lu-EDTMP in dogs with spontaneous tumor involving bone: pharmacokinetics, dosimetry and therapeutic efficacy. Curr Radiopharm. doi:CRP-EPUB-65891 [pii]
9. De Saint-Hubert M et al (2011) Preclinical imaging of therapy response using metabolic and apoptosis molecular imaging. Mol Imaging Biol 13:995–1002. doi:10.1007/s11307-010-0412-z

10. Flexner C (2007) HIV drug development: the next 25 years. Nat Rev Drug Discov 6:959–966. doi:10.1038/nrd2336
11. Wlodawer A (2002) Rational approach to AIDS drug design through structural biology. Annu Rev Med 53:595–614. doi:10.1146/annurev.med.53.052901.131947
12. Lele DRD (2009) Chapter 7: Small animal PET and SPECT for basic research and drug development. In Principles and practice of nuclear medicine and correlative medical imaging. Jaypee Brothers Medical Publishers Pvt. Ltd., New Delhi, INDIA. ISBN 9788184484816
13. Matthews KA, Kaufman TC, Gelbart WM (2005) Research resources for Drosophila: the expanding universe. Nat Rev Genet 6:179–193. doi:10.1038/nrg1554nrg1554 [pii]
14. Kavanagh K, Reeves EP (2004) Exploiting the potential of insects for in vivo pathogenicity testing of microbial pathogens. FEMS Microbiol Rev 28:101–112. doi:10.1016/j.femsre.2003.09.002
15. Antunes LC, Imperi F, Carattoli A, Visca P (2011) Deciphering the multifactorial nature of Acinetobacter baumannii pathogenicity. PLoS One 6:e22674. doi:10.1371/journal.pone.0022674
16. Aperis G et al (2007) Galleria mellonella as a model host to study infection by the Francisella tularensis live vaccine strain. Microbes Infect 9:729–734. doi:10.1016/j.micinf.2007.02.016
17. Waterfield NR et al (2008) Rapid Virulence Annotation (RVA): identification of virulence factors using a bacterial genome library and multiple invertebrate hosts. Proc Natl Acad Sci U S A 105:15967–15972. doi:10.1073/pnas.0711114105
18. Hajar R (2011) Animal testing and medicine. Heart Views 12:42. doi:10.4103/1995-705X.81548
19. Guerrini A (ed) (2003) Experimenting with humans and animals: from Galen to animal rights. Johns Hopkins introductory studies in the history of science. John Hopkins University Press, New York, ISBN-13: 978-0801871979
20. Vanhooren V, Libert C (2013) The mouse as a model organism in aging research: usefulness, pitfalls and possibilities. Ageing Res Rev 12:8–21. doi:10.1016/j.arr.2012.03.010
21. Beck JA et al (2000) Genealogies of mouse inbred strains. Nat Genet 24:23–25. doi:10.1038/71641
22. Smetana K Jr, Holub M, Slavcev A (1989) Foreign body reaction against cellophane in the athymic nude mice. J Biomed Mater Res 23:947–951. doi:10.1002/jbm.820230810
23. Hansen CT, Fogh J, Giovanella B (1978). In: Fogh J (ed) The nude mouse in experimental and clinical research, vol 1, Ch. 1. Academic Press, New York, pp 1–35
24. Cespedes MV, Casanova I, Parreno M, Mangues R (2006) Mouse models in oncogenesis and cancer therapy. Clin Transl Oncol 8:318–329. doi:10.1007/s12094-006-0177-7
25. Bosma MJ, Carroll AM (1991) The SCID mouse mutant: definition, characterization, and potential uses. Annu Rev Immunol 9:323–350. doi:10.1146/annurev.iy.09.040191.001543
26. Blunt T et al (1995) Defective DNA-dependent protein kinase activity is linked to V(D)J recombination and DNA repair defects associated with the murine scid mutation. Cell 80:813–823. doi:10.1016/0092-8674(95)90360-7
27. Bankert RB, Hess SD, Egilmez NK (2002) SCID mouse models to study human cancer pathogenesis and approaches to therapy: potential, limitations, and future directions. Front Biosci 7:c44–c62
28. Bankert RB, Egilmez NK, Hess SD (2001) Human-SCID mouse chimeric models for the evaluation of anti-cancer therapies. Trends Immunol 22:386–393. doi:10.1016/S1471-4906(01)01943-3
29. Bankert RB et al (1989) Human lung tumors, patients' peripheral blood lymphocytes and tumor infiltrating lymphocytes propagated in scid mice. Curr Top Microbiol Immunol 152:201–210
30. Kamel-Reid S et al (1989) A model of human acute lymphoblastic leukemia in immune-deficient SCID mice. Science 246:1597–1600
31. Shultz LD, Ishikawa F, Greiner DL (2007) Humanized mice in translational biomedical research. Nat Rev Immunol 7:118–130. doi:10.1038/nri2017

32. Shultz LD et al (2005) Human lymphoid and myeloid cell development in NOD/LtSz-scid IL2R gamma null mice engrafted with mobilized human hemopoietic stem cells. J Immunol 174:6477–6489. doi:10.4049/jimmunol.174.10.6477
33. Shultz LD et al (1995) Multiple defects in innate and adaptive immunologic function in NOD/LtSz-scid mice. J Immunol 154:180–191
34. Wang J, Li W (2014) Discovery of novel second mitochondria-derived activator of caspase mimetics as selective inhibitor of apoptosis protein inhibitors. J Pharmacol Exp Ther 349:319–329. doi:10.1124/jpet.113.212019
35. Uribe V et al (2012) Rescue from excitotoxicity and axonal degeneration accompanied by age-dependent behavioral and neuroanatomical alterations in caspase-6-deficient mice. Hum Mol Genet 21:1954–1967. doi:10.1093/hmg/dds005
36. Nagy A, Gertsenstein M, Vintersten K, Behringer R (2003) Manipulating the mouse embryo: a laboratory manual. Cold Spring Harbor Laboratory, Cold Spring Harbor
37. Palmiter RD, Brinster RL (1986) Germ-line transformation of mice. Annu Rev Genet 20:465–499. doi:10.1146/annurev.ge.20.120186.002341
38. Brinster RL, Palmiter RD (1984) Introduction of genes into the germ line of animals. Harvey Lect 80:1–38
39. Smithies O, Gregg RG, Boggs SS, Koralewski MA, Kucherlapati RS (1985) Insertion of DNA sequences into the human chromosomal beta-globin locus by homologous recombination. Nature 317:230–234
40. Doetschman T et al (1987) Targetted correction of a mutant HPRT gene in mouse embryonic stem cells. Nature 330:576–578. doi:10.1038/330576a0
41. Gossler A, Doetschman T, Korn R, Serfling E, Kemler R (1986) Transgenesis by means of blastocyst-derived embryonic stem cell lines. Proc Natl Acad Sci U S A 83:9065–9069
42. Robertson E, Bradley A, Kuehn M, Evans M (1986) Germ-line transmission of genes introduced into cultured pluripotential cells by retroviral vector. Nature 323:445–448. doi:10.1038/323445a0
43. Mitchell AS et al (2015) Functional, morphological, and apoptotic alterations in skeletal muscle of ARC deficient mice. Apoptosis 20:310–326. doi:10.1007/s10495-014-1078-9
44. Hanahan D, Wagner EF, Palmiter RD (2007) The origins of oncomice: a history of the first transgenic mice genetically engineered to develop cancer. Genes Dev 21:2258–2270. doi:10.1101/gad.1583307
45. Turk B (2006) Targeting proteases: successes, failures and future prospects. Nat Rev Drug Discov 5:785–799. doi:10.1038/nrd2092
46. Abbenante G, Fairlie DP (2005) Protease inhibitors in the clinic. Med Chem 1:71–104
47. Zaman MA, Oparil S, Calhoun DA (2002) Drugs targeting the renin-angiotensin-aldosterone system. Nat Rev Drug Discov 1:621–636. doi:10.1038/nrd873
48. Schimmer AD et al (2004) Small-molecule antagonists of apoptosis suppressor XIAP exhibit broad antitumor activity. Cancer Cell 5:25–35. doi:10.1016/S1535-6108(03)00332-5
49. Vlasuk GP (1993) Structural and functional characterization of tick anticoagulant peptide (TAP): a potent and selective inhibitor of blood coagulation factor Xa. Thromb Haemost 70:212–216
50. Premzl A, Zavasnik-Bergant V, Turk V, Kos J (2003) Intracellular and extracellular cathepsin B facilitate invasion of MCF-10A neoT cells through reconstituted extracellular matrix in vitro. Exp Cell Res 283:206–214. doi:10.1016/S0014-4827(02)00055-1
51. McGovern SL, Helfand BT, Feng B, Shoichet BK (2003) A specific mechanism of nonspecific inhibition. J Med Chem 46:4265–4272. doi:10.1021/jm030266r
52. De Clercq E (2004) Antiviral drugs in current clinical use. J Clin Virol 30:115–133. doi:10.1016/j.jcv.2004.02.009
53. Stanton A (2003) Therapeutic potential of renin inhibitors in the management of cardiovascular disorders. Am J Cardiovasc Drugs 3:389–394. doi:10.2165/00129784-200303060-00002
54. Mervaala E et al (2000) Blood pressure-independent effects in rats with human renin and angiotensinogen genes. Hypertension 35:587–594

55. Mentlein R, Gallwitz B, Schmidt WE (1993) Dipeptidyl-peptidase IV hydrolyses gastric inhibitory polypeptide, glucagon-like peptide-1(7-36)amide, peptide histidine methionine and is responsible for their degradation in human serum. Eur J Biochem 214:829–835
56. Kieffer TJ, McIntosh CH, Pederson RA (1995) Degradation of glucose-dependent insulinotropic polypeptide and truncated glucagon-like peptide 1 in vitro and in vivo by dipeptidyl peptidase IV. Endocrinology 136:3585–3596. doi:10.1210/endo.136.8.7628397
57. Marguet D et al (2000) Enhanced insulin secretion and improved glucose tolerance in mice lacking CD26. Proc Natl Acad Sci U S A 97:6874–6879. doi:10.1073/pnas.120069197
58. Nagakura T et al (2001) Improved glucose tolerance via enhanced glucose-dependent insulin secretion in dipeptidyl peptidase IV-deficient Fischer rats. Biochem Biophys Res Commun 284:501–506. doi:10.1006/bbrc.2001.4999
59. Demuth HU, McIntosh CH, Pederson RA (2005) Type 2 diabetes – therapy with dipeptidyl peptidase IV inhibitors. Biochim Biophys Acta 1751:33–44. doi:10.1016/j.bbapap.2005.05.010
60. Thurmond RL et al (2004) Identification of a potent and selective noncovalent cathepsin S inhibitor. J Pharmacol Exp Ther 308:268–276. doi:10.1124/jpet.103.056879
61. Altmann E, Green J, Tintelnot-Blomley M (2003) Arylaminoethyl amides as inhibitors of the cysteine protease cathepsin K-investigating P1' substituents. Bioorg Med Chem Lett 13:1997–2001. doi:10.1016/S0960-894X(03)00344-5
62. Hardy JA, Wells JA (2004) Searching for new allosteric sites in enzymes. Curr Opin Struct Biol 14:706–715. doi:10.1016/j.sbi.2004.10.009
63. Hardy JA, Lam J, Nguyen JT, O'Brien T, Wells JA (2004) Discovery of an allosteric site in the caspases. Proc Natl Acad Sci U S A 101:12461–12466. doi:10.1073/pnas.0404781101
64. Harris JL, Peterson EP, Hudig D, Thornberry NA, Craik CS (1998) Definition and redesign of the extended substrate specificity of granzyme B. J Biol Chem 273:27364–27373
65. Thornberry NA et al (1997) A combinatorial approach defines specificities of members of the caspase family and granzyme B. Functional relationships established for key mediators of apoptosis. J Biol Chem 272:17907–17911
66. Grabowskal U, Chambers TJ, Shiroo M (2005) Recent developments in cathepsin K inhibitor design. Curr Opin Drug Discov Devel 8:619–630
67. Gocheva V et al (2006) Distinct roles for cysteine cathepsin genes in multistage tumorigenesis. Genes Dev 20:543–556. doi:10.1101/gad.1407406
68. Du C, Fang M, Li Y, Li L, Wang, X (2000) Smac, a mitochondrial protein that promotes cytochrome c-dependent caspase activation by eliminating IAP inhibition. Cell 102:33–42. doi:10.1016/S0092-8674(00)00008-8
69. Verhagen AM et al (2000) Identification of DIABLO, a mammalian protein that promotes apoptosis by binding to and antagonizing IAP proteins. Cell 102:43–53. doi:10.1016/S0092-8674(00)00009-X
70. Li M, Song T, Yin ZF, Na YQ (2007) XIAP as a prognostic marker of early recurrence of nonmuscular invasive bladder cancer. Chin Med J (Engl) 120:469–473
71. Mizutani Y et al (2007) Overexpression of XIAP expression in renal cell carcinoma predicts a worse prognosis. Int J Oncol 30:919–925
72. Tamm I et al (2000) Expression and prognostic significance of IAP-family genes in human cancers and myeloid leukemias. Clin Cancer Res 6:1796–1803
73. Foster FM et al (2009) Targeting inhibitor of apoptosis proteins in combination with ErbB antagonists in breast cancer. Breast Cancer Res 11:R41. doi:10.1186/bcr2328
74. Glinsky GV (2006) Genomic models of metastatic cancer: functional analysis of death-from-cancer signature genes reveals aneuploid, anoikis-resistant, metastasis-enabling phenotype with altered cell cycle control and activated Polycomb Group (PcG) protein chromatin silencing pathway. Cell Cycle 5:1208–1216. doi:10.4161/cc.5.11.2796
75. Berezovskaya O et al (2005) Increased expression of apoptosis inhibitor protein XIAP contributes to anoikis resistance of circulating human prostate cancer metastasis precursor cells. Cancer Res 65:2378–2386. doi:10.1158/0008-5472.CAN-04-2649 [pii]

76. Cossu F et al (2009) Structural basis for bivalent Smac-mimetics recognition in the IAP protein family. J Mol Biol 392:630–644. doi:10.1016/j.jmb.2009.04.033-S0022-2836(09)00478-1 [pii]
77. Sun H et al (2004) Structure-based design of potent, conformationally constrained Smac mimetics. J Am Chem Soc 126:16686–16687. doi:10.1021/ja047438+
78. Lu J et al (2008) SM-164: a novel, bivalent Smac mimetic that induces apoptosis and tumor regression by concurrent removal of the blockade of cIAP-1/2 and XIAP. Cancer Res 68:9384–9393. doi:10.1158/0008-5472.CAN-08-265568/22/9384 [pii]
79. Sun H et al (2008) Structure-based design, synthesis, evaluation, and crystallographic studies of conformationally constrained Smac mimetics as inhibitors of the X-linked inhibitor of apoptosis protein (XIAP). J Med Chem 51:7169–7180. doi:10.1021/jm8006849
80. Sun H et al (2007) Design, synthesis, and characterization of a potent, nonpeptide, cell-permeable, bivalent Smac mimetic that concurrently targets both the BIR2 and BIR3 domains in XIAP. J Am Chem Soc 129:15279–15294. doi:10.1021/ja074725f
81. Flygare JA et al (2012) Discovery of a potent small-molecule antagonist of inhibitor of apoptosis (IAP) proteins and clinical candidate for the treatment of cancer (GDC-0152). J Med Chem 55:4101–4113. doi:10.1021/jm300060k
82. Peng Y et al (2012) Bivalent Smac mimetics with a diazabicyclic core as highly potent antagonists of XIAP and cIAP1/2 and novel anticancer agents. J Med Chem 55:106–114. doi:10.1021/jm201072x
83. Wu G et al (2000) Structural basis of IAP recognition by Smac/DIABLO. Nature 408:1008–1012. doi:10.1038/35050012
84. Vucic D et al (2005) Engineering ML-IAP to produce an extraordinarily potent caspase 9 inhibitor: implications for Smac-dependent anti-apoptotic activity of ML-IAP. Biochem J 385:11–20. doi:10.1042/BJ20041108
85. Wang XJ et al (2010) Crystal structures of human caspase 6 reveal a new mechanism for intramolecular cleavage self-activation. EMBO Rep 11:841–847. doi:10.1038/embor.2010.141
86. Albrecht S et al (2007) Activation of caspase-6 in aging and mild cognitive impairment. Am J Pathol 170:1200–1209. doi:10.2353/ajpath.2007.060974
87. Graham RK et al (2010) Cleavage at the 586 amino acid caspase-6 site in mutant huntingtin influences caspase-6 activation in vivo. J Neurosci 30:15019–15029. doi:10.1523/JNEUROSCI.2071-10.201030/45/15019 [pii]
88. Hermel E et al (2004) Specific caspase interactions and amplification are involved in selective neuronal vulnerability in Huntington's disease. Cell Death Differ 11:424–438. doi:10.1038/sj.cdd.4401358
89. Slee EA et al (1999) Ordering the cytochrome c-initiated caspase cascade: hierarchical activation of caspases-2, -3, -6, -7, -8, and -10 in a caspase-9-dependent manner. J Cell Biol 144:281–292
90. Zhang J, Gorostiza OF, Tang H, Bredesen DE, Galvan V (2010) Reversal of learning deficits in hAPP transgenic mice carrying a mutation at Asp664: a role for early experience. Behav Brain Res 206:202–207. doi:10.1016/j.bbr.2009.09.013-S0166-4328(09)00528-2 [pii]
91. Galvan V et al (2006) Reversal of Alzheimer's-like pathology and behavior in human APP transgenic mice by mutation of Asp664. Proc Natl Acad Sci U S A 103:7130–7135. doi:10.1073/pnas.0509695103
92. Galvan V et al (2008) Long-term prevention of Alzheimer's disease-like behavioral deficits in PDAPP mice carrying a mutation in Asp664. Behav Brain Res 191:246–255. doi:10.1016/j.bbr.2008.03.035
93. Saganich MJ et al (2006) Deficits in synaptic transmission and learning in amyloid precursor protein (APP) transgenic mice require C-terminal cleavage of APP. J Neurosci 26:13428–13436. doi:10.1523/JNEUROSCI.4180-06.2006
94. Nguyen TV et al (2008) Signal transduction in Alzheimer disease: p21-activated kinase signaling requires C-terminal cleavage of APP at Asp664. J Neurochem 104:1065–1080. doi:10.1111/j.1471-4159.2007.05031.xJNC5031 [pii]

95. Banwait S et al (2008) C-terminal cleavage of the amyloid-beta protein precursor at Asp664: a switch associated with Alzheimer's disease. J Alzheimers Dis 13:1–16
96. Bredesen DE, John V, Galvan V (2010) Importance of the caspase cleavage site in amyloid-beta protein precursor. J Alzheimers Dis 22:57–63. doi:10.3233/JAD-2010-100537
97. Harris JA et al (2010) Many neuronal and behavioral impairments in transgenic mouse models of Alzheimer's disease are independent of caspase cleavage of the amyloid precursor protein. J Neurosci 30:372–381. doi:10.1523/JNEUROSCI.5341-09.2010
98. Nikolaev A, McLaughlin T, O'Leary DD, Tessier-Lavigne M (2009) APP binds DR6 to trigger axon pruning and neuron death via distinct caspases. Nature 457:981–989. doi:10.1038/nature07767
99. Zheng TS, Hunot S, Kuida K, Flavell RA (1999) Caspase knockouts: matters of life and death. Cell Death Differ 6:1043–1053. doi:10.1038/sj.cdd.4400593
100. Graham RK et al (2006) Cleavage at the caspase-6 site is required for neuronal dysfunction and degeneration due to mutant huntingtin. Cell 125:1179–1191. doi:10.1016/j.cell.2006.04.026
101. Pouladi MA et al (2009) Prevention of depressive behaviour in the YAC128 mouse model of Huntington disease by mutation at residue 586 of huntingtin. Brain 132:919–932. doi:10.1093/brain/awp006
102. Milnerwood AJ et al (2010) Early increase in extrasynaptic NMDA receptor signaling and expression contributes to phenotype onset in Huntington's disease mice. Neuron 65:178–190. doi:10.1016/j.neuron.2010.01.008
103. Metzler M et al (2010) Phosphorylation of huntingtin at Ser421 in YAC128 neurons is associated with protection of YAC128 neurons from NMDA-mediated excitotoxicity and is modulated by PP1 and PP2A. J Neurosci 30:14318–14329. doi:10.1523/JNEUROSCI.1589-10.2010
104. Adams DJ et al (2005) A genome-wide, end-sequenced 129Sv BAC library resource for targeting vector construction. Genomics 86:753–758. doi:10.1016/j.ygeno.2005.08.003
105. van der Weyden L, Adams DJ, Bradley A (2002) Tools for targeted manipulation of the mouse genome. Physiol Genomics 11:133–164. doi:10.1152/physiolgenomics.00074.2002
106. Elmore S (2007) Apoptosis: a review of programmed cell death. Toxicol Pathol 35:495–516. doi:10.1080/01926230701320337
107. Niers JM, Kerami M, Pike L, Lewandrowski G, Tannous BA (2011) Multimodal in vivo imaging and blood monitoring of intrinsic and extrinsic apoptosis. Mol Ther 19:1090–1096. doi:10.1038/mt.2011.17
108. Schambach SJ, Bag S, Schilling L, Groden C, Brockmann MA (2010) Application of micro-CT in small animal imaging. Methods 50:2–13. doi:10.1016/j.ymeth.2009.08.007
109. Buscombe JR, Wong B (2013) PET a tool for assessing the in vivo tumour cell and its microenvironment? Br Med Bull 105:157–167. doi:10.1093/bmb/lds041
110. Jang BS (2013) MicroSPECT and MicroPET imaging of small animals for drug development. Toxicol Res 29:1–6. doi:10.5487/TR.2013.29.1.001
111. de Kemp RA, Epstein FH, Catana C, Tsui BM, Ritman EL (2010) Small-animal molecular imaging methods. J Nucl Med 51(Suppl 1):18S–32S. doi:10.2967/jnumed.109.068148
112. Thorek D et al (2012) Cerenkov imaging – a new modality for molecular imaging. Am J Nucl Med Mol Imaging 2:163–173
113. Galban CJ et al (2010) Applications of molecular imaging. Prog Mol Biol Transl Sci 95:237–298. doi:10.1016/B978-0-12-385071-3.00009-5
114. Lee JS, Kim JH (2014) Recent advances in hybrid molecular imaging systems. Semin Musculoskelet Radiol 18:103–122. doi:10.1055/s-0034-1371014
115. Zhu A, Shim H (2011) Current molecular imaging positron emitting radiotracers in oncology. Nucl Med Mol Imaging 45:1–14. doi:10.1007/s13139-011-0075-y75 [pii]
116. Meikle SR, Kench P, Kassiou M, Banati RB (2005) Small animal SPECT and its place in the matrix of molecular imaging technologies. Phys Med Biol 50:R45–R61. doi:10.1088/0031-9155/50/22/R01 [pii]

117. Li J, Chen L, Du L, Li M (2013) Cage the firefly luciferin! – a strategy for developing bioluminescent probes. Chem Soc Rev 42:662–676. doi:10.1039/c2cs35249d
118. Sato A, Klaunberg B, Tolwani R (2004) In vivo bioluminescence imaging. Comp Med 54:631–634
119. Miyoshi S (2014) Challenges of imaging as a biomarker in drug research and drug. Yakugaku Zasshi 134:465–472. doi:10.1248/yakushi.13-00248-2
120. Stout DB, Zaidi H (2008) Preclinical multimodality imaging in vivo. PET Clin 3:251–273. doi:10.1016/j.cpet.2009.03.001
121. Dandekar M, Tseng JR, Gambhir SS (2007) Reproducibility of ^{18}F-FDG microPET studies in mouse tumor xenografts. J Nucl Med 48:602–607. doi:10.2967/jnumed.106.036608
122. Herschman HR (2003) Micro-PET imaging and small animal models of disease. Curr Opin Immunol 15:378–384. doi:10.1016/S0952-7915(03)00066-9
123. Haubner R, Beer AJ, Wang H, Chen X (2010) Positron emission tomography tracers for imaging angiogenesis. Eur J Nucl Med Mol Imaging 37(Suppl 1):S86–S103. doi:10.1007/s00259-010-1503-4
124. Knapp FF Jr, Mirzadeh S (1994) The continuing important role of radionuclide generator systems for nuclear medicine. Eur J Nucl Med 21:1151–1165
125. Kung MP, Kung HF (2005) Mass effect of injected dose in small rodent imaging by SPECT and PET. Nucl Med Biol 32:673–678. doi:10.1016/j.nucmedbio.2005.04.002
126. Willmann JK, van Bruggen N, Dinkelborg LM, Gambhir SS (2008) Molecular imaging in drug development. Nat Rev Drug Discov 7:591–607. doi:10.1038/nrd2290
127. Kowalsky A, Cohn M (1964) Application of nuclear magnetic resonance in biochemistry. Annu Rev Biochem 33:481–518. doi:10.1146/annurev.bi.33.070164.002405
128. Hall LD (1964) Nuclear magnetic resonance. Adv Carbohydr Chem 19:51–93
129. Andersson-Engels S, Johansson J, Svanberg K, Svanberg S (1991) Fluorescence imaging and point measurements of tissue: applications to the demarcation of malignant tumors and atherosclerotic lesions from normal tissue. Photochem Photobiol 53:807–814
130. Flecknell PA (1993) Anesthesia and perioperative care. Methods Enzymol 225:16–33
131. Kreissl MC et al (2006) Noninvasive measurement of cardiovascular function in mice with high-temporal-resolution small-animal PET. J Nucl Med 47:974–980. PMID:16741307 [pii]
132. Yang T, Haimovitz-Friedman A, Verheij M (2012) Anticancer therapy and apoptosis imaging. Exp Oncol 34:269–276
133. Smith C, Gibson DF, Tait JF (2009) Transmembrane voltage regulates binding of annexin V and lactadherin to cells with exposed phosphatidylserine. BMC Biochem 10:5. doi:10.1186/1471-2091-10-5
134. Tait JF (2008) Imaging of apoptosis. J Nucl Med 49:1573–1576. doi:10.2967/jnumed.108.052803
135. Blankenberg FG et al (1999) Imaging of apoptosis (programmed cell death) with 99mTc annexin V. J Nucl Med 40:184–191
136. Lahorte CM et al (2004) Apoptosis-detecting radioligands: current state of the art and future perspectives. Eur J Nucl Med Mol Imaging 31:887–919. doi:10.1007/s00259-004-1555-4
137. Blankenberg FG et al (1998) In vivo detection and imaging of phosphatidylserine expression during programmed cell death. Proc Natl Acad Sci U S A 95:6349–6354
138. Beekman CA et al (2011) Questioning the value of (99m)Tc-HYNIC-annexin V based response monitoring after docetaxel treatment in a mouse model for hereditary breast cancer. Appl Radiat Isot 69:656–662. doi:10.1016/j.apradiso.2010.12.012
139. Kartachova M et al (2004) In vivo imaging of apoptosis by 99mTc-Annexin V scintigraphy: visual analysis in relation to treatment response. Radiother Oncol 72:333–339. doi:10.1016/j.radonc.2004.07.008
140. Wang F et al (2008) Imaging paclitaxel (chemotherapy)-induced tumor apoptosis with 99mTc C2A, a domain of synaptotagmin I: a preliminary study. Nucl Med Biol 35:359–364. doi:10.1016/j.nucmedbio.2007.12.007
141. Thapa N et al (2008) Discovery of a phosphatidylserine-recognizing peptide and its utility in molecular imaging of tumour apoptosis. J Cell Mol Med 12:1649–1660. doi:10.1111/j.1582-4934.2008.00305.xJCMM305 [pii]

142. Damianovich M et al (2006) ApoSense: a novel technology for functional molecular imaging of cell death in models of acute renal tubular necrosis. Eur J Nucl Med Mol Imaging 33:281–291. doi:10.1007/s00259-005-1905-x
143. Cohen A et al (2009) From the Gla domain to a novel small-molecule detector of apoptosis. Cell Res 19:625–637. doi:10.1038/cr.2009.17
144. Cohen A et al (2007) Monitoring of chemotherapy-induced cell death in melanoma tumors by N,N′-Didansyl-L-cystine. Technol Cancer Res Treat 6:221–234. doi:10.1177/153303460700600310 [pii]
145. Aloya R et al (2006) Molecular imaging of cell death in vivo by a novel small molecule probe. Apoptosis 11:2089–2101. doi:10.1007/s10495-006-0282-7
146. Podichetty AK et al (2009) Fluorinated isatin derivatives. Part 2. New N-substituted 5-pyrrolidinylsulfonyl isatins as potential tools for molecular imaging of caspases in apoptosis. J Med Chem 52:3484–3495. doi:10.1021/jm8015014
147. Zhou D et al (2006) Synthesis, radiolabeling, and in vivo evaluation of an ^{18}F-labeled isatin analog for imaging caspase-3 activation in apoptosis. Bioorg Med Chem Lett 16:5041–5046. doi:10.1016/j.bmcl.2006.07.045
148. De Saint-Hubert M, Prinsen K, Mortelmans L, Verbruggen A, Mottaghy FM (2009) Molecular imaging of cell death. Methods 48:178–187. doi:10.1016/j.ymeth.2009.03.022
149. Edgington LE et al (2009) Noninvasive optical imaging of apoptosis by caspase-targeted activity-based probes. Nat Med 15:967–973. doi:10.1038/nm.1938
150. Niu G, Chen X (2010) Apoptosis imaging: beyond annexin V. J Nucl Med 51:1659–1662. doi:10.2967/jnumed.110.078584
151. Bullok KE et al (2007) Biochemical and in vivo characterization of a small, membrane-permeant, caspase-activatable far-red fluorescent peptide for imaging apoptosis. Biochemistry 46:4055–4065. doi:10.1021/bi061959n
152. Kim K et al (2006) Cell-permeable and biocompatible polymeric nanoparticles for apoptosis imaging. J Am Chem Soc 128:3490–3491. doi:10.1021/ja057712f
153. Stefflova K, Chen J, Li H, Zheng G (2006) Targeted photodynamic therapy agent with a built-in apoptosis sensor for in vivo near-infrared imaging of tumor apoptosis triggered by its photosensitization in situ. Mol Imaging 5:520–532
154. Liu JJ, Wang W, Dicker DT, El-Deiry WS (2005) Bioluminescent imaging of TRAIL-induced apoptosis through detection of caspase activation following cleavage of DEVD-aminoluciferin. Cancer Biol Ther 4:885–892. doi:10.4161/cbt.4.8.2133
155. Hickson J et al (2010) Noninvasive molecular imaging of apoptosis in vivo using a modified firefly luciferase substrate, Z-DEVD-aminoluciferin. Cell Death Differ 17:1003–1010. doi:10.1038/cdd.2009.205

Index

K. Bose (ed.), *Proteases in Apoptosis: Pathways, Protocols and Translational Advances*, DOI 10.1007/978-3-319-19497-4

Q

R

S

Printed by Printforce, the Netherlands